Metal–Organic and Organic Molecular Magnets

Metal–Organic and Organic Molecular Magnets

Edited by

P. Day
Fullerian Professor of Chemistry, The Royal Institution of Great Britain, London, UK

A. E. Underhill
Pro-vice Chancellor, University College of North Wales, Bangor, Gwynedd, UK

The proceedings of The Royal Society Discussion Meeting on Metal–Organic and Organic Molecular Magnets held on 24–25 March 1999 at The Royal Society, London

This work was first published in *Phil. Trans. R. Soc. Lond. A*, 1999, **357**, 2849–3184.

Special publication No. 252

ISBN 0-85404-764-6

A catalogue record for this book is available from the British Library

Published by The Royal Society of Chemistry,
Thomas Graham House, Science Park, Milton Road,
Cambridge CB4 0WF, UK

For further information see the RSC web site at www.rsc.org

Typeset by Paston PrePress Ltd
Printed and bound by MPG Books Ltd, Bodmin, Cornwall

Preface

One of the defining features of chemical science over the last decade has been an increasing preoccupation with the properties of matter in bulk, whether it be the flow of molten polymers or (as in this volume) designing and synthesizing magnetic materials from purely molecular components. The reasons for trying to make magnets from molecules are numerous: processing from solution at low temperatures, coupling magnetism with optical properties, looking for new lattice topologies and interaction mechanisms and finally, it must be said, the thrill of making architectures of molecules that never existed before and finding that they have quite unlooked for properties.

The many disparate strands of current work on molecular-based magnetic compounds were brought together recently in a Discussion Meeting held by The Royal Society of which this book is an outcome. Here the reader will find accounts of the discovery of the first purely organic ferromagnet, of metal–organic compounds with quite exceptionally high coercivities, of unprecedented photomagnetic effects, of novel physical methods like muon spin rotation, of room temperature magnets formed from aqueous solution under ambient conditions, and much more. As a pendant to the main theme, we also include the text of the 1999 Bakerian Lecture, The Royal Society's premier award lecture in the physical sciences, which coincidentally takes up several themes explored in the Discussion Meeting.

The collection of contributions from so many of the leading practitioners in this newly emerging subject will be of interest not only to synthetic and physical chemists but also to materials scientists and condensed matter physicists.

We thank the staff of The Royal Society and the publication departments of The Royal Society and the Royal Society of Chemistry for their help in bringing this volume to fruition.

P. Day
A. E. Underhill

Contents

Molecular-based magnets: setting the scene   1
P. Day and A. E. Underhill

p-Nitrophenyl nitronyl nitroxide: the first organic ferromagnet   4
M. Kinoshita
Discussion: P. Day   19

Crystal architectures of organic molecular-based magnets   22
D. B. Amabilino, J. Cirujeda and J. Veciana
Discussion: P. Day   40

Unusual crystal structures and properties of nitronylnitroxide radicals.
Possible RVB states in molecule-based magnets   41
K. Awaga, N. Wada, I. Watanabe and T. Inabe
Discussion: M. Verdaguer   70

Muon-spin-rotation studies of organic magnets   71
S. J. Blundell
Discussion: M. Verdaguer   84

High-spin polymeric arylamines   86
R. J. Bushby, D. Gooding and M. E. Vale

Room-temperature molecule-based magnets   105
M. Verdaguer, A. Bleuzen, C. Train, R. Garde, F. Fabrizi de Biani and
C. Desplanches

Design of novel magnets using Prussian blue analogues   123
K. Hashimoto and S. Ohkoshi

Magnetic anisotropy in molecule-based magnets   150
O. Kahn

Multifunctional coordination compounds: design and properties   169
S. Decurtins
Discussion: M. Verdaguer   184

Ferrimagnetic and metamagnetic layered cobalt(II)-hydroxides: first
observation of a coercive field greater than 5 T   185
M. Kurmoo
Discussion: B. J. Bushby, M. Verdaguer   205

Towards magnetic liquid crystals   206
K. Binnemans, D. W. Bruce, S. R. Collinson, R. Van Deun,
Yu. G. Galyametdinov and F. Martin

Quantum size effects in molecular magnets   221
D. Gatteschi

Large metal clusters and lattices with analogues to biology   240
D. J. Price, F. Lionti, R. Ballou, P. T. Wood and A. K. Powell

New high-spin clusters featuring transition metals   260
E. K. Brechin, A. Graham, P. E. Y. Milne, M. Murrie, S. Parsons and
R. E. P. Winpenny

From ferromagnets to high-spin molecules: the role of the organic ligands   279
T. Mallah, A. Marvilliers and E. Rivière
Discussion: P. Day   298

Molecular-based magnets: an epilogue   299
J. S. Miller

The Bakerian Lecture, 1999

*Lecture held at The Royal Institution 10 February 1999 and
University of Oxford 15 February 1999*

The Bakerian Lecture, 1999. The molecular chemistry of magnets and
superconductors   303
P. Day

Molecular-based magnets: setting the scene

By Peter Day[1] and Allan E. Underhill[2]

[1] Davy Faraday Research Laboratory, The Royal Institution of Great Britain, 21 Albemarle Street, London W1X 4BS, UK
[2] School of Chemistry, University College of North Wales, Bangor, Gwynedd LL57 2DG, UK

In 1839, when Michael Faraday published the first picture of the lines of magnetic flux around a magnet (Faraday 1832), shown in figure 1, the cylinder of material in the centre of the figure could only have been one material, iron. Over the succeeding 160 years, the number of substances showing spontaneous magnetization has increased enormously, while their variety has broadened dramatically. Yet till quite recently, the field of magnetic materials has been traditionally confined to metals, among which the current 'market leaders' such as lanthanide–cobalt and Nd–Fe–B have achieved large technological significance. Among non-metallic phases, transition metal oxides made an early appearance in the years immediately before and after World War II, and the technologically driven need to understand and optimize their properties led to the phenomenological theories of Néel, complemented later by the microscopic models of Mott (1949), Anderson (1963) and Goodenough (1955). The latter, in particular, set out the orbital symmetry rules that brought the subject of cooperative magnetism firmly within the ambit of the solid-state chemist.

Halides, chalcogenides and pnicnides are for the most part continuous lattice compounds, and, apart from isolated instances that could be regarded more or less as curiosities (such as diethyldithiocarbamato–Fe(III) chloride (Wickman *et al.* 1967*a, b*) and Mn phthalocyanine (Barraclough *et al.* 1970)), magnetic solids built up from molecular coordination or organometallic complexes only arrived on the scene quite recently. Early work by the Dutch school (de Jongh & Miedema 1974) had shown that layer perovskite halide salts of Cu(II) were useful models for ferromagnetism in insulating two-dimensional lattices, while we found that by replacing Cu ($S = 1/2$) with Cr ($S = 3/2$) organic intercalated insulating ferromagnets could be synthesized with Curie temperatures up to 50 K (Bellitto & Day 1976; for reviews see Day (1986) and Bellitto & Day (1992)). These Cu(II) and Cr(II) salts were also excellent realizations of the ferromagnetic exchange arising from orthogonality between orbitals on neighbouring metal centres containing the unpaired electrons, of the kind pointed out by Goodenough (1963) and Kanamori (1959) for continuous lattice oxides. Organometallic charge transfer salts and one-dimensional ferrimagnetic coordination polymers followed in the 1980s (Miller *et al.* 1988; Nakatami *et al.* 1991), while the early 1990s saw the first ferromagnets made from purely organic molecular building blocks, without any metal atoms at all, heralding a new field of p-electron magnetism to complement that of d- and f-electron materials (Tamura *et al.* 1991).

Given the truly enormous number of magnetic solids prepared, characterized and exploited in the last century or so, it is legitimate to ask what new features, either

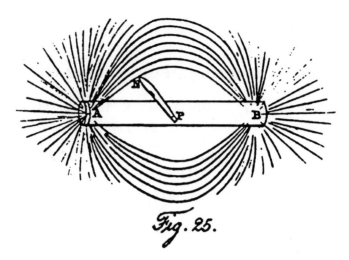

Figure 1. The first published picture of lines of magnetic flux around a magnet (Faraday 1832).

experimental or theoretical, the molecular-based materials have brought to this field. There are many. From the point of view of synthesis and processing, the contrast with conventional magnetic materials could not be more stark: substances made at (or close to) ambient temperature, usually from solution compared with high temperature metallurgical or ceramic processes. A real chance exists to make a soluble magnet! Correlating magnets with other properties, it should be pointed out that all known molecular-based magnetic compounds are insulators, the precise inverse of the situation for continuous lattice materials. That simple fact has consequences for many of the accompanying properties, of which the most striking (and potentially one of the most useful) is optical. The molecular-based materials are frequently transparent to infrared and visible light. More than 20 years have passed since we first demonstrated the striking colour change occurring in an insulating ferromagnetic transition metal salt on passing through the Curie temperature as a result of coupling excitons to spin waves (Bellitto & Day 1978; Bellitto *et al.* 1980). Combining magnetism with properties only found in the molecular solid state, such as mesomorphism or chirality, would appear to be another potentially fruitful source of novel physics, as is the construction of unusual lattice topologies, such as the Kagomé lattice, with which to test statistical thermodynamic models of critical behaviour. One apparent drawback of molecular-based magnetic materials, especially in the realm of information storage, is that the large majority are very soft magnets, i.e. they have quite low coercivities. However, even this limitation is now being breached, as two contributions to this issue demonstrate (Kahn, this issue; Kurmoo, this issue).

In all, the design, synthesis and study of molecular-based metal-organic and organic magnets has brought supramolecular and coordination chemistry, as well as purely organic synthetic chemistry to bear on a field previously dominated by condensed matter physics. In doing so, it has given new impetus to these fields and provided physics with new objects for study. While technological spin-off is in its infancy, it can confidently be predicted that new applications not replacing, but complementing, existing materials are not far away. It is to be hoped that the present collection of contributions, stemming from the lively Discussion Meeting organized under the auspices of the Royal Society and held on 24–25 March 1999, will be of wide interest to chemists and physicists, and will serve to broaden appreciation of this fascinating multidisciplinary topic.

References

Anderson, P. W. 1963 *Sol. State Phys.* **14**, 99.

Barraclough, C. C., Martin, R. L., Mitra, S. & Sherwood, R. C. 1970 *J. Chem. Phys.* **53**, 1638.

Bellitto, C. & Day, P. 1976 *J. Chem. Soc. Chem. Commun.*, p. 870.

Bellitto, C. & Day, P. 1978 *J. Chem. Soc. Chem. Commun.*, p. 511.

Bellitto, C. & Day, P. 1992 *J. Mater. Chem.* **2**, 265.

Bellitto, C., Fair, M. J., Wood, T. E. & Day, P. 1980 *J. Phys.* C **13**, L627.

Day, P. 1986 *J. Mag. Mag. Mat.* **54–57**, 1273.

de Jongh, L. J. & Miedema, A. R. 1974 *Adv. Phys.* **23**, 1.

Faraday, M. 1832 *Phil. Trans. R. Soc. Lond.* **122**, 125.

Goodenough, J. B. 1955 *Phys. Rev.* **100**, 564.

Goodenough, J. B. 1963 *Magnetism and the chemical bond.* Intrascience: New York.

Kanamori, J. 1959 *J. Phys. Chem. Solids* **10**, 87.

Miller, J. S., Epstein, A. J. & Reiff, W. M. 1988 *Chem. Rev.* **88**, 201.

Mott, N. F. 1949 *Proc. Phys. Soc.* A **62**, 416.

Nakatami, K., Bergerat, P., Codjovi, E., Mathionière, C., Pei, Y. & Kahn, O. 1991 *Inorg. Chem.* **30**, 3977.

Tamura, M., Nakazawa, Y., Shiomi, D., Nozawa, K., Hosokoshi, Y., Ishikawa, M., Takahashi, M. & Kinoshita, M. 1991 *Chem. Phys. Lett.* **186**, 401.

Wickman, H. H., Trozzolo, A. M., Williams, H. J., Hull, G. W. & Merritt, F. R. 1967a *Phys. Rev.* **155**, 563.

Wickman, H. H., Trozzolo, A. M., Williams, H. J., Hull, G. W. & Merritt, F. R. 1967b *Phys. Rev.* **163**, 526.

p-Nitrophenyl nitronyl nitroxide: the first organic ferromagnet

BY MINORU KINOSHITA

Faculty of Science and Engineering, Science University of Tokyo,
Daigaku-dori 1-1-1, Onoda-shi, Yamaguchi 756-0884, Japan

The transition to ferromagnetic order was realized in 1991 with the discovery of a p-nitrophenyl nitronyl nitroxide ($C_{13}H_{16}N_3O_4$) crystal. This was the first example of a ferromagnet without metal elements. Its ferromagnetism below the transition temperature of 0.6 K has been established by various experiments such as susceptibility, magnetization, heat capacity, zero-field muon spin rotation, neutron diffraction and ferromagnetic resonance measurements. Details of the results of these experiments are described in this paper.

Keywords: p-nitrophenyl nitronyl nitroxide; p-NPNN; muon spin rotation;
pressure effect on ferromagnetism; ferromagnetic to antiferromagnetic transition;
charge-transfer mechanism

1. Introduction

Solid-state properties of organic compounds have been extensively studied for several decades and it has been revealed that organic solids possess the potential ability to exhibit various interesting properties. The development of organic conductors and superconductors is one such example. In contrast to these advances, the absence of an organic ferromagnet was one of the most conspicuous problems about 10 years ago. The first explicit theoretical approach for an organic ferromagnet was proposed as early as 1963 by McConnell (1963). However, no purely organic crystal was found to be a three-dimensional 'bulk' ferromagnet, even after extensive studies on the magnetic properties of organic solids.

About 15 years ago, only a few organic radical crystals were known to exhibit an intermolecular ferromagnetic interaction. One of them was the galvinoxyl radical (Mukai *et al.* 1967; Mukai 1969). We then initiated extensive studies on this compound to search for conditions favouring the ferromagnetic interaction in organic solids. After experimental and theoretical studies over a couple of years, we derived the following conclusion (Awaga *et al.* 1986*a*–*c*, 1987*a*, *b*, 1988; Hosokoshi *et al.* 1997; Kinoshita 1991, 1993*a*). The requirement for the ferromagnetic intermolecular interaction is twofold, namely:

(a) large spin polarization within a radical molecule; and

(b) small SOMO–SOMO overlap and large SOMO–FOMO overlap between neighbouring radicals.

Here SOMO stands for the singly occupied molecular orbital, and FOMO the fully occupied (or unoccupied) molecular orbitals. Condition (a) states the requirement a radical molecule has to fulfil. The concept of spin polarization was well studied in the 1960s, particularly in odd-alternate organic compounds such as galvinoxyl and nitronyl nitroxide radicals. The spin polarization originates from an exchange interaction in a radical molecule. On the other hand, condition (b) is related to intermolecular interactions and to the relative location of the neighbouring radicals in a crystal. According to these conditions, the ferromagnetic intermolecular interaction originates in the exchange interaction within a molecule, which is always ferromagnetic. If the latter interaction is strong enough, it spreads out over a crystal through intermolecular charge-transfer interaction between SOMOs and FOMOs. The first radical employed fulfilling these conditions was *p*-nitrophenyl nitronyl nitroxide and it became the first example of an organic ferromagnet (Tamura *et al.* 1991; Nakazawa *et al.* 1992; Kinoshita 1993*b*, 1995, 1996).

2. Molecular and electronic structure of *p*-nitrophenyl nitronyl nitroxide

Figure 1*a* shows the molecular structure of *p*-nitrophenyl nitronyl nitroxide (abbreviated as *p*-NPNN hereafter). The dot near the NO group denotes the unpaired electron that is responsible for the magnetism. The unpaired electron is mobile over the whole molecule, but mostly localizes on the ONCNO moiety and resides to a very small extent on the other parts of the molecule. The highly localized nature of the unpaired electron on the ONCNO moiety assures the molecule of large spin polarization as a result of strong exchange interactions between the unpaired electron and the lone-pair electrons on the hetero atoms. This qualitative prediction is also supported by an unrestricted Hartree–Fock (UHF) calculation of the molecular orbitals. As shown in figure 1*b*, the energy of the molecular orbital containing the unpaired electron is much more stabilized than that of the fourth FOMO containing an electron of the opposite spin direction.

3. Crystal structure and magnetic interactions

There are four polymorphic forms, α-, β-, γ- and δ-phases, known for *p*-NPNN (Allemand *et al.* 1991†; Awaga *et al.* 1989*a*; Turek *et al.* 1991). The crystallographic data of these phases are summarized in table 1. Among them the orthorhombic β-phase is most stable, and the other phases gradually transform into the β-phase when maintained around room temperature. The molecular arrangement on the *ac*-plane of the β-phase crystal is shown in figure 2*a*. The molecules on the *ac*-plane are arranged in a parallel manner with the long molecular axis along the *a*-axis. Since the crystal belongs to the $F2dd$ space group, the lattice can be divided into two face-centred orthorhombic sublattices, each deviating by $\frac{1}{4}a$, $\frac{1}{4}b$ and $\frac{1}{4}c$. Thus the crystal structure is similar to that of diamond or, more precisely, zincblende, as shown schematically in figure 2*b*, where the radical is denoted by an ellipsoid. All the molecules on the *ac*-plane at $y = 0$ are tilted in one way and those at $y = \frac{1}{4}b$ are tilted in the other way with respect to the *ac*-plane. The best-fit planes of the ONCNO moiety are tilted by

† The δ-phase was initially denoted as the β_h-phase in this article. However, it was renamed with permission from Wudl, because its crystal structure is closely related to that of the β- and γ-phases.

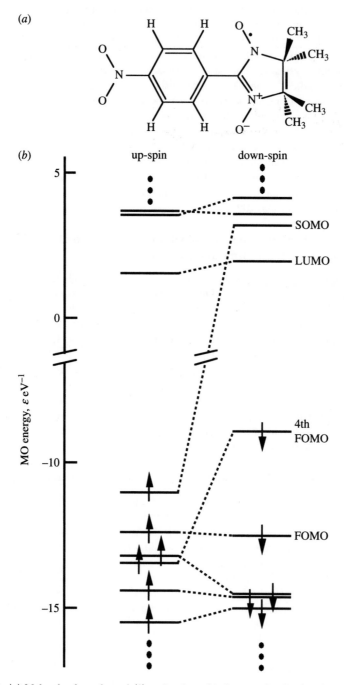

Figure 1. (*a*) Molecular formula and (*b*) molecular orbital energy levels of *p*-nitrophenyl nitronyl nitroxide. The energy levels near the singly occupied orbital (SOMO) are shown.

Table 1. *Crystal structures of p-nitrophenyl nitronyl nitroxide*

	α-phase	β-phase	γ-phase	δ-phase
system	monoclinic	orthorhombic	triclinic	monoclinic
space group	$P2_1/c$	$F2dd$	$P\bar{1}$	$P2_1/c$
a (Å)	7.302	12.347	9.193	8.963
b (Å)	7.617	19.350	12.105	23.804
c (Å)	24.677	10.960	6.471	6.728
α (deg)			97.35	
β (deg)	93.62		104.44	104.25
γ (deg)			82.22	
Z	4	8	2	4
V (Å^3)	1369.7	2618.5	687.6	1391.3
density, ρ (g cm^{-3})	1.354	1.416	1.349	1.333

$\pm 18.40°$ from the ac-plane, those of the phenyl rings by $\pm 68.45°$, and those of the nitro groups by $\pm 84.70°$.

The crystal structure at 6 K is also known from the neutron diffraction measurements (Zheludev *et al.* 1994*b*). The crystal contracts thermally maintaining its crystal symmetry, and the lattice constants of $a = 12.16$, $b = 19.01$ and $c = 10.71$ Å are reported. The contraction is largest along the c-axis (2.24%). As a result, the molecules are tilted at a greater angle than at room temperature. The tilt angles are $\pm 21.7°$, $\pm 71.2°$ and $\pm 89.1°$, respectively. These tilt angles are summarized in figure 2*c*. These changes in the tilt angle indicate that the molecules, when the crystal is cooled to 6 K, undergo librational rotation by $\pm 3.3°$ about the a-axis in preservation of the molecular shape, only the nitro groups being further rotated internally by $\pm 1°$.

It is to be noted that the density of the β-phase crystal is, as shown in table 1, the largest of the four polymorphic phases. The density increases to $\rho_c = 1.498$ g cm^{-3} at 6 K.

The expected dominant exchange paths, J_{12}, J_{13} and J_{14}, are also shown in figure 2*b* by the broken, full and dotted lines, respectively. Theoretical calculation indicates that the first two paths are ferromagnetic and the third is slightly antiferromagnetic (Okumura *et al.* 1993).

4. Ferromagnetic interactions

The paramagnetic susceptibility of the β-phase crystal was first measured in 1989 (Awaga & Maruyama 1989). The temperature dependence of magnetic susceptibility is shown in figure 3 for the field direction along the b-axis. The low field susceptibility obeys the Curie–Weiss law ($\chi = C/(T - \Theta)$) with a Weiss constant of $\Theta = +1$ K in the temperature range above *ca.* 4 K, suggesting the presence of weak ferromagnetic intermolecular interactions. Since the Weiss constant is very small, the ferromagnetic interaction is checked by measuring the field dependence of the magnetization at low temperatures. As shown in figure 4, the magnetization grows more steeply at lower temperature (Tamura *et al.* 1991). This indicates that the spins are connected by means of ferromagnetic interaction. In addition, this experiment ensures that the sample is not contaminated with a ferromagnetic impurity.

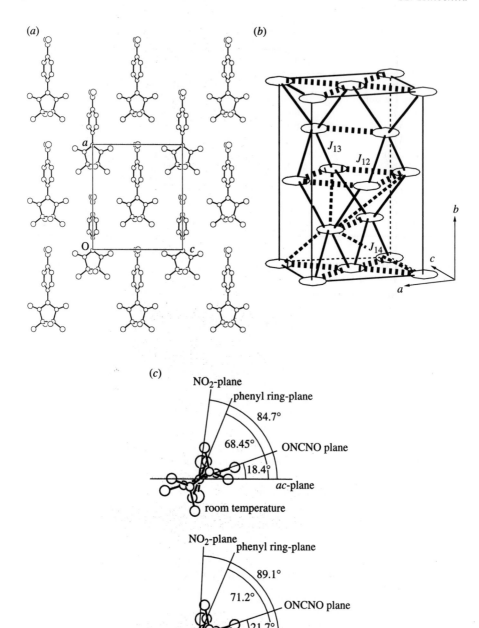

Figure 2. (*a*) The molecular arrangement on the *ac*-plane in the β-phase crystal of *p*-nitrophenyl nitronyl nitroxide. (*b*) A schematic bird's-eye-view of the crystal structure of the β-phase crystal. Each radical molecule is given by the ellipsoid. (*c*) The molecular shape viewed along the long axis at room temperature and at 6 K. The tilt angles of the best fit planes are noted with respect to the *ac*-plane.

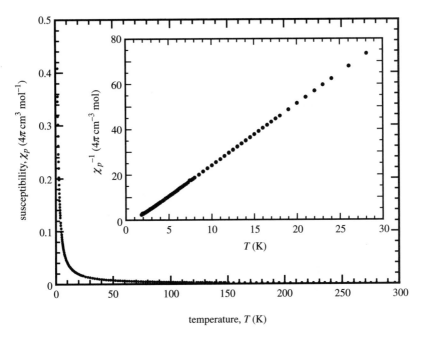

Figure 3. The temperature dependence of the static susceptibility of the β-phase p-nitrophenyl nitronyl nitroxide. The inset is the plot of reciprocal susceptibility against temperature.

5. Transition to the ferromagnetic ordered state

In 1991, the transition towards the ferromagnetic ordered state was discovered in a β-phase crystal by the measurements of AC susceptibility and heat capacity (Tamura *et al.* 1991; Nakazawa *et al.* 1992). The results of these measurements are shown in figure 5. The heat capacity has a λ-type sharp peak at the critical temperature of $T_c = 0.60$ K and reveals the existence of a transition. The corresponding entropy amounts to 85% of $R \ln 2$ in the range up to 2 K, and the transition is magnetic and bulk in nature.

As the AC susceptibility diverges at around T_c, the ordered state is, without doubt, a ferromagnetic state. In fact, the magnetization curve at 0.44 K traces a hysteresis loop characteristic of ferromagnetism, as shown in figure 6. The magnetization is almost saturated at an applied field as low as *ca.* 5 mT and the coercive force is quite small. The rapid saturation suggests a small magnetic anisotropy in the system. The g-factors observed in the paramagnetic resonance experiments are $g_a = 2.0070$, $g_b = 2.0030$ and $g_c = 2.0106$. The linewidth is also almost independent of field direction at room temperature.

6. Evidence for ferromagnetism

Further evidence for ferromagnetism has been provided by various experiments such as the measurements of the temperature dependence of heat capacity in applied magnetic fields of various strengths (Nakazawa *et al.* 1992), the zero-field muon spin

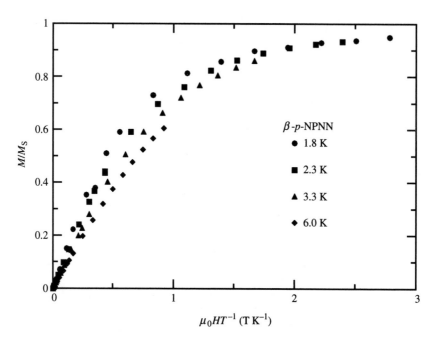

Figure 4. The plots of the magnetization of β-phase p-NPNN against $\mu_0 H / T$ measured at
$T = 1.8$, 2.3, 3.3 and 6.0 K.

rotation (Uemura *et al.* 1993; Le *et al.* 1993; Blundell *et al.* 1995), the ferromagnetic
resonance (Oshima *et al.* 1995*a*, *b*), the neutron diffraction (Zheludev *et al.* 1994*a*, *b*)
and the pressure effect on the magnetic properties (Takeda *et al.* 1995, 1996; Mito
et al. 1997).

(*a*) *Heat capacity in a magnetic field*

The heat-capacity temperature dependences at various magnetic field strengths are
illustrated in figure 7 (Nakazawa *et al.* 1992). The sharp peak in the zero field is
slightly rounded, and is shifted to the higher-temperature side as the magnetic field is
increased. This is a feature of ferromagnetic materials. For ferromagnetic substances,
the critical temperature cannot be defined in a finite magnetic field. When there is a
ferromagnetic interaction among the spins, they have a tendency to align themselves
in parallel along the magnetic field at very low temperatures and the spin system is
ordered by a weak field slightly above the ferromagnetic transition temperature. Thus
the sharp peak of the heat capacity shifts and becomes rounded as in the
paramagnetic region. In the case of antiferromagnetic order, the peak remains, up to
a certain field strength. Therefore, this experiment ensures the ferromagnetism of the
β-phase crystal below 0.6 K.

(*b*) *Zero-field muon spin rotation*

Another piece of evidence for ferromagnetism was obtained by the measurements of
zero-field muon spin rotation (ZF-μSR). Figure 8 shows some of the results of ZF-μSR
experiments performed with the initial muon spin polarization perpendicular to the

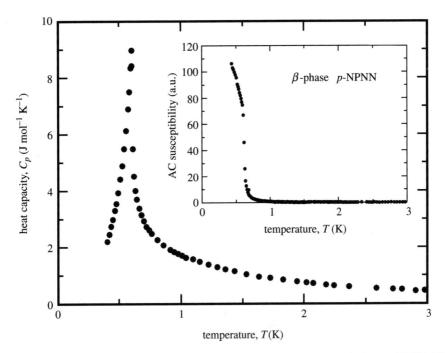

Figure 5. Temperature dependence of the magnetic heat capacity of the β-phase p-NPNN. The inset shows the temperature dependence of the AC susceptibility.

b-axis (Uemura *et al.* 1993; Le *et al.* 1993). The oscillating signals observed at lower temperatures are due to the precession of the muons implanted in the crystal. Since there is no applied field, it is obvious that the precession is caused by the internal field coming from the spontaneous magnetization. The long-lasting oscillations indicate that the muons experience a rather homogeneous local field, which requires the ferromagnetic spin network to be commensurate with the crystallographic structure. Thus, the results of ZF-μSR experiments clearly demonstrate the appearance of spontaneous magnetic order in the β-phase crystal.

The oscillation frequency is approximately related to the internal field by $\nu_\mu = (\gamma_\mu/2\pi)B_{int}$, where the muon gyromagnetic ratio $\gamma_\mu/2\pi = 135.53$ MHz T^{-1}. In figure 9, the frequency is plotted against temperature. The frequency extrapolated to 0 K corresponds approximately to the local field of 15.5 mT. The solid line in figure 9 shows a fit of $M(T) \propto M(0)[1 - (T/T_c)^\alpha]^\beta$ with $\alpha = 1.86$ and $\beta = 0.32$ ($\alpha = 1.74$ and $\beta = 0.36$ are reported by Blundell *et al.* (1995)). The agreement between the μSR results and the solid line is remarkably good, which allows us to discuss the results in two interesting regions, namely $T \to 0$ K and $T \to T_c$. At temperatures well below T_c, M decreases with increasing temperature as $[M(0) - M(T)] \propto T^\alpha$, close to the magnon-like behaviour of $[M(0) - M(T)] \propto T^{1.5}$. Near T_c, $M(T) \propto (T_c - T)^\beta$ with the critical magnetization exponent $\beta = 0.32$, in agreement with a value of $\frac{1}{3}$ expected for a three-dimensional Heisenberg system. The temperature dependence of $M(T)$ in p-NPNN is thus consistent with that of three-dimensional Heisenberg systems both at low temperature and near T_c.

The amplitude of the oscillations diminishes to *ca.* 20% when the initial muon spin

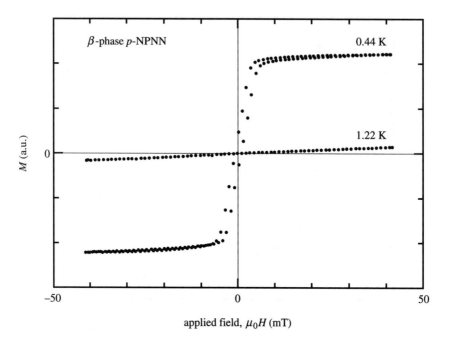

Figure 6. The magnetization curves of the β-phase p-NPNN measured at the temperatures above and below the transition temperature, T_c.

polarization is parallel to the b-axis. This suggests that the spin orientation in different domains is not aligned randomly and is most likely along the b-axis. Recent ferromagnetic resonance experiments (Oshima *et al.* 1995a, b) and neutron diffraction measurements by Schweizer's group (Zheludev *et al.* 1994a, b) also show that the magnetic easy axis is along the b-axis in the β-phase.

(c) Neutron diffraction

The spin density on each atom has been determined by neutron diffraction experiments (Zheludev *et al.* 1994a). This allows us to calculate the magnetic dipole–dipole interaction, D, in the ordered state. According to our calculation, it is $D_a/k = -0.016$ K, $D_b/k = -0.029$ K and $D_c/k = +0.045$ K (Kinoshita 1994; see also Blundell *et al.* 1995). The spin system is most stable when the spin alignment is along the b-axis. This agrees with the easy axis assignment mentioned above.

(d) Pressure effect

When the pressure is applied to the crystal, the AC susceptibility exhibits remarkable changes. Figure 10 shows the pressure dependence of the AC susceptibility of the polycrystalline sample of β-phase p-NPNN up to $p = 1.04$ GPa measured under the AC field $H_{AC}(\nu) < 1$ Oe ($= 10^3/4\pi$ A m^{-1}) with the frequency $\nu = 15.9$ Hz. It is found that the critical temperature, T_c, defined by the crossing point of the extrapolated lines from above and below T_c, approximately agrees with that determined from the heat capacity peak. As shown in figure 10, T_c shifts towards the

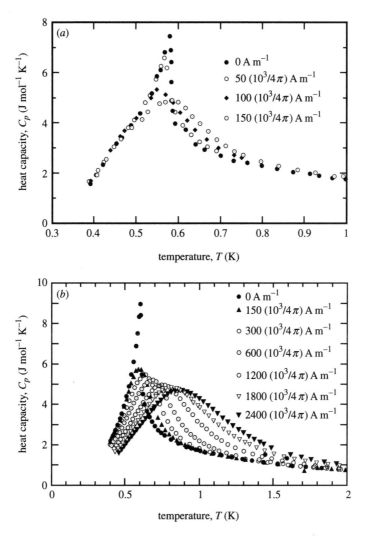

Figure 7. Temperature dependence of the heat capacity of the β-phase p-NPNN in the different applied magnetic field from 0 to 2400 Oe ($= 2400(10^3/4\pi)$ A m^{-1}).

lower temperature side with the initial gradient $\mathrm{d}[T_c(p)/T_c(p_0)]/\mathrm{d}p = -0.48$ GPa^{-1}, and the magnitude of the susceptibility decreases gradually as the pressure increases from p_0 ($= 0$ MPa). In the low-pressure region below $p \approx 650$ MPa, however, the ferromagnetic behaviour is still preserved below $T_c(p)$, as characterized by the shape of χ_{AC}.

In the high-pressure region above 650 MPa, on the contrary, the magnitude of χ_{AC} becomes quite small, the susceptibility in the ordered state decreases sensitively with the pressure increase, and the shoulder-like curve of χ_{AC} around $T_c(p)$ changes into a cusp, as shown in the inset of figure 10. Furthermore, $T_c(p)$ turns to increase as the pressure increases with a gradient of $\mathrm{d}[T_c(p)/T_c(p_c)]/\mathrm{d}p = +0.04$ GPa^{-1}, where

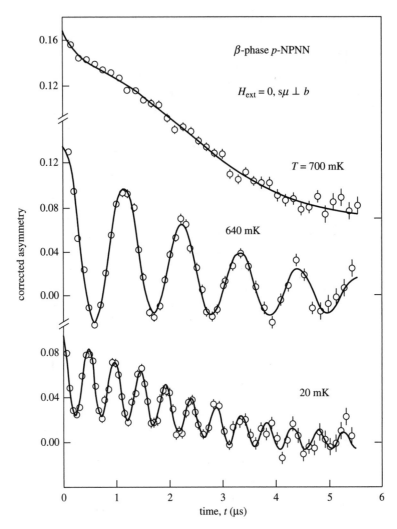

Figure 8. The ZF-μSR time spectra observed in the β-phase single crystals of p-NPNN with
initial muon spin polarization perpendicular to the b-axis.

$p_c = 650 \pm 50$ MPa is the critical pressure. These results suggest that the magnetic
order below T_c is of an antiferromagnet under high pressure.

The antiferromagnetic behaviour is also recognized in the external field dependence
of χ_{AC} at constant pressure, $p = 690$ MPa, as shown in figure 11. $T_c(p)$ shifts to a
lower temperature as the field increases, contrary to the case of a ferromagnet. Thus,
we can conclude that pressurization induces a ferromagnetic-to-antiferromagnetic
transition in the β-phase crystal.

7. Charge-transfer mechanism

These experimental results could be explained in terms of a charge-transfer
mechanism by a competition between the ferromagnetic and antiferromagnetic

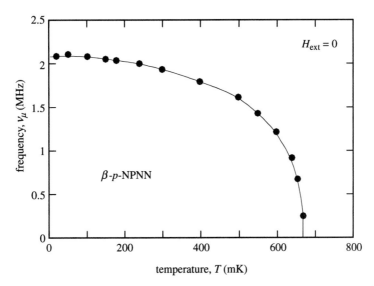

Figure 9. The temperature dependence of the muon spin precession frequency in the β-phase *p*-NPNN below T_c in a zero applied field. The frequency is proportional to the spontaneous magnetization, $M(T)$. The solid curve is a fit to the mean field magnetization $M(T) \propto [1 - (T/T_c)^\alpha]^\beta$, with $\alpha = 1.86$ and $\beta = 0.32$.

interactions. The effective exchange interaction between molecules A and B contains the kinetic (J_{AB}^{K}) and the potential (J_{AB}^{P}) terms as

$$J_{AB} = J_{AB}^{K} + J_{AB}^{P}. \qquad (7.1)$$

Both terms depend on the overlap of molecular orbitals (MOs) on molecules A and B. The essential point of the charge-transfer mechanism is that the kinetic exchange interaction is described by a sum of terms contributing to antiferromagnetic and ferromagnetic interactions:

$$J_{AB}^{K} = -\frac{t_{SS}^{2}}{U} + \frac{t_{SF}^{2}}{U^{2}} J_{in} + (\text{terms related to other paths}), \qquad (7.2)$$

where t_{SS} denotes the transfer integral between the SOMOs of molecules A and B, t_{SF} the transfer integral between SOMO and other FOMOs, U is the on-site Coulomb repulsion, and J_{in} is the intramolecular exchange integral. Then interplay or frustration among these contributions may result in $J_{AB} \approx 0$ for a certain condition, giving $T_c(p) \approx 0$ K. In the case of the β-phase of *p*-NPNN, we must take at least twelve interacting molecules adjacent to a central molecule in its zincblende-like structure of figure 2b. The corresponding exchange integrals are classified into three types, J_{12}, J_{13} and J_{14} (strictly speaking, J_{14} is further divided into two types) from the symmetry of the lattice. As mentioned before, J_{12} and J_{13} are, from theoretical calculation, known to be ferromagnetic and J_{14} is only weakly antiferromagnetic. On the other hand, reduction of the dimensionality from the three- to two-dimensional ferromagnetic system has been pointed out from the appearance of the short-range order effect by the heat capacity measurements under pressure (Takeda *et al.* 1995, 1996). From the crystal symmetry, it is obvious that only the exchange integral J_{12} is

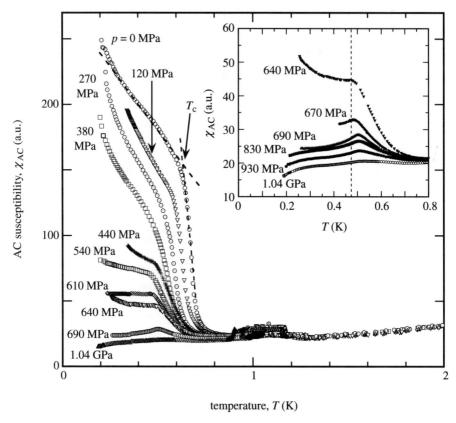

Figure 10. The temperature dependence of AC susceptibility of the polycrystalline β-phase p-NPNN under various pressures up to 1.04 GPa.

responsible for the two-dimensional ferromagnetic interaction. The transition temperature, T_{3D}, in such a reduced system can be written in terms of the mean field theory as

$$kT_{3D} \propto S^2 \xi_{2D}^2 (J_{12}, T_{3D})\{|J_{13} + J_{14}|\}, \tag{7.3}$$

where ξ_{2D} is the spin correlation length in the ac-plane in which $J_{12}/k \approx 0.8$ K is estimated from the heat capacity curve. Therefore, the antiferromagnetic behaviour at $p > p_c = 650 \pm 50$ MPa can be ascribed to a change in the sign of J_{13}. This means that the relative importance of the first and second terms in equation (7.2) relieve each other for J_{13} under high pressure.

8. Lattice constants under high pressure

The change of the magnetic interactions discussed above would be closely related to changes in molecular packing and deformation in the molecular shape. The lattice constants under various pressures have been determined by the Riedvelt method from the X-ray powder patterns obtained on the polycrystalline sample of β-phase p-NPNN by the use of imaging plates. The powder patterns observed are similar to one

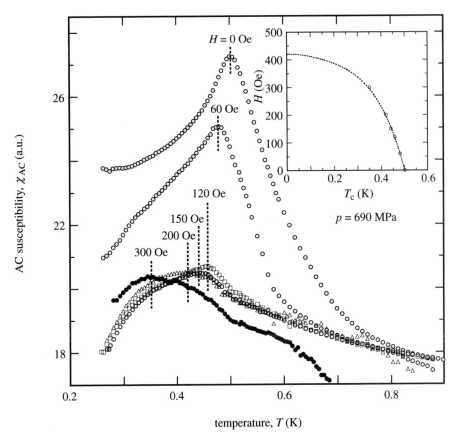

Figure 11. The temperature dependence of AC susceptibility of the β-phase p-NPNN at various external fields under the constant pressure of 690 MPa. The inset shows the T_c versus H phase boundary.

another, indicating that the crystal symmetry is maintained under pressure up to 1.26 GPa (Takeda *et al.* 1998). The peak positions, of course, shift towards the direction of wider angles or to the direction of lattice contractions as the pressure increases. The lattice constants are plotted against pressure in figure 12. The crystal changes in two steps. The linear and volume compressibilities are summarized in table 2. The biggest contraction (*ca.* 4.5%) is found again along the c-axis as in the case of thermal contraction. The crystal density increases up to as large as 1.58 g cm^{-3} at 1.26 GPa.

The two-step contraction observed can be understood in the following way. The thermal contraction at 6 K nearly corresponds to the compressive contraction at *ca.* 400 MPa, although the thermal and compressive contraction of the lattice may or may not be the same. From this, we can expect that the molecules are librationally rotated about the a-axis to some extent more under pressures up to *ca.* 550 MPa. However, the plane of the nitro group is already in the upright orientation at 6 K (or at *ca.* 400 MPa) with respect to the ac-plane, and further pressurization beyond

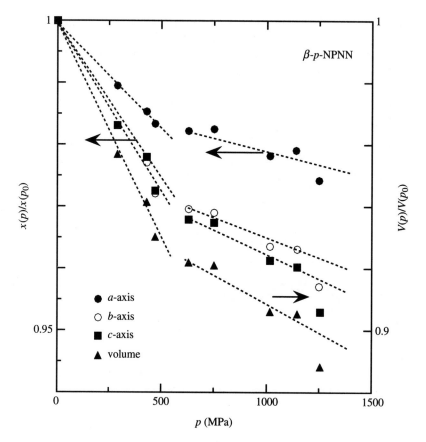

Figure 12. The pressure dependence of the lattice parameters and the unit-cell volume of the β-phase p-NPNN at room temperature.

Table 2. *Compressibility of the β-phase p-nitrophenyl nitronyl nitroxide crystal in units of* Pa^{-1}

	$\kappa_a = -\dfrac{1}{a}\dfrac{\partial a}{\partial p}$	$\kappa_b = -\dfrac{1}{b}\dfrac{\partial b}{\partial p}$	$\kappa_c = -\dfrac{1}{c}\dfrac{\partial c}{\partial p}$	$\kappa = -\dfrac{1}{V}\dfrac{\partial V}{\partial p}$
$p < 550$ MPa	34.4×10^{-11}	56.6×10^{-11}	55.1×10^{-11}	143×10^{-11}
$p > 550$ MPa	11.5×10^{-11}	18.4×10^{-11}	22.3×10^{-11}	49.0×10^{-11}

550 MPa will not increase the tilt angle further for the nitro group. Then, the other parts of the molecule can rotate internally about the long molecular axis under higher pressures. Internal rotations of the five-membered ring with the bulky tetra-methylethylene group would be the most probable candidates to take place in order for the molecule to become more and more planar and for the crystal to become more compact.

Such molecular deformation would cause a change in the electronic structure of the molecule and in the intermolecular magnetic interactions. At this moment we cannot conclude definitely, because the crystal structure at very low temperature and under high pressure is not available. However, it is natural to expect that the molecules take a more planar form and a more parallel arrangement in the crystal on compression at low temperature. The intermolecular charge-transfer interaction between the SOMOs on adjacent molecules would become more efficient, resulting in interchange of the importance of ferromagnetic and antiferromagnetic interactions, as discussed in the preceding section.

9. Other purely organic ferromagnets

Following the finding of the first example of an organic ferromagnet in the β-phase *p*-NPNN, a dozen of purely organic ferromagnets have been found. Some of them has been compiled in the literature (Kinoshita 1997). The highest ferromagnetic Curie temperature of 1.48 K was obtained for diazaadamantane dinitroxide (Chiarelli *et al.* 1993).

References

Allemand, P.-M., Fite, C., Canfield, P., Srdanov, G., Keder, N., Wudl, F. & Canfield, P. 1991 On the complexities of short range ferromagnetic exchange in a nitronyl nitroxide. *Synth. Met.* **41–43**, 3291–3295.

Awaga, K. & Maruyama, Y. 1989 Ferromagnetic intermolecular interaction of the organic radical, 2-(4-nitrophenyl)-4,4,5,5-tetramethyl-4,5-dihydro-1H-imidazolyl-1-oxyl 3-oxide. *Chem. Phys. Lett.* **158**, 556–558.

Awaga, K., Sugano, T. & Kinoshita, M. 1986*a* Organic radical clusters with ferromagnetic intermolecular interactions. *Solid State Commun.* **57**, 453–456.

Awaga, K., Sugano, T. & Kinoshita, M. 1986*b* Ferromagnetic intermolecular interactions in a series of organic mixed crystals of galvinoxyl radical and its precursory closed shell compound. *J. Chem. Phys.* **85**, 2211–2218.

Awaga, K., Sugano, T. & Kinoshita, M. 1986*c* ESR evidence for the ferro- and antiferro-magnetic intermolecular interaction in pure and dilute mixed crystals of galvinoxyl. *Chem. Phys. Lett.* **128**, 587–590.

Awaga, K., Sugano, T. & Kinoshita, M. 1987*a* Thermodynamic properties of the mixed crystals of galvinoxyl radical and its precursory closed shell compound: the large entropy cooperating with the spin system. *J. Chem. Phys.* **87**, 3062–3068.

Awaga, K., Sugano, T. & Kinoshita, M. 1987*b* Ferromagnetic intermolecular interaction of the galvinoxyl radical: cooperation of spin polarization and charge-transfer interaction. *Chem. Phys. Lett.* **141**, 540–544.

Awaga, K., Sugano, T. & Kinoshita, M. 1988 Ferromagnetic intermolecular interaction of organic radical, galvinoxyl. *Synth. Met.* **27**, B631–B638.

Awaga, K., Inabe, T., Nagashima, U. & Maruyama, Y. 1989*a* Two-dimensional network of the ferromagnetic radical, 2-(4-nitrophenyl)-4,4,5,5-tetramethyl-4,5-dihydro-1H-imidazol-1-oxyl 3-N-oxide. *J. Chem. Soc. Chem. Commun.*, pp. 1617–1618.

Blundell, S. J., Pattenden, P. A., Pratt, F. L., Valladares, R. M., Sugano, T. & Hayes, W. 1995 μ^+SR of the organic ferromagnet *p*-NPNN: diamagnetic and paramagnetic states. *Europhys. Lett.* **31**, 573–578.

Chiarelli, R., Novak, M. A., Rassat, A. & Tholence, J. L. 1993 A ferromagnetic transition at 1.48 K in an organic nitroxide. *Nature* **363**, 147–149.

Hosokoshi, Y., Tamura, M. & Kinoshita, M. 1997 Pressure effect on organic radicals. *Mol. Cryst. Liq. Cryst.* **306**, 423–430.

Kinoshita, M. 1991 Intermolecular ferromagnetic coupling in organic radical crystals. In *Magnetic molecular materials* (ed. D. Gatteschi, O. Kahn, J. S. Miller & F. Palacio). NATO ASI Series E, vol. 198, pp. 87–103.

Kinoshita, M. 1993*a* Ferromagnetism in organic radical crystal. *Mol. Cryst. Liq. Cryst.* **232**, 1–12.

Kinoshita, M. 1993*b* Bulk ferromagnetism of the crystal of the organic radical, *p*-NPNN. *Synth. Met.* **56**, 3279–3284.

Kinoshita, M. 1994 Ferromagnetism of organic radical crystals. *Jap. J. Appl. Phys.* **33**, 5718–5733.

Kinoshita, M. 1995 Ferromagnetism and antiferromagnetism in organic radical crystals. *Physica* B **213** /**214**, 257–261.

Kinoshita, M. 1996 Organic magnetic materials with cooperative magnetic properties. In *Molecular magnetism. From molecular assemblies to the devices* (ed. P. Coronado, D. Delhaes, D. Gatteschi & J. S. Miller). NATO ASI Series E, vol. 321, pp. 449–472.

Kinoshita, M. 1997 Magnetic properties of organic solids. In *Organic molecular solids. Properties and applications* (ed. W. Jones), pp. 379–414. Boca Raton, FL: CRC.

Le, L. P., Keren, A., Luke, G. M., Wu, W. D., Uemura, Y. J., Tamura, M., Ishikawa, M. & Kinoshita, M. 1993 Searching for spontaneous magnetic order in an organic ferromagnet. μSR studies of β-phase *p*-NPNN. *Chem. Phys. Lett.* **206**, 405–408.

McConnell, H. M. 1963 Ferromagnetism in solid free radicals. *J. Chem. Phys.* **39**, 1910.

Mito, M., Kawae, T., Takumi, M., Nagata, K., Tamura, M., Kinoshita, M. & Takeda, K. 1997 Pressure-induced ferro- to antiferromagnetic transition in a purely organic compound, β-phase *p*-nitrophenyl nitronyl nitroxide. *Phys. Rev.* B **56**, 14 255–14 258.

Mukai, K. 1969 Anomalous magnetic properties of stable crystalline phenoxyl radicals. *Bull. Chem. Soc. Jap.* **42**, 40–46.

Mukai, K., Nishiguchi, H. & Deguchi, Y. 1967 Anomaly in the χ–T curve of galvinoxyl radical. *J. Phys. Soc. Jap.* **23**, 125.

Nakazawa, Y., Tamura, M., Shirakawa, N., Shiomi, D., Takahashi, M., Kinoshita, M. & Ishikawa, M. 1992 Low-temperature magnetic properties of ferromagnetic organic radical, *p*-nitrophenyl nitronyl nitroxide. *Phys. Rev.* B **46**, 8906–8914.

Okumura, M., Mori, W. & Yamaguchi, K. 1993 *Mol. Cryst. Liq. Cryst.* **232**, 35–44.

Oshima, K., Kawanoue, H., Haibara, Y., Yamazaki, H., Awaga, K., Tamura, M., Ishikawa, M. & Kinoshita, M. 1995*a* Ferromagnetic resonance and nonlinear absorption in *p*-NPNN. *Synth. Met.* **71**, 1821–1822.

Oshima, H., Haibara, Y., Yamazaki, H., Awaga, K., Tamura, M. & Kinoshita, M. 1995*b* Ferromagnetic resonance in *p*-NPNN below 1 K. *Mol. Cryst. Liq. Cryst.* **271**, 29–34.

Takeda, K., Konishi, K., Tamura, M. & Kinoshita, M. 1995 Magnetism of the β-phase *p*-nitrophenyl nitronyl nitroxide crystal. *Mol. Cryst. Liq. Cryst.* **273**, 57–66.

Takeda, K., Konishi, K., Tamura, M. & Kinoshita, M. 1996 Pressure effects on intermolecular interactions of the organic ferromagnetic crystalline β-phase *p*-nitrophenyl nitronyl nitroxide. *Phys. Rev.* B **53**, 3374–3380.

Takeda, K., Mito, M., Kawae, T., Takumi, M., Nagata, K., Tamura, M. & Kinoshita, M. 1998 Pressure dependence of intermolecular interactions in the genuine organic β-phase *p*-nitrophenyl nitronyl nitroxide crystal accompanying a ferro- to antiferromagnetic transition. *J. Phys. Chem.* B **102**, 671–676.

Tamura, M., Nakazawa, Y., Shiomi, D., Nozawa, K., Hosokoshi, Y., Ishikawa, M., Takahashi, M. & Kinoshita, M. 1991 Bulk ferromagnetism in the β-phase crystal of the *p*-nitrophenyl nitronyl nitroxide radical. *Chem. Phys. Lett.* **186**, 401–404.

Turek, P., Nozawa, K., Shiomi, D., Awaga, K., Inabe, T., Maruyama, Y. & Kinoshita, M. 1991 Ferromagnetic coupling in a new phase of the *p*-nitrophenyl nitronyl nitroxide radical. *Chem. Phys. Lett.* **180**, 327–331.

Uemura, Y. J., Le, L. P. & Luke, G. M. 1993 Muon spin relaxation studies in organic superconductors and organic magnets. *Synth. Met.* **56**, 2845–2850.

Zheludev, A., Ressouche, M. & Schweizer, E. 1994a Neutron diffraction of a ferromagnetic phase transition in a purely organic crystal. *Solid State Commun.* **90**, 233–235.

Zheludev, A., Bonnet, M., Ressouche, E., Schweizer, J., Wan, M. & Wang, H. 1994b Experimental spin density in the first purely organic ferromagnet: the β-para-nitrophenyl nitronyl nitroxide. *J. Mag. Mag. Mater.* **135**, 147–160.

Discussion

P. DAY (*The Royal Institution, London, UK*). In view of the large size of the crystals of NPNN that are available, have any coherent inelastic neutron scattering experiments been carried out to define the spin wave dispersion along the various crystallographic directions? I ask because the crystal structure suggests that the magnetic correlations may be quite strongly two dimensional in character.

M. KINOSHITA. No, we have not done such experiments. However, the dominant two-dimensional character of this compound was suggested from heat capacity analysis. The temperature dependence of the magnetic heat capacity is well explained theoretically by introducing a small correction to a square lattice model.

Crystal architectures of organic molecular-based magnets

BY DAVID B. AMABILINO, JOAN CIRUJEDA AND JAUME VECIANA

Institut de Ciència de Materials de Barcelona (CSIC),
Campus Universitari, 08193-Bellaterra, Spain

The crystal architectures of molecular materials play deciding roles when it comes to the determination of magnetic properties. In turn, the build-up of these arrangements is shown to depend on the molecular features present in the compounds employed. An analysis of the molecular and supramolecular aspects of the α-nitronyl aminoxyl radicals is presented, and a critical appraisal of the state of the art in interpretation of magnetic interactions and ability to crystallize predetermined relative orientations between radicals is presented.

Keywords: organic magnets; magnetic interactions; McConnell I; hydrogen bonds; supramolecular chemistry; structure–magnetism correlation

1. Introduction

An organic ferromagnet—this elusive target has spurred the efforts of many research groups over the last decade, and few have attained it (Takahashi *et al.* 1991; Stephens *et al.* 1992; Chiarelli *et al.* 1993; Cirujeda *et al.* 1995*a*; Nogami *et al.* 1995*a*; Banister *et al.* 1996; Matsushita *et al.* 1997). The group of free radicals which have attracted most attention in this regard are the α-nitronyl aminoxyl family (figure 1). Their appeal derives from their relative stability, and (normally) relatively easy synthesis which allows the introduction of a wide variety of functional groups. While the weak magnetic interactions between the unpaired electrons occasion transition temperatures which are extremely low (always below 2 K), these molecules are ideal for studies on magnetism, owing to the well-defined location of the unpaired electron.

Experiment (Zheludev *et al.* 1994; Bonnet *et al.* 1995) and *ab initio* calculations (Novoa *et al.* 1995; Yamanaka *et al.* 1995) agree that in the α-nitronyl aminoxyl unit the spin density in the singly occupied molecular orbital (SOMO) is distributed mainly over the two oxygen and two nitrogen atoms of the ONCNO conjugated system. The central carbon atom of this moiety is a node in the SOMO. However, extrapolating the knowledge of the electronic nature of the molecules to understanding and predicting magnetic behaviours in these organic compounds is an abstruse quandary.

In an effort to rationalize and understand the magnetic properties of these compounds, we seek to exploit the intimate relationship between the structural and electronic characteristics of the molecules. In particular, analysis of the bank of structural data concerning the α-nitronyl aminoxyl radicals in the solid state permits access to a gamut of information which in principle could provide insight into the driving forces that decide the magnetic interactions between the molecules. For this reason, a

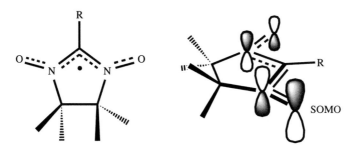

Figure 1. General chemical structure of the family of α-nitronyl aminoxyl radicals and a representation of their singly occupied molecular orbital (SOMO).

data base has been constructed (Cirujeda 1997; Deumal *et al.* 1998*a*) which contains the X-ray structures of the majority of the α-nitronyl aminoxyl radicals whose magnetic properties have been measured. With this tool in hand it is possible to attempt correlation between molecular or supramolecular structures and the bulk properties. Before looking at the crystal architecture, it is instructive to scrutinize the bricks of which it is made. The discussion here will start from the bottom—the molecule—and work up to considering the way in which the units aggregate and interact.

2. Statistical analysis of the geometry of the radicals in the solid state

A data base used for the statistical analysis of the radicals was constructed (Cirujeda 1997) from crystal structures in the Cambridge Crystallographic Structure Data Base as well as from personal communications. It comprised (after removal of structures with disorder or of poor resolution, and of structures whose distances within the imidazolyl ring were greater by three times the standard deviation of the mean) 110 crystal structures incorporating 135 crystallographically independent α-nitronyl aminoxyl units of purely organic radicals and their related coordination compounds.

Firstly, we will only consider the molecular aspects of the imidazolyl ring of the radicals, with the covalently attached oxygen atoms, which carry mainly the unpaired electron, and the pendant methyl groups at the 4 and 5-positions, which protect the radical centre sterically, but may also prove important in the propagation of magnetic interactions. This common unit to all the α-nitronyl aminoxyl radicals in the data base is shown in figure 2 together with the atom numbering which has been employed in the subsequent discussion. The structural correlation was carried out according to the method described by Bürgi & Dunitz (1994). All the interatomic distances have been analysed, along with the angles formed by the covalent bonds between any three of the atoms, as well as the torsion angles present. The results are presented in table 1, and the mean values are presented in figure 3.

In the case of the distances and angles—which show unimodal distributions (figure 4)—the statistical treatment of the unweighted sample mean is trivial, and most of the values are as might be expected based on the known averages for each type of bond (Bürgi & Dunitz 1994). For example, the C4—C5 bond length is 1.545 Å, compared with 1.543 Å for the C_{sp^3}—C_{sp^3} bond in the structure correlation analysis of cyclopentane. The distances between C4 and C5 and their neighbouring nitrogen atoms are 1.499 and 1.500 Å, identical (within the standard deviation) to the overall mean of all C_{sp^3}—N^+ distances.

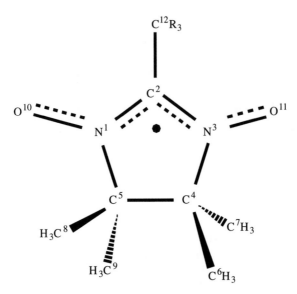

Figure 2. Atom numbering used in the discussion of the statistical analysis of the geometries of the α-nitronyl aminoxyl radicals.

Interestingly, the values of the C—N distances between the two nitrogen atoms in the ring and the carbon atom between them are not statistically different, and have a mean value of approximately 1.344 Å, in accord with the conjugated nature of the system. The distances in imidazole for the C_{sp^2}—N bond (N1—C2) and the C_{sp^2}=N bond (C2=N3) are 1.349 and 1.313 Å, respectively. This observation implies that little of the π-electron density is located in the C—N bond in the α-nitronyl aminoxyl radicals. The N—O distances are statistically identical, being approximately 1.281 Å long, slightly more than the N—O distance in pyridine N-oxides (1.304 Å).

The distances, angles and torsion angles on either 'side' of the α-nitronyl aminoxyl unit are the same within the error involved in the measurement, supporting the notion that the ONCNO group is fully conjugated. Indeed, the torsion angles associated with the conjugated unit are extremely small.

In the case of the few bond angles and the torsion angles which exhibit bimodal distributions, the standard deviation is, as would be expected intuitively, large. It is noteworthy that the smallest standard deviations pertain to the conjugated parts of the molecules, described by four of the torsion angles to be encountered in table 1. The bimodal nature of the torsion angle N1—C5—C4—N3, which describes the puckering of the imidazolyl moiety, is displayed in figure 5 (Minguet *et al.* 1999*a*). Analysis of the puckering angle using a principal component analysis reveals that the conformation of the imidazole moiety can be described to a level of significance of 97% by only *one* of *any* of the torsion angles, for example N1—C5—C4—N3, and all the other angles are defined by this one. This result is caused by the conjugation of three of the five atoms in the ring. Therefore, there are two predominant conformations, the $^5T_4$ and $^4T_5$, which are depicted in figure 5. The statistically most probable angles for the torsion, referred to from now on as T_{IM}, are +24° and −24°, as determined by statistical analysis. The most favoured torsion angles are found between 16 and 32°, and angles greater than the latter are highly unlikely, marking a significant increase in the

Table 1. *Mean distances and angles, along with their standard deviations, in the α-nitronyl aminoxyl radicals in their crystals*

variable	distance (Å) and standard deviation (Å)	variable	angle (deg) and standard deviation (Å)	variable[b]	torsion angle (deg) and standard deviation (Å)
N1–C2	1.340 (0.016)	N1–C2–N3	108.3 (1.2)	N1–C2–N3–C4	−0.5 (8.0)
C2–N3	1.347 (0.015)	C2–N3–C4	112.3 (1.2)	C2–N3–C4–C5	1.5 (18.7)
N3–C4	1.500 (0.013)	N3–C4–C5	101.1 (1.2)	N3–C4–C5–N1	−1.7 (20.4)
C4–C5	1.545 (0.019)	C4–C5–N1	100.8 (1.0)	C4–C5–N1–C2	1.7 (18.6)
C5–N1	1.499 (0.014)	C5–N1–C2	112.7 (1.3)	C5–N1–C2–N3	−0.8 (7.8)
N1–O10	1.285 (0.014)	O10–N1–C2	125.7 (1.2)	O10–N1–C2–N3	0.7 (3.7)
N3–O11	1.278 (0.014)	C2–N3–O11	125.7 (1.2)	N1–C2–N3–O11	0.3 (3.4)
C4–C6	1.521 (0.030)	O11–N3–C4	121.7 (1.2)	O11–N3–C4–C5	−1.3 (13.3)
C4–C7	1.518 (0.033)	C5–N1–O10	121.1 (1.4)	O11–N3–C4–C6	61.0 (15.6)
C5–C8	1.514 (0.034)	N1–C5–C8	107.8 (2.5)[a]	O11–N3–C4–C7	−57.7 (15.9)
C5–C9	1.522 (0.028)	N1–C5–C9	107.5 (2.6)[a]	N3–C4–C5–C8	−117.6 (23.5)
		N3–C4–C6	107.7 (2.5)[a]	N3–C4–C5–C9	113.5 (23.7)
		N3–C4–C7	107.8 (2.5)[a]	C4–C5–N1–O10	−1.64 (12.8)
		C4–C5–C8	114.9 (2.3)[a]	C6–C4–N3–C2	−118.9 (21.1)
		C4–C5–C9	114.5 (2.6)[a]	C6–C4–C5–N1	113.7 (23.4)
		C5–C4–C6	114.4 (2.5)[a]	C6–C4–C5–C8	−2.1 (26.7)
		C5–C4–C7	114.8 (2.6)[a]	C6–C4–C5–C9	−131.1 (26.8)
		C6–C4–C7	110.1 (2.0)	C6–C4–N3–C2	122.4 (21.3)
		C8–C5–C9	110.1 (2.1)	C7–C4–C5–N1	−117.6 (23.6)
				C7–C4–C5–C8	126.6 (26.9)
				C7–C4–C5–C9	−2.4 (26.9)
				C8–C5–N1–C2	122.6 (20.9)
				C8–C5–N1–O10	−57.5 (15.1)
				C9–C5–N1–C2	−118.7 (21.1)
				C9–C5–N1–O10	61.2 (15.3)
				O10–N1–C2–C12	−0.6 (4.6)
				O11–N3–C2–C12	−0.3 (4.4)

[a]These values have bimodal distributions.
[b]All the torsion angles present bimodal distributions.

potential energy curve at this point. In contrast, the slope of the curve towards a planar arrangement is much shallower.

The vast majority of the α-nitronyl aminoxyl radicals which have been prepared and studied to date are of the phenyl variety, with a melange of substituents scattered around the positions of the aromatic ring, whose nature has been determined by commercial or synthetic accessibility to the aldehyde precursors. The introduction of this group brings with it an additional stereochemical aspect to the conformation of the molecules. When the angle between the plane formed by the ONCNO unit and the phenyl substituent, A_{PNN}, is non-zero, and considering that there are two possible enantiomeric conformations for the imidazolyl ring, four diastereomeric conformations exist, comprising two enantiomeric pairs (figure 6). The twist angles can be named

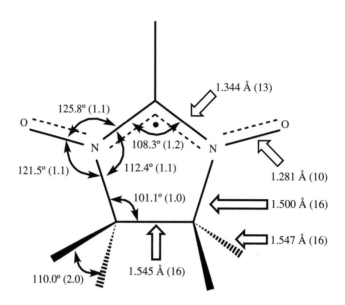

Figure 3. Average bond distances and angles from the statistical analysis of crystals containing
α-nitronyl aminoxyl radicals.

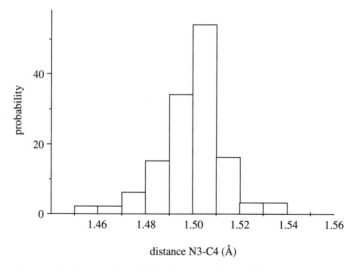

Figure 4. An example of the unimodal distributions obtained for the bond distances, in this case
for the N3—C4 bond.

using the helical denominators for left- and right-handed helices, M and P,
respectively. Therefore, the two enantiomeric pairs are MP and PM and MM and
PP. In the former enantiomers, the twists of the imidazolyl unit and the angle between
the two rings are opposite, and the global conformation is more or less flat, with the
atoms C4 and C5 eclipsed approximately with the plane of the phenyl group. Therefore
the MP and PM enantiomeric conformations are referred to as *pseudo-eclipsed*. The

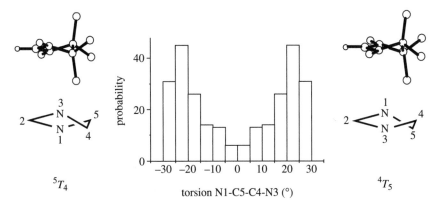

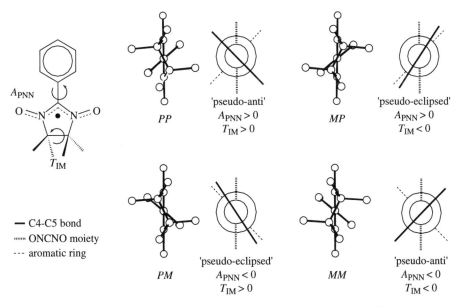

Figure 5. The bimodal distribution obtained from the statistical study for the torsion angle formed by the atoms N1—C5—C4—N3, corresponding to the puckering of the imidazolyl moiety in the radicals, which gives rise to the two enanatiomers depicted.

Figure 6. Definition of the angles A_{PNN} and T_{IM} and their signs in the four diastereomeric conformations that phenyl α-nitronyl aminoxyl radicals can adopt.

remaining diastereomeric conformers have a single-handed twist down the length of the molecule, and are referred to as *pseudo-anti*.

Let as first deal with the angles T_{IM} and A_{PNN} independently. As we have seen, the analysis of the crystal structures shows that the torsion angle formed by the atoms N1, C5, C4 and N3 present in the imidazolyl ring must always be present. Indeed, AM1 semiempirical calculations (Cirujeda 1997) reveal that the planar conformation is a local energy maximum, the minima being located at $+25°$ and $-25°$. It is likely that the interconversion between the two enantiomers is an extremely rapid equilibrium at room temperature in solution, since the barrier to the process is a mere 0.6 kcal mol^{-1}. As for the angle between the imidazolyl and aromatic rings, A_{PNN}, in the crystal structures in the data base the average of its absolute value is 27.5°, with a standard deviation of 8.5°. The angles are located principally between 10 and 45°. The value is surely determined by various factors, among them: (i) the tendency to form a planar conformation to encourage π-electron conjugation; (ii) the presence of a hydrogen bond between the *ortho* hydrogen atoms of the phenyl group and the oxygen atoms of the radical; (iii) the presence of a bulky group in the *ortho* position of the phenyl ring which induces a high angle; (iv) non-covalent interactions between the molecules, i.e. hydrogen bonds; and (v) other crystal packing forces. We shall return to the latter aspect subsequently.

In an effort to detect a relation between T_{IM} and A_{PNN} in the crystals of the phenyl α-nitronyl aminoxyl radicals, we have plotted the calculated probability of the global conformation of the molecules in terms of the two angles, the result of which is shown in figure 7. Clearly, the most favoured conformers are the pseudo-eclipsed ones, *MP* and *PM*, which point to an overall flat conformation. The preference for this latter conformation is statistically significant, since it was ratified in all the statistical tests tried. There is no excess of either of the two enantiomers, since none of the sample contained chiral molecules. Therefore, although some of the molecules crystallize in chiral space groups, there are, in principle at least, an equal number of dextro- and levo-rotatory crystals. It should be pointed out that the introduction of a chiral substituent to the phenyl nitronyl nitroxide induces the preferential formation of only one of the enantiomers of one of the diastereomers (Minguet *et al.* 1999*b*).

One of the important questions related to our analysis of the crystal structures of these radicals is: are the conformations observed in the solid a result of intrinsic preference for one conformation, or are they bound by crystal packing forces? The answer to this question will have an influence on how the crystals of new radicals are forged. In this regard, it is interesting to remark that in our statistical analysis, the conformation with $A_{PNN} = 0$ is clearly disfavoured. However, in the radical 2-phenylbenzimidazol-1-yl N, N′-dioxide (figure 8), in which the –CMe$_2$CMe$_2$– group is replaced by an *ortho* -substituted benzene ring, this angle is only 10.3°, and the molecule, whose crystals have dominantly antiferromagnetic interactions present, has a practically planar shape (Kusaba *et al.* 1997). The absence of a twisting force in the imidazolyl unit in this structure would seem to favour a low A_{PNN} angle, and indicates that the reason for the favoured pseudo-eclipsed geometry is the presence of a molecule with a flat shape in the crystal, in line with the ideas put forward by Kitaigorodskii (1961).

3. Crystal architecture and magnetism–structure correlations

Naturally, the structural study tells us little about the electronic distribution of the molecules in the crystals, for which neutron diffraction studies are necessary (Zheludev *et al.* 1994), which in turn relies on the formation of relatively large single crystals, a requisite that cannot always be fulfilled. An alternative procedure for studying the spin distribution of molecules in crystals is the analysis of solid state ^{1}H and ^{13}C nuclear magnetic resonance spectroscopy (Heisse *et al.* 1999), combined with *ab initio* calculations and with other spectroscopic methods such as UV–visible and electron paramagnetic resonance (EPR) spectroscopies. The latter technique provides the

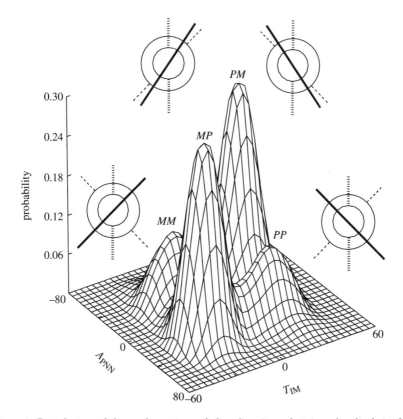

Figure 7. Distribution of the conformations of phenyl α-nitronyl aminoxyl radicals in their crystals, as defined by the angles A_{PNN} and T_{IM}.

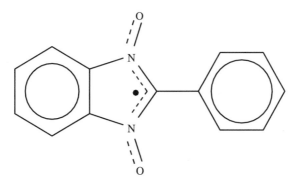

Figure 8. The radical 2-phenylbenzimidazol-1-yl N, N'-dioxide.

interesting opportunity of studying the spin distribution in the isolated molecules in solution. For example, studies of the conformations of phenyl α-nitronyl aminoxyl radicals in solution using UV–vis spectroscopy—aided by applying a linear solvation energy relationship multiparametric analysis—shows that the angle A_{PNN} varies dramatically with the polarity of the solvent (Cirujeda *et al.* 1997). Generally, as the polarity of the solvent increases so does the angle A_{PNN}, an effect which has also been followed by EPR spectroscopy (Cirujeda *et al.* 1999). The spin density of the free electron on the aromatic moiety decreases dramatically as the torsion angle increases. A similar situation might be expected in the solid state, which in turn will influence the interactions between the delocalized free electrons and therefore sway dramatically the bulk properties.

So far, we have seen that it is possible to establish the conformational and electronic structure of α-nitronyl aminoxyl compounds based on a combination of solid-state structural data and solution-state spectroscopic data. However, the identification of the magnetic interaction pathways is elusive. Indeed, perhaps one of the most important unanswered questions remaining in the area of molecular magnetism of systems incorporating organic molecules is: what are the molecular arrangements— what one might call *magnetic synthons* (not necessarily the same as the supramolecular synthons)—which favour ferromagnetic interactions?

Currently, the analyses of interactions based on structural data use mainly the first model described by McConnell (1963), which is based on a Heisenberg spin Hamiltonian and is proposed to explain through space magnetic interactions and the dependence of distance and relative disposition of the orbitals in which the unpaired electron(s) move. Ferromagnetic interactions are predicted only when 'touching' molecules have spin populations of opposite sign close to one another. Therefore, one would predict that if the single occupied molecular orbitals of the oxygen atoms come into close proximity (figure 9), an antiferromagnetic interaction will result, while if the oxygen atom of one molecule approaches one or more of the four methyl groups located above the carbon atom between the two nitrogen atoms of the imidazolyl moiety of another molecule, the interaction will generally be ferromagnetic.

If the scenario painted by the McConnell (1963) I model is correct, one might expect that there would be spatial segregation of ferromagnetic and antiferromagnetic relative dispositions of radical moieties. We employed in collaboration with J. J. Novoa (Universitat de Barcelona) our structural database to test this hypothesis (Deumal *et al.* 1998a, 1999). A group of 47 purely organic α-nitronyl aminoxyl radicals was chosen which showed either dominant ferromagnetic (FM, 23 of them) or dominant antiferromagnetic (AFM, 24 of them) interactions. In order to describe the relative orientations of the ONCNO units in the radicals, six parameters are necessary (figure 10): the closest oxygen–oxygen distance, two N—O···O angles, two C—N—O···O torsion angles and one N—O···O—N torsion angle.

Only considering the closest O···O distances (below 10 Å, of which there were 1312), the shortest distances in the FM and AFM subsets were 3.158 and 3.159 Å, respectively, and is a general state of affairs across the sample. In fact the distribution of subsets is independent of the distance cut-off, be it $3, 4, 5, 6, 7, 8$ or 9 Å. Therefore, short N—O···O—N distances *per se* do not rule out ferromagnetic interactions. At all these distances, the angular distributions are averaged. In principle, one should expect clustering at certain angles for FM or AFM subsets. This situation was not observed. Therefore, given that looking at one distance and one torsion angle does not show division of subsets, the common application of these parameters to describe magnetic interactions must be questioned. Even allowance for hydrogen bonding did not separate the subsets. The situation is even more disparaging when one considers that a factor analysis of all six parameters showed interpenetration of the FM and AFM subsets. So how should the magnetic interactions be evaluated? As only the ONCNO

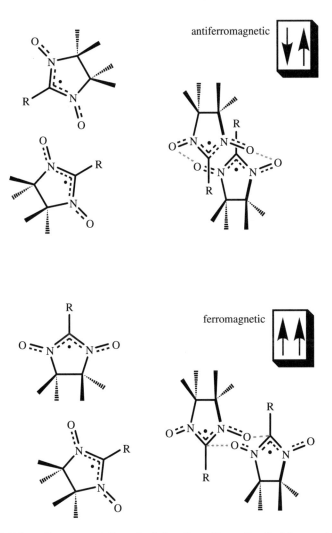

Figure 9. Schematic representation of relative dispositions of radical fragments and the interactions that they would be expected to show according to the McConnell (1963) I theory.

unit was considered in this statistical analysis, it seems likely that consideration of the whole molecule is necessary to describe the magnetic interaction. A small spin density could be combined with a large exchange coupling between the free electrons to produce a significant magnetic interaction. The testing of this hypothesis is being realized presently. Another important issue in this field is the scope and limitations of the McConnell (1963) I model, whose validity has been analysed recently (Deumal *et al.* 1998*b*)

4. Crystal architectures engineered

In the face of the uncertain significance of certain packing arrangements for the propagation of ferromagnetic interactions, a necessity still exists to 'engineer' crystals

R
O
N
N
O
D
A1
A2
T1
T2
T3
O
N
N
R
O

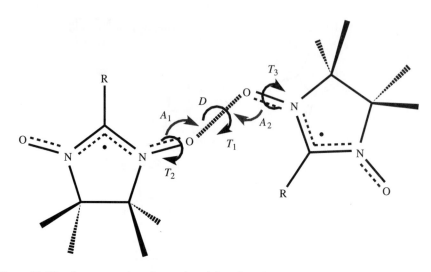

Figure 10. The distances and angles used to define the relative orientations of radical fragments in the crystals of purely organic α-nitronyl aminoxyl radicals.

Csp3-H···O-N
Csp2-H···O-N
RCOOH···O-N
RXH···O-N
X = O, N, S

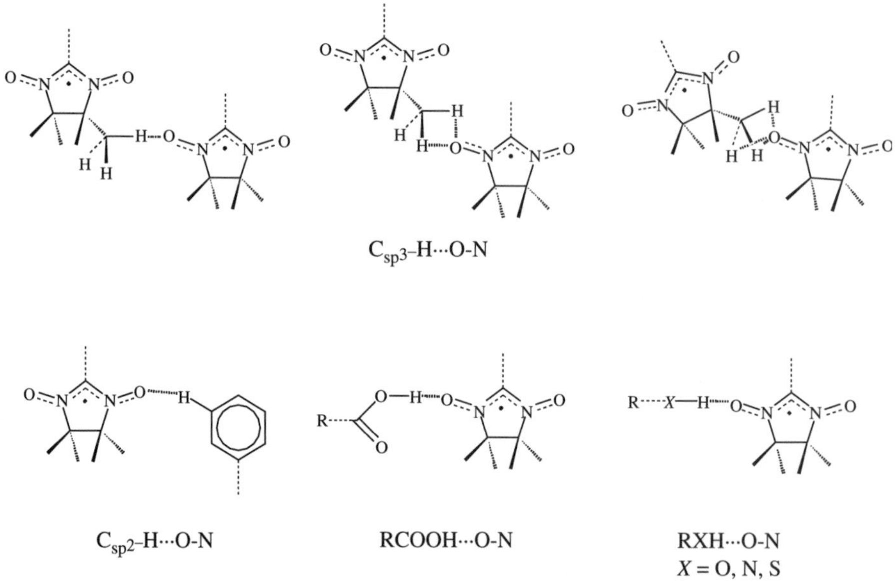

Figure 11. Supramolecular synthons relevant to the crystal design of α-nitronyl aminoxyl radicals.

such that the relative dispositions of the radicals and the dimensionality of the interactions between them is controlled. In order to control relative dispositions, dominion over non-covalent interactions is necessary, and the concept of 'supramolecular synthons' (Desiraju 1995) is a useful one. The α-nitronyl aminoxyl radicals have recurring patterns in their crystal structures which can be regarded as 'supramolecular synthons'. These synthons include (figure 11): C_{sp^3}—H···O hydrogen bonds (Desiraju

1996; Steiner 1997, 1999) from the methyl groups attached to the radical ring to the oxygen atoms bearing the unpaired electron; C_{sp^2}—H$\cdots$O hydrogen bonds from the aromatic ring to the same oxygen atoms; and X—H$\cdots$O hydrogen bonds, where X is an electronegative atom.

The weak C_{sp^3}—H$\cdots$O hydrogen bonds are a frequently encountered cement between the α-nitronyl aminoxyl radicals (Novoa *et al.* 1997; Novoa & Deumal 1997), and may be involved in the propagation of magnetic interactions within them, as has been implied in the case of TEMPO-derived magnets (Nogami *et al.* 1994, 1995b; Maruta *et al.* 1999), most notably in 2OHNN (Cirujeda *et al.* 1995a) in which a three-dimensional network of hydrogen bonds results in crystals which show ferromagnetic ordering. Additional types of weak forces which have been used recently (Jürgens *et al.* 1997) in an attempt to influence crystal packing are C—H$\cdots$Cl (Taylor & Kennard 1982) and Cl$\cdots$Cl interactions. In this case, while the crystal packing was modified, the magnetic interactions were antiferromagnetic in all the examples studied. Stronger intermolecular hydrogen bonds have proven a particularly versatile tool for the preparation of organic molecular magnets, and have also been invoked to explain transmission of spin–spin magnetic interactions (Pontillon *et al.* 1997).

A particularly interesting clan of radicals which are substituted with hydrogen-bond donors are those in which the aromatic moiety has a pendant hydroxyl group. The phenolic hydrogen atom is an extremely good hydrogen-bond donor. For example, the radical 4OHNN (Hernàndez *et al.* 1993; Cirujeda *et al.* 1995b), in which the molecules form chains in which hydrogen bonds are given by the hydroxyl group at the 4-position of the phenyl ring and received by one of the oxygen atoms of the radical moiety (figure 12). In turn, these chains are bonded by two C_{sp^3}—H$\cdots$O hydrogen bonds from diametrically opposite methyl groups in one radical moiety and the remaining NO group of a molecule in the adjacent chain. The resulting two-dimensional sheet is virtually flat. Indeed, the ferromagnetic interactions are quasi-two-dimensional, as revealed by EPR studies on oriented single crystals.

In contrast to its *para*-substituted cousin, the radical 3OHNN forms dimers with itself in the solid state, as depicted schematically in figure 13. These dimers pack together in the crystals in manners which are determined by weaker C—H$\cdots$O hydrogen bonds, forming a herringbone-type pattern two molecules thick (Cirujeda *et al.* 1995a) to give a solid which has dominantly antiferromagnetic interactions between the spins.

It is possible to use the 3OHNN synthon for the preparation of solids with higher structural dimensionality as far as the strong interactions are concerned, as in the case of the dihydroxyphenyl radicals 34OHNN and 35OHNN shown in figure 14. The dimers of the type shown in 3OHNN are also present in the crystals formed by both these radicals. In the case of 34OHNN (Cirujeda *et al.* 1995c) the second OH group forms a hydrogen bond to the NO group not involved in formation of the head-to-tail dimer, so as to generate a chain. In these crystals there are very apparent competing ferro- and antiferromagnetic interactions. Meanwhile, 35OHNN (Cirujeda *et al.* 1996; Matsushita *et al.* 1997) forms one-dimensional chains containing the 3OHNN synthon in polymeric tapes. In these crystals, which show dominant antiferromagnetic interactions, the tapes come together thanks to C—H$\cdots$O hydrogen bonds from the methyl groups in one tape to the phenolic oxygen atoms in another. Remarkably, one of the phases of the related 25OHNN, which also has structures which are replete with hydrogen bonds, is a bulk ferromagnet (Matsushita *et al.* 1997).

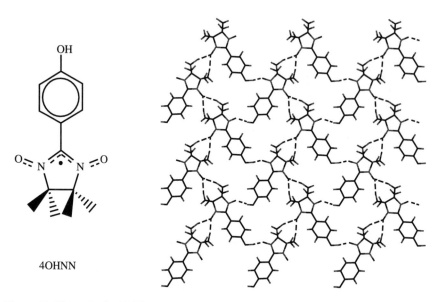

Figure 12. The radical 4OHNN and the two-dimensional sheet which it forms in its crystals.

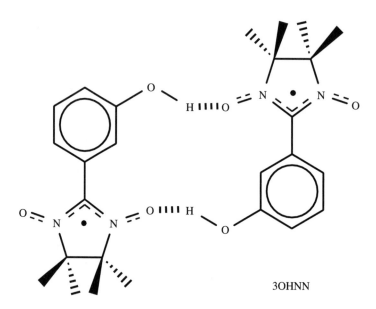

Figure 13. The hydrogen-bonded dimer formed by 3OHNN.

A quite intriguing example of an α-nitronyl aminoxyl radical which is hydrogen bonded in the solid state is that derived from 5-methyl-1,2,4-triazole (5MTNN, Lang *et al.* 1996). The molecules form chains which are sustained by N—H···O hydrogen bonds (shown schematically in figure 15) from N3 of the triazole ring to the nitroxide oxygen atom. The ferromagnetic interactions in the crystals are extremely strong for

34OHNN

35OHNN

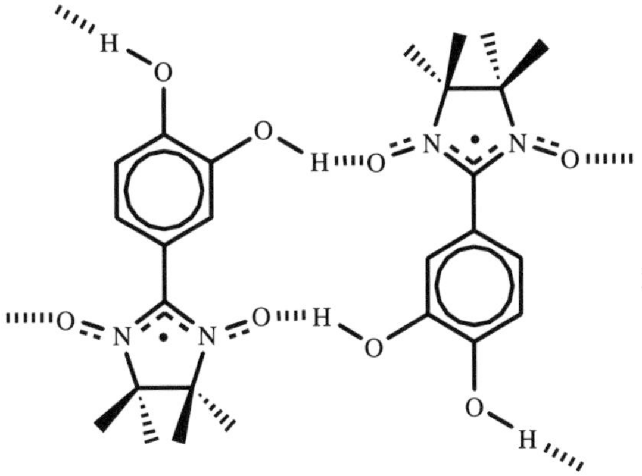

Figure 14. Schematic representation of the packing patterns of the radicals 34OHNN and 35OHNN which contain the supramolecular synthon discovered in 3OHNN.

5MTNN

45MTNN

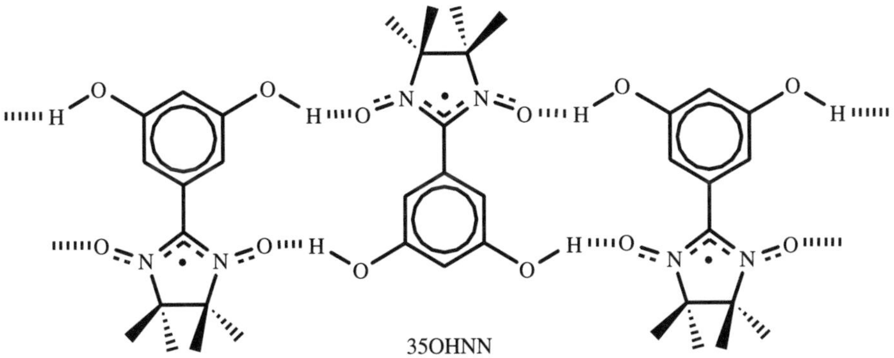

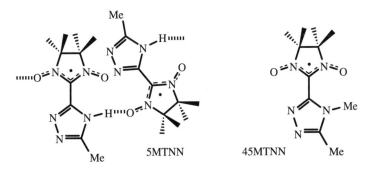

Figure 15. The radical 5MTNN whose crystals contain chains of molecules linked through strong hydrogen bonds and the radical 45MTNN whose packing is determined by weaker intermolecular forces.

an organic molecule, with $J = 14.8\ \mathrm{cm}^{-1}$. The 4,5-dimethyl triazole derivative (45MTNN), which does not have the strong hydrogen bonds between the radicals, shows metamagnetic behaviour (Sutter *et al.* 1997).

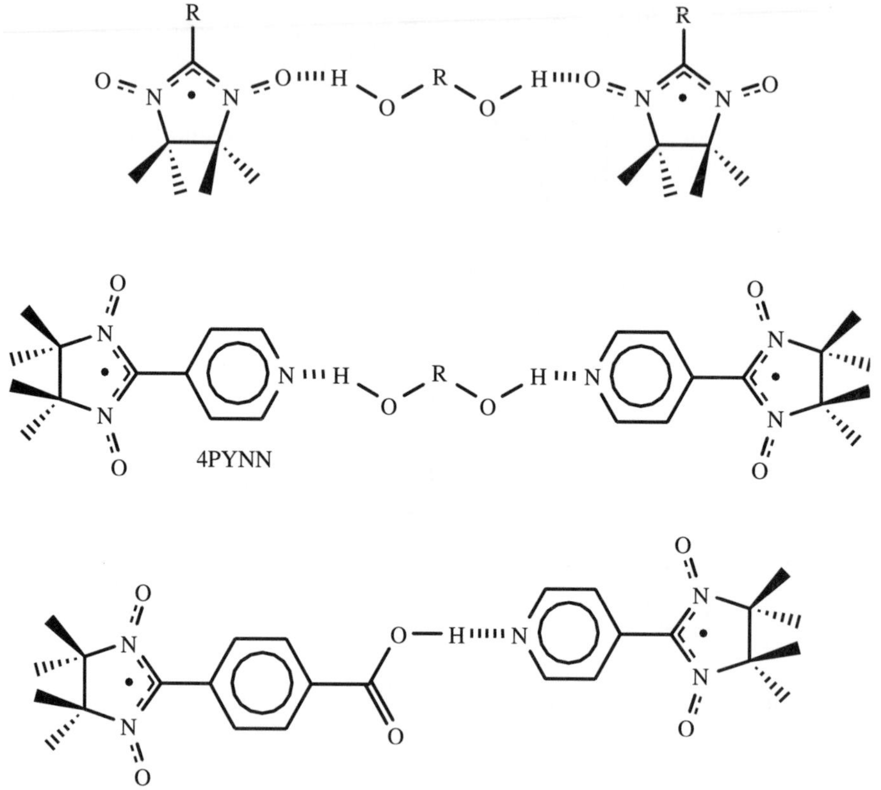

Figure 16. Examples of zero-dimensional trimeric and dimeric aggregates incorporating α-nitronyl aminoxyl radicals.

The discussion has centred so far on radicals which crystallize in a way which is dominated by non-covalent interactions. An attractive possibility which is available to chemists, and which appeals more to the idea of crystal engineering, is to cocrystallize radicals, either with neutral molecules or with other radicals. For example, cocrystallization of 4PYNN with three diacids has produced zero-dimensional 2:1 radical:acid hydrogen-bonded complexes (Otsuka et al. 1998). Similar heteroradical complexes have been prepared by crystallizing 4PYNN or 3PYNN with 4COOHNN (Otsuka et al. 1997). All these crystalline complexes show antiferromagnetic coupling between the radical units. Attempts have also been made to cocrystallize di- or tri-α-nitronyl aminoxyl radicals—which have intramolecular ferromagnetic interactions between free electrons—into stacks with diamagnetic or radical species in order to try and extend the magnetic interactions beyond the molecule (Izuoka et al. 1995). However, the sought after ferro- or ferrimagnetic were not forthcoming, perhaps because of antiferromagnetic interactions between stacks.

One complex which does show ferromagnetic interactions is the one formed between the phenyl α-nitronyl aminoxyl radical and phenylboronic acid, which consists of helical chains comprising alternating components, linked by strong hydrogen bonds between the acid groups of the diamagnetic species and the oxygen atoms of the radical

(Akita *et al.* 1995). At this stage, it is unclear whether the magnetic interaction results from coupling through the boronic acid or by close approaches of radicals between helical chains.

In all these examples, although control has been demonstrated in the formation of the zero- or one-dimensional complexes, the next dimension of crystal does not possess controlling elements, and this situation reflects a general obstacle that exists in the field of crystal engineering. While crystal engineers have achieved interesting results with zero-dimensional dimers, one-dimensional, and a few two-dimensional packing patterns that can be more or less designed, it is at present impossible to contrive *a priori* even simple molecules that will come together in a predetermined way in three dimensions, let alone molecules with a function. In particular, the 'grammar of crystal packing' (Pratt Brock & Dunitz 1994) for the group of compounds discussed here has yet to be deciphered, although, as demonstrated, there are many letters and words! In the realm of molecular magnets, interlayer packing often rules, and holds sway over the nature of the bulk properties.

Perhaps a way forward in unravelling the supramolecular synthons which are appropriate for the propagation of ferromagnetic interactions is to generate small discrete zero- or one-dimensional aggregates in solution and to study their magnetic properties. The radical 3OHNN and some derivatives related to it are presently being studied in collaboration with J. J. Novoa using various techniques to test the viability of this approach.

5. Conclusions and outlook

The analysis of the molecular and supramolecular aspects of α-nitronyl aminoxyl radicals in their crystals using structure correlation methods gives the mean molecular geometries of the molecules, and has allowed a structure–magnetism relationship which does not show any clear-cut regions of ferromagnetic or antiferromagnetic interactions. While the analysis presented in this paper is applied to the family of α-nitronyl aminoxyl radicals, it is also useful in the analysis of other organic radicals. It seems likely that appreciation of interactions between molecules as a whole, and not only the parts which bear mainly the spin, is necessary to account for the magnetic behaviour of the materials. This assertion implies that even small spin densities in certain regions of the molecules can have profound effects on their properties. For instance, hydrogen bonds between the oxygen atoms bearing the majority of the spin density in the radicals to hydrogen atoms with very small spin densities have proven to be very significant in the transmittance of magnetic interactions. The analysis of spin densities in the solution and solid states, and their correlation with structure and properties, will help to confirm or repeal this hypothesis.

We thank all our collaborators whose names are cited in the references and who have contributed to this work over the years, especially Ms Maria Minguet for her recent contribution to the study of conformations in the phenyl α-nitronyl aminoxyl radical derivatives, and also to Professor Juan J. Novoa and Dr Mercè Deumal (Universitat de Barcelona) for their continuous eagerness to perform calculations and valuable comments. We also thank the Fundación Ramón Areces for generous financial support of a project, and to DGES (proyecto PB96-0862-C02-C01), CIRIT (1998 SGR-96-00106) and to the 3MD Network of the EU TMR programme (contract ERBFMRXCT980181).

References

Akita, T., Mazaki, Y. & Kobayashi, K. & Kobayashi, K. 1995 Ferromagnetic spin interaction in a crystalline molecular complex formed by inter-heteromolecular hydrogen bonding: a 1:1 complex of phenyl nitronyl nitroxide radical and phenylboronic acid. *J. Chem. Soc. Chem. Commun.*, pp. 1861–1862.

Banister, A. J., Bricklebank, N., Lavender, I., Rawson, J. M., Gregory, C. I., Tanner, B. K., Clegg, W., Elsegood, M. R. J. & Palacio, F. 1996 Spontaneous magnetisation in a sulphur–nitrogen radical at 36 K. *Angew. Chem. Int. Ed. Engl.* **35**, 2533–2535.

Bonnet, M., Luneau, D., Ressouche, E., Rey, P., Schweizer, J., Wan, M., Wang, H. & Zheludev, A. 1995 The experimental spin density of two nitrophenyl nitroxides: a nitronyl nitroxide and an imino nitroxide. *Mol. Cryst. Liq. Cryst.* **271**, 35–53.

Bürgi, H.-B. & Dunitz, J. D. (eds) 1994 *Structure correlation*, vols 1 and 2. Weinheim: VCH.

Chiarelli, R., Novak, M. A., Rassat, A. & Tholence, J. L. 1993 A ferromagnetic transition at 1.48 K in an organic nitroxide. *Nature* **363**, 147–149.

Cirujeda, J. 1997 La influència dels ponts d'hidrogen sobre l'agregació supramolecular de radicals alfa-nitronil nitròxid en solució i en estat sòlid. Obtenció i estudi de materials magnètics moleculars. PhD thesis, Universitat Ramon Llull, Barcelona.

Cirujeda, J., Mas, M., Molins, E., Lanfranc de Panthou, F., Laugier, J., Park, J. G., Paulsen, C., Rey, P., Rovira, C. & Veciana, J. 1995*a* Control of the structural dimensionalty in hydrogen-bonded self-assemblies of open-shell molecules. Extension of intermolecular ferromagnetic interactions in α-phenyl nitronyl nitroxide radicals into three dimensions. *J. Chem. Soc. Chem. Commun.*, pp. 709–710.

Cirujeda, J., Hernàndez-Gasió, E., Rovira, C., Stanger, J.-L., Turek, P. & Veciana, J. 1995*b* Role of hydrogen bonds in the propogation of ferromagnetic interactions in organic solids. I. The *p*-hydroxyphenyl α-nitronyl aminoxyl radical case. *J. Mater. Chem.* **5**, 243–252.

Cirujeda, J., Ochando, L. E., Amigó, J. M., Rovira, C., Rius, J. & Veciana, J. 1995*c* Structure determination from powder X-ray diffraction of a hydrogen-bonded molecular solid with competing ferromagnetic and antiferromagnetic interactions: the 2-(3,4-dihydroxyphenyl)-α-nitronyl nitroxide radical. *Angew. Chem. Int. Ed. Engl.* **34**, 55–57.

Cirujeda, J., Rovira, C., Stanger, J.-L., Turek, P. & Veciana, J. 1996 The self-assembly of hydroxylated phenyl α-nitronyl nitroxide radicals. In *Magnetism: a supramolecular function* (ed. O. Kahn), pp. 219–248. Dordrecht: Kluwer.

Cirujeda, J., Jürgens, O., Vidal-Gancedo, J., Rovira, C., Turek, P. & Veciana, J. 1997 The influence of chemical surroundings on the properties of α-nitronyl nitroxide radicals in solution. Its relation with the magnetic properties in the solid state. *Mol. Cryst. Liq. Cryst.* **305**, 367–384.

Cirujeda, J., Vidal-Gancedo, J., Jürgens, O., Mota, F., Novoa, J. J., Rovira, C. & Veciana, J. 1999 Spin density distribution of α-nitronyl nitroxide radicals from experimental and *ab initio* calculated ESR isotropic hyperfine coupling constants. (Submitted.)

Deumal, M., Cirujeda, J., Veciana, J. & Novoa, J. J. 1998*a* Structure–magnetism relationships in α-nitronyl nitroxide radicals: pitfalls and lessons to be learned. *Adv. Mater.* **10**, 1461–1466.

Deumal, M., Novoa, J. J., Bearpark, M. J., Celani, P., Olivucci, M. & Robb, M. A. 1998*b* On the validity of the McConnell I model of ferromagnetic interactions: The [2.2]paracyclophane example. *J. Phys. Chem.* A **102**, 8404–8412.

Deumal, M., Cirujeda, J., Veciana, J. & Novoa, J. J. 1999 Structure–magnetism relationships in α-nitronyl nitroxide radicals. *Chem. Eur. J.* **5**, 1631–1642.

Desiraju, G. R. 1995 Supramolecular synthons in crystal engineering: a new organic synthesis. *Angew. Chem. Int. Ed. Engl.* **34**, 2311–2327.

Desiraju, G. R. 1996 The C—H···O hydrogen bond: structural implications and supramolecular design. *Acc. Chem. Res.* **29**, 441–449.

Heisse, H., Köhler, F. H., Mota, F., Novoa, J. J. & Veciana, J. 1999 Determination of the spin distribution in α-nitronyl nitroxides by solid state ^{1}H, ^{2}H and ^{13}C NMR spectroscopy. *J. Am. Chem. Soc.* (In the press.)

Hernàndez, E., Mas, M., Molins, E., Rovira, C. & Veciana, J. 1993 Hydrogen bonds as a crystal design element for organic molecular solids with intermolecular ferromagnetic interactions. *Angew. Chem. Int. Ed. Engl.* **32**, 882–884.

Izuoka, A., Kumai, R. & Sugawara, T. 1995 Crystal designing organic ferrimagnets. *Adv. Mater.* **7**, 672–674.

Jürgens, O., Cirujeda, J., Mas, M., Mata, I., Cabrero, A., Vidal-Gancedo, J., Rovira, C., Molins, E. & Veciana, J. 1997 Magnetostructural study of substituted α-nitronyl aminoxyl radicals with chlorine and hydroxy groups as crystalline design elements. *J. Mater. Chem.* **7**, 1723–1730.

Kitaigorodskii, A. I. 1961 *Organic chemical crystallography*. New York: Consultants Bureau.

Kusaba, Y, Tamura, M., Hosokoshi, Y., Kinoshita, M., Sawa, H., Kato, R. & Kobayashi, H. 1997 Isolation of crystals of a planar nitronyl nitroxide radical: 2-phenylbenzimidazol-1-yl N, N'-dioxide. *J. Mater. Chem.* **7**, 1377–1382.

Lang, A., Pei, Y., Ouahab, L. & Kahn, O. 1996 Synthesis, crystal structure, and magnetic properties of 5-methyl-1,2,4-triazole nitronyl nitroxide: a one-dimensional compound with unusually large ferromagnetic intermolecular interactions. *Adv. Mater.* **8**, 60–62.

McConnell, H. M. 1963 Ferromagnetism in solid free radicals. *J. Chem. Phys.* **39**, 1910.

Maruta, G., Takeda, S., Imachi, R., Ishida, T., Nogami, T. & Yamaguchi, K. 1999 Solid-state high-resolution 1H and 2D NMR study of the electron spin density distribution of the hydrogen-bonded organic ferromagnetic compound 4-hydroxylimino-TEMPO. *J. Am. Chem. Soc.* **121**, 424–431.

Matsushita, M. M., Izuoka, A., Sugawara, T., Kobayashi, T., Wada, N., Takeda, N. & Ishikawa, M. 1997 Hydrogen-bonded organic ferromagnet. *J. Am. Chem. Soc.* **119**, 4369–4379.

Minguet, M., Amabilino, D. B., Cirujeda, J., Wurst, K., Mata, I., Molins, E., Novoa, J. J. & Veciana, J. 1999*a* Stereochemistry in phenyl α-nitronyl nitroxide molecular materials. (Submitted.)

Minguet, M., Amabilino, D. B., Mata, I., Molins, E. & Veciana, J. 1999*b* Crystal engineering and magnetism of hydrogen-bonded phenyl nitronyl nitroxides. *Synth. Met.* **103**, 2253–2256.

Nogami, T., Tomioka, K., Ishida, T., Yoshikawa, H., Yasui, M., Iwasaki, F., Iwamura, H., Takeda, N. & Ishikawa, M. 1994 A new organic ferromagnet: 4-benzylideneamino-2,2,6,6-tetramethylpiperidin-1-oxyl. *Chem. Lett.*, pp. 29–32.

Nogami, T., Togashi, K., Tsuboi, H., Ishida, T., Yoshikawa, H., Yasui, M., Iwasaki, F., Iwamura, H., Takeda, N. & Ishikawa, M. 1995*a* New organic ferromagnets: 4-Arylmethyleneamino-2,2,6,6-tetramethylpiperidin-1-oxyl (aryl = phenyl, 4-biphenylyl, 4-chlorophenyl, and 4-phenoxyphenyl). *Synth. Met.* **71**, 1813–1814.

Nogami, T., Ishida, T., Tsuboi, H., Yoshikawa, H., Yamamoto, H., Yasui, M., Iwasaki, F., Iwamura, H., Takeda, N. & Ishikawa, M. 1995*b* Ferromagnetism in organic radical crystal of 4-(*p*-chlorobenzylideneamino)-2,2,6,6-tetramethylpiperidin-1-oxyl. *Chem. Lett.*, pp. 635–636.

Novoa, J. J. & Deumal, M. 1997 Theoretical analysis of the packing and polymorphism of molecular crystals using quantum mechanical methods: The packing of the 2-hydro nitronyl nitroxide. *Mol. Cryst. Liq. Cryst.* **305**, 143–156.

Novoa, J. J., Mota, F., Veciana, J. & Cirujeda, J. 1995 *Ab initio* computation of the spin population of substituted α-nitronyl nitroxide radicals. *Mol. Cryst. Liq. Cryst.* **271**, 79–90.

Novoa, J. J., Deumal, M., Kinoshita, M., Hosokoshi, Y., Veciana, J. & Cirujeda, J. 1997 A theoretical analysis of the packing and polymorphism of the 2-hydro nitronyl nitroxide crystal. *Mol. Cryst. Liq. Cryst.* **305**, 129–141.

Otsuka, T., Okuno, T., Ohkawa, M., Inabe, T. & Awaga, K. 1997 Hydrogen-bonded acid-base molecular complexes of nitronyl nitroxides. *Mol. Cryst. Liq. Cryst.* **306**, 285–292.

Otsuka, T., Okuno, T., Awaga, K. & Inabe, T. 1998 Crystal structures and magnetic properties of acid-base molecular complexes, (p-pyridyl nitronylnitroxide)2X (X = hydroquinone, fumaric acid and squaric acid). *J. Mater. Chem.* **8**, 1157–1163.

Pontillon, Y., Ressouche, E., Romero, F., Schweizer, J. & Ziessel, R. 1997 Spin density in the free radical NitPy(CCH). *Physica* B **234**–**236**, 788–789.

Pratt Brock, C. & Dunitz, J. D. 1994 Towards a grammar of crystal packing. *Chem. Mater.* **6**, 1118–1127.

Steiner, T. 1997 Unrolling the hdrogen bond properties of C—H⋯O interactions. *Chem. Commun.*, pp. 727–734.

Steiner, T. 1999 Not all short C—H⋯O contacts are hydrogen bonds: the prototypical example of contacts to C=O⁺—H. *Chem. Commun.*, pp. 313–314.

Stephens, P. W., Cox, D., Lauher, J. W., Mihaly, L., Wiley, J. B., Allemand, P.-M., Hirsch, A., Holczer, K., Li, Q., Thompson, J. D. & Wudl, F. 1992 Lattice structure of the fullerene ferromagnet TDAE-C_{60}. *Nature* **355**, 331–332.

Sutter, J. P., Lang, A., Kahn, O., Paulsen, C., Ouahab, L. & Pei, Y. 1997 Ferromagnetic interactions, and metamagnetic behaviour of 4,5-dimethyl-1,2,4-triazole-nitronyl-nitroxide. *J. Magn. Magn. Mater.* **171**, 147–152.

Takahashi, M., Turek, P., Nakazawa, Y., Tamura, M., Nozawa, K., Shiomi, D., Ishikawa, M. & Kinoshita, M. 1991 Discovery of a quasi-1D organic ferromagnet, p-NPNN. *Phys. Rev. Lett.* **67**, 746–748.

Taylor, R. & Kennard, O. 1982 Crystallographic evidence for the existence of C—H⋯O, C—H⋯N and C—H⋯Cl hydrogen bonds. *J. Am. Chem. Soc.* **104**, 5063–5070.

Yamanaka, S., Kawakami, T., Nagao, H. & Yamaguchi, K. 1995 Theoretical studies of spin populations on nitronyl nitroxide, phenyl nitronyl nitroxide and P-NPNN. *Mol. Cryst. Liq. Cryst.* **271**, 19–28.

Zheludev, A., Barone, V., Bonnet, M., Delley, B., Grand, A., Ressouche, E., Rey, P., Subra, R. & Schweizer, J. 1994 Spin density in a nitronyl nitroxide free radical. Polarized neutron diffraction investigation and *ab initio* calculations. *J. Am. Chem. Soc.* **116**, 2019–2027.

Discussion

P. Day (*The Royal Institution, London, UK*). What are the implications of the task of correlation between mean neighbour intermolecular spacing and the magnetic exchange for the mechanism of the exchange?

J. Veciana. The magneto-structural correlation performed with this family of free radicals has important implications relative to the mechanism of magnetic exchange between neighbouring molecules. The fact that no direct relationship exists between the sign of dominant magnetic interactions and the intermolecular distances and disposition of the groups carrying the vast majority of the free electron spin density could be due to two different reasons. Firstly, the McConnell I model might not be valid and could be just an oversimplification of the actual operative mechanism. Secondly, dominant magnetic interactions could originate from the relative disposition of all the magnetically active sites with little or large spin density instead of local contacts between 'touching' molecules. So far, there is no experimental evidence that rules out either of the two options. It is therefore necessary to perform further experimental and theoretical work in order to clarify this issue. This is a crucial point in molecular magnetism since most molecular solids and open-shell molecules are designed under considerations based on the McConnell I model.

Unusual crystal structures and properties of nitronylnitroxide radicals. Possible RVB states in molecule-based magnets

By Kunio Awaga[1,2], Nobuo Wada[1], Isao Watanabe[3]
and Tamotsu Inabe[4]

[1]*Department of Basic Science, Graduate School of Arts and Sciences,
The University of Tokyo, Komaba, Meguro, Tokyo 153-8902, Japan*
[2]*PRESTO, Japan Science and Technology Corporation, Kudan-Minami,
Chiyoda, Tokyo 102-0074, Japan*
[3]*Muon Science Laboratory, The Institute of Physical and Chemical Research
(RIKEN), Hirosawa, Wako, Saitama 351-0198, Japan*
[4]*Division of Chemistry, Graduate School of Science, Hokkaido University,
Sapporo 060-0810, Japan*

We studied two kinds of novel low-dimensional magnetic systems found in the crystals of nitronylnitroxide molecular compounds: Kagome lattice and spin ladder. (i) In the crystals of m-MPYNN $\cdot$ X (m-MPYNN is m-N-methylpyridinium nitronylnitroxide and X = I, BF_4, ClO_4, etc.), m-MPYNN ($S = 1/2$) exists as a ferromagnetic dimer and the dimers form a triangular lattice with a weak interdimer antiferromagnetic coupling that brings about spin frustration. The magnetic system can be regarded as a spin-1 Kagome antiferromagnet at low temperatures. The low-temperature magnetic measurements revealed a spin-gap ground state, which was possibly identical with the RVB state. The temperature dependence of the heat capacity did not show long-range magnetic ordering down to 0.1 K, but short-range ordering caused by the interdimer interaction. The μ^+SR exhibited the temperature-independent depolarization behaviour down to 30 mK, supporting no long-range ordering and a non-magnetic ground state. By replacing the N-methyl group in m-MPYNN with longer alkyl chains, slightly distorted Kagome lattices were obtained, in which the spin gap state collapsed. (ii) The crystal of p-EPYNN $\cdot$ [Ni(dmit)$_2$] (p-EPYNN is p-N-ethylpyridinium nitronylnitroxide and [Ni(dmit)$_2$] is nickel bis(4,5-dithiolato-1,3-dithiole-2-thione)) includes a ladder structure of [Ni(dmit)$_2$] ($S = 1/2$) sandwiched by ferromagnetic chains of p-EPYNN ($S = 1/2$). The magnetic measurements clearly demonstrated a spin gap of *ca.* 1000 K for the molecular spin ladder. Impurity effects on the ladder were studied by making a solid solution system, p-EPYNN $\cdot$ [Ni(dmit)$_2$]$_{1-x}$ $\cdot$ [Au(dmit)$_2$]$_x$ with $0 \leqslant x \leqslant 0.5$. The doping of the non-magnetic [Au(dmit)$_2$] increased Curie defects in the ladder, and resulted in antiferromagnetic behaviour of p-EPYNN at low temperatures.

Keywords: Kagome antiferromagnet; spin ladder; nitronylnitroxide; spin gap

1. Introduction

Magnetic properties of a large number of transition metal complexes, organic radicals and metal–organic radical complexes have been studied so far, and the search for

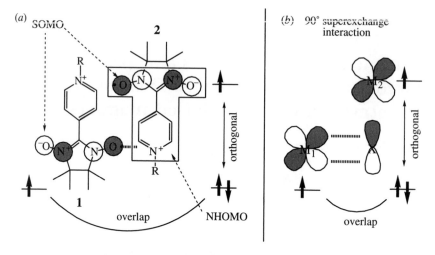

Scheme 1.

molecule-based magnetic materials has intensified in recent years (Day 1993). In this field, a stable organic radical family, nitronylnitroxide, has attracted much interest, because of potential ferromagnetic properties. Various nitronylnitroxide derivatives have been found to exhibit ferromagnetic intermolecular interactions in their bulk crystals (Awaga & Maruyama 1989; Wang *et al.* 1993; Turek *et al.* 1991; Sugano *et al.* 1992; Awaga *et al.* 1992*a*, 1994*a*, *b*; Hernàndez *et al.* 1993; Tamura *et al.* 1993; Panthou *et al.* 1993; Inoue & Iwamura 1993; Sugawara *et al.* 1994; Matsushita *et al.* 1997; Akita *et al.* 1995; Cirujeda *et al.* 1995; Okuno *et al.* 1995) since the discovery of the first pure organic ferromagnet, *p*-nitrophenyl nitronylnitroxide (Kinoshita *et al.* 1991). The nitronylnitroxides have also been known as a bidentate ligand for various transition and rare-earth metal ions. Ferromagnetic ground states have been observed in these complexes (Gatteschi *et al.* 1987; Caneschi *et al.* 1988, 1989; Benelli *et al.* 1993). Besides the ferromagnetic properties, it is notable that the solid-state nitronylnitroxides often exhibit unusual crystal structures.

The electronic structures of the nitronylnitroxides have been examined by means of electron paramagnetic resonance (Davis *et al.* 1972; Takui *et al.* 1995; D'Anna & Wharton 1970), ultraviolet photoelectron spectroscopy (Awaga *et al.* 1993), neutron diffraction (Ressouche *et al.* 1993; Zheludev *et al.* 1994) and NMR (Maruta *et al.* 1999). This radical family possesses a strong spin polarization effect, mainly because of the spatial closeness between the unpaired π electron and the non-bonding electrons (n–π exchange interaction). It is believed that the spin polarization effect stabilizes triplet charge transfer (CT) excited states, and the admixture of these states to the ground state results in a ferromagnetic intermolecular interaction (Awaga *et al.* 1987). This mechanism was originally proposed by McConnell (1967), and the theoretical importance of the spin polarization effect was pointed out by Yamaguchi *et al.* (1986).

While chemical modifications for the substituent at the α-carbon in nitronylnitroxide have been carried out extensively by many groups, we prepared the molecular compounds of pyridyl and *N*-alkylpyridinium nitronylnitroxides, combining them with various acids and anions. The principal aim of this project was to expand the chemistry of nitronylnitroxide and to discover new organic ferromagnets and more unusual crystal structures and properties. Another aim of the *N*-alkylpyridinium nitronylnitroxide cation project was to control the intermolecular arrangement for ferromagnetic interaction. It is known that the oxygen atoms in the nitronylnitroxide moiety possess a large negative polarized charge (Awaga *et al.* 1989). Therefore, the intermolecular arrangement caused by an electrostatic interaction between the oxygen and the pyridinium ring is naturally to be expected, as shown in scheme 1. The singly occupied molecular orbital (SOMO) of nitronylnitroxide is localized on the two NO groups, making a node on the middle α-carbon, and has little population in the aromatic substituent, while the other frontier closed-shell orbitals (NLUMO, NHOMO, etc.) are distributed on both the nitronylnitroxide moiety and the substituent (Awaga *et al.* 1989). Therefore, the distance between the two SOMOs is so long in this geometry that the antiferromagnetic interaction would be suppressed. The contact between the NO group and the pyridinium ring means an intermolecular interaction between the SOMO in molecule **1** and the NHOMO, NLUMO, etc., in **2**. The latter orbitals are orthogonal to the SOMO in *2* because they belong to the same molecule. This situation can be compared with the 90° superexchange interaction that is well known to be ferromagnetic in the field of inorganic chemistry and physics (Anderson 1963). A ferromagnetic interaction is to be expected in the geometry depicted in scheme 1.

The molecular compounds of the pyridyl and pyridinium nitronylnitroxides prepared by our group are summarized in scheme 2. The *N*-protonation of pyridyl nitronylnitroxide was carried out by the reaction with HBr. The obtained materials were (pyridyl nitronylnitroxide)$_2$HBr in which the proton bridged two pyridyl ring with a [NHN]$^+$ hydrogen bond (Okuno *et al.* 1995). We also prepared various acid–base complexes of pyridyl nitronylnitroxide (Otsuka *et al.* 1998, 1997). The *N*-alkylpyridinium nitronylnitroxides were combined with various anions, such as, I$^-$, ClO$_4^-$, BF$_4^-$, MCl$_4^{2-}$ (M = Mn^{2+} ($S = 5/2$) and Co^{2+} ($S = 3/2$)), [Ni(dmit)$_2$]$^-$, [Mn$_{12}$O$_{12}$(O$_2$CPh)$_{16}$(H$_2$O)$_4$]$^-$, etc. (Awaga *et al.* 1992*b*, 1994*a*; Yamaguchi *et al.* 1996; Imai *et al.* 1996; Takeda & Awaga 1997). Besides these works, a pyridinium nitronylnitroxide was used as a counter cation for a Cu^{2+}–Mn^{2+} bimetallic complex anion by Kahn and co-workers (Stump *et al.* 1993). In the course of our study, we obtained quite unusual spin systems: Kagome lattice and spin ladder, which are topics of the research on low-dimensional magnetic materials. In this paper we will describe the crystal structures and properties of the two systems found in the molecular compounds of nitronylnitroxide.

2. Organic Kagome antiferromagnets, *m*-MPYNN · X

Geometrical frustration in antiferromagnetic systems with triangular coordination symmetry is of interest to recent physics. In such a triangle, two nearest neighbours to a given spin are themselves nearest neighbours and antiferromagnetic couplings among them cannot be completely satisfied. The frustration prevents long-range magnetic order from being established and allows novel kinds of low-temperature magnetic states to develop (Fazekas & Anderson 1974; Wen *et al.* 1989; Chandra & Coleman 1991). The Heisenberg Kagome antiferromagnet, whose lattice is shown in scheme 3, is one of the most interesting of the frustrated systems (Syozi 1951). While the number of the nearest neighbours is six in the simple triangular lattice, it is

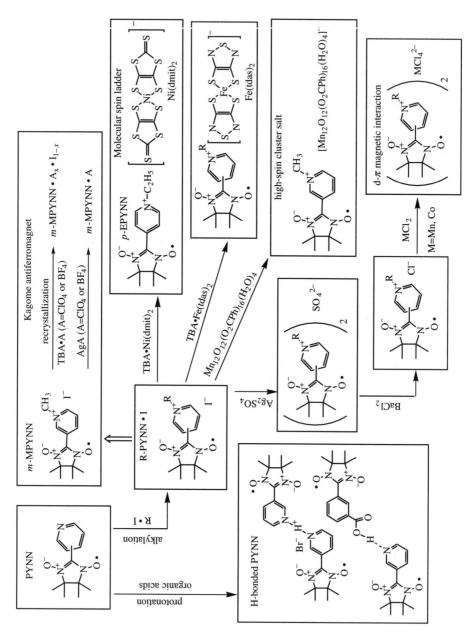

Scheme 2.

decreased to be four in the Kagome lattice. This fact permits more freedom for the alignment of the magnetic moments on the Kagome lattice, so that the Kagome antiferromagnets are predicted to exhibit rich non-trivial ground-state degeneracy. The actual ground state may be governed by subtle effects, such as a quantum

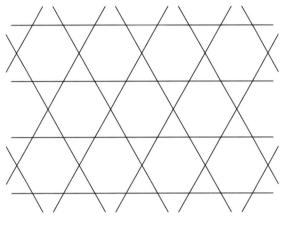

Scheme 3.

effect, a single-ion magnetic anisotropy, next-nearest-neighbour interactions, and so on.

From the viewpoint of materials, the spin-frustrated systems which have been experimentally studied so far are limited in inorganic materials (Ramirez 1991). Among them, the mineral hydronium jarosite (Wills & Harrison 1996) and the two-dimensional solid ^{3}He adsorbed on graphite (Estner *et al.* 1993; Siqueira *et al.* 1997) are known as Kagome magnets.

(*a*) *Unusual crystal structure of* m-MPYNN · X

Recently we found that the crystals of *m*-*N*-methylpyridinium nitronylnitroxide (abbreviated as *m*-MPYNN) involve an antiferromagnetic Kagome lattice (Awaga *et al.* 1992*b*, 1994*a*). As shown in scheme 2, *m*-MPYNN · I was obtained by the reaction of *m*-pyridyl nitronylnitroxide and methyl iodide. Recrystallization of *m*-MPYNN · I with the presence of excess TBA · A (TBA = tetrabutylammonium, and A = BF$_4$, ClO$_4$) gave a crystalline solid solution, *m*-MPYNN · A$_x$ · I$_{1-x}$. The reaction of equivalent amounts of *m*-MPYNN · I and Ag · A resulted in immediate precipitation of Ag · I, leaving iodide-free *m*-MPYNN · A in the solution. Recrystallization of *m*-MPYNN · I, *m*-MPYNN · A$_x$ · I$_{1-x}$ and *m*-MPYNN · A from their acetone solutions resulted in hexagonal-shaped single crystals, including one acetone molecule per three *m*-MPYNN. The crystals of the simple iodide salt, *m*-MPYNN · I · $(\frac{1}{3})$(acetone), were not stable: in the air they immediately turned into mosaic by evaporation of the crystal solvent, but the other crystals including BF$_4^-$ or ClO$_4^-$ were stable.

The structure of *m*-MPYNN · X · $(\frac{1}{3})$(acetone) (X = I, BF$_4$, ClO$_4$, etc.) belongs to a trigonal space group. The *m*-MPYNN molecules exist as a dimer, and the dimer units form a two-dimensional triangular lattice parallel to the *ab*-plane. Figure 1*a* shows a projection of the organic layer of *m*-MPYNN onto the *ab*-plane. The radical dimer is located on each side of the triangles. In other words, the *m*-MPYNN molecules form a bond-alternated hexagonal lattice, as shown schematically in figure 1*b*. In the intradimer arrangement there is a very short intermolecular interatomic distance of less than 3 Å between the NO group and the pyridinium

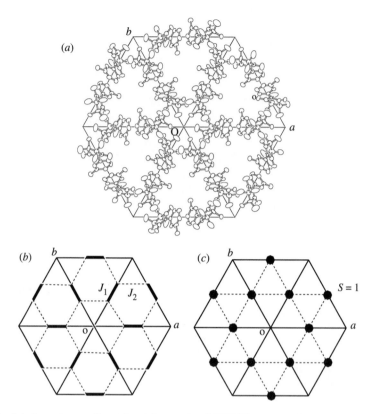

Figure 1. (a) Organic two-dimensional layer of m-MPYNN projected onto the ab-plane. (b) Bond-alternated hexagonal lattice. The parameters, J_1 and J_2, are intradimer and interdimer magnetic interactions, respectively. (c) Kagome lattice.

ring. This short contact is probably caused by an electrostatic interaction between the positive charge on the pyridinium ring and the negative charge polarized on the O atom. As explained before, this is a wanted arrangement in which a ferromagnetic coupling can be expected. In the interdimer arrangement, on the other hand, there is a weak contact between the NO groups. The NO···NO contact means an overlap between SOMOs, which always contributes to antiferromagnetic coupling.

Figure 2 shows a side view of the trigonal lattice, where nine units of the m-MPYNN dimers on the surface of the hexagonal prism are drawn. The unit cell includes two organic layers at the heights of $z = 0$ and $z = \frac{1}{2}$, between which there is a large separation. A third of the anions are in the organic layer, joining the m-MPYNN molecules, and the remainder are between the layers, compensating the excess of positive charge in the organic layers. The crystal solvent, acetone, is located between the organic layers at the centre of the triangle.

(b) Magnetic properties of m-MPYNN · X

X-band EPR measurements were performed on the m-MPYNN · ClO_4 · $(\frac{1}{3})$(acetone) single crystal (Awaga et al. 1994a; Hasegawa et al. 1995). The lineshape at room

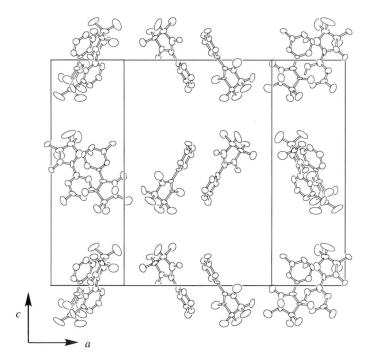

Figure 2. Side view of the organic two-dimensional layers. Nine m-MPYNN dimers on the surface of the hexagonal prism are drawn.

temperature was close to a Lorentzian shape, while it depended slightly on direction of the external field. Whereas one-dimensional magnetic systems exhibit EPR lines with large deviation from the Lorentzian, the deviation of a two-dimensional system is very small (Richards & Salamon 1974). From angular dependence experiments, the principal g values were obtained to be $g_{\parallel} = 2.0060$ and $g_{\perp} = 2.0058$, where $g_{\parallel}$ and $g_{\perp}$ are those values of g which are parallel and perpendicular to the c-axis, respectively. The g-value anisotropy in m-MPYNN $\cdot$ ClO$_4$ $\cdot$ ($\frac{1}{3}$)(acetone) was small, as well as those in most of the organic radicals. We examined the temperature dependence of $g_{\parallel}$ and $g_{\perp}$ and found a g-value shift below 100 K that depended seriously on the crystal shape. It was well explained in terms of the demagnetizing effect expressed by the Kittel's equation for the ferromagnetic resonance (Kittel 1947, 1948).

Figure 3 shows the angular dependence of the peak-to-peak linewidth ΔH_{pp} in the ac (open circles) and ab (closed circles) planes at room temperature. θ is the angle between the external field and the c-axis in the ac-plane and that between the field and the a-axis in the ab-plane, respectively. When the field is in the ab-plane the linewidth shows no dependence on the field direction, but when the field is in the ac-plane it makes minimums at the magic angles. The angular dependence can be well fitted to the theoretical equation (Richards & Salamon 1974)

$$\Delta H_{\text{pp}} = A(3\cos^2\theta - 1)^2 + B, \qquad (2.1)$$

where θ is the angle between the field and the c-axis. The solid curves in figure 3 show the best fits, obtained with $A = 0.094$ mT and $B = 0.259$ mT. Since the EPR

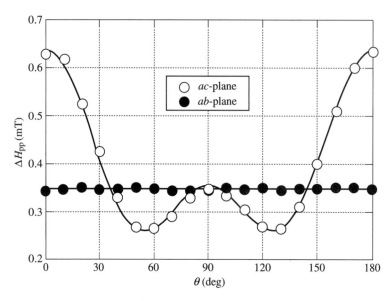

Figure 3. Angular dependence of the peak-to-peak linewidth of the m-MPYNN·ClO$_4$·($\frac{1}{3}$)(acetone) single crystal, in the ac (open circles) and ab (closed circles) planes. θ is the angle between the field and the c-axis in the ac-plane, and that between the field and the a-axis in the ab-plane, respectively.

measurements concluded the two-dimensional Heisenberg-spin character, the magnetic properties of the material would be governed by the two parameters: the intradimer interaction J_1 and the interdimer interaction J_2, shown in figure 1b.

The temperature dependence of the paramagnetic susceptibilities χ_p for the m-MPYNN · BF$_4$ · ($\frac{1}{3}$)(acetone) is shown in figure 4, where $\chi_p T$ is plotted as a function of temperature. The value of $\chi_p T$ increases as the temperature is decreased from room temperature down to $ca.$ 10 K, indicating a ferromagnetic interaction. The intradimer ferromagnetic coupling J_1 is confirmed. After passing through a maximum near 10 K, $\chi_p T$ shows a quick decrease which suggests that the interdimer magnetic interaction J_2 is antiferromagnetic. The observed behaviour is quite consistent with the intradimer and the interdimer molecular arrangements. The observed temperature dependence can be interpreted well in terms of the ferromagnetic J_1 and the antiferromagnetic J_2, using

$$\chi = \frac{4C}{T\{3 + \exp(-2J_1/k_\mathrm{B}\,T)\} - 4J_2/k_\mathrm{B}\,T}, \tag{2.2}$$

where C is the Curie constant and k_B is the Boltzmann constant. The derivation of equation (2.2) is described elsewhere (Awaga *et al.* 1994a). The solid curve in figure 4 is the theoretical best fit to the data, obtained with $J_1/k_\mathrm{B} = 11.6$ K and $J_2/k_\mathrm{B} = -1.6$ K. Below $|J_1|/k_\mathrm{B}$ K, the radical dimer can be regarded as a spin-1 Heisenberg spin which is located at the midpoint of each side of the triangle, as shown in figure 1c. It is expected that the interdimer antiferromagnetic coupling J_2 gives rise to spin frustration among the triplet spins. The spin lattice in figure 1c is exactly coincident with the Kagome lattice. Therefore, the magnetic system in this organic

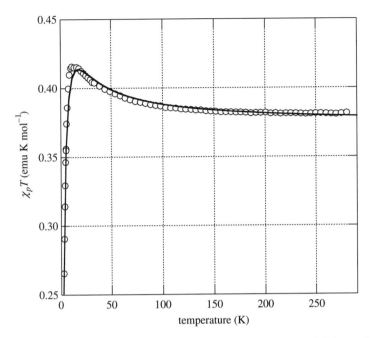

Figure 4. Temperature dependence of the paramagnetic susceptibilities χ_p for
$m\text{-MPYNN} \cdot \text{BF}_4 \cdot (\frac{1}{3})(\text{acetone})$.

material will be characterized as a spin-1 Kagome antiferromagnet in the temperature range below $|J_2|/k_B$ K.

Figure 5 shows the temperature dependence of the AC magnetic susceptibilities χ_{AC} for the oriented single crystals of $m\text{-MPYNN} \cdot \text{BF}_4 \cdot (\frac{1}{3})(\text{acetone})$ below 0.8 K (Wada *et al.* 1997). The magnetic field was parallel to the *c*-axis. The value increases with decreasing temperature down to 0.24 K, and, after passing through a maximum, it approaches to zero at the absolute zero. We have confirmed no magnetic anisotropy in this behaviour. This clearly indicates that the ground state is not an antiferromagnetic ordered state but a spin gap state. In fact the low-temperature data can be well fitted to the gap equation

$$\chi = A\left(\frac{\Delta}{k_B T}\right)^f \exp\left(-\frac{\Delta}{k_B T}\right), \tag{2.3}$$

where A is a constant and Δ is the magnetic gap. The parameter f depends on the density of the excited states against the excitation energy, but is fixed to be unity in the analyses (Bulaevskii 1969). The solid curve in figure 5 is the theoretical one obtained with $A = 0.52$ and $\Delta/k_B = 0.25$ K. The spin gap states have been observed in the spin Peierls systems (Jacobs *et al.* 1976) and the Haldene gap systems (Haldene 1983), although these precedents were one-dimensional magnetic systems. To our knowledge, this is the first example of a spin gap state found in the two-dimensional magnetic materials. It is worth noting here that Anderson predicted the so-called resonating valence bond (RVB) state on a triangular antiferromagnetic lattice, which

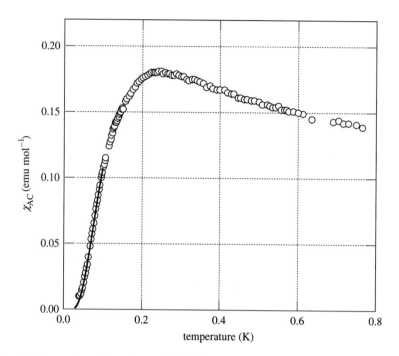

Figure 5. Temperature dependence of the AC magnetic susceptibilities χ_{AC} for the oriented single crystals of m-MPYNN $\cdot$ BF$_4$ $\cdot$ $(\frac{1}{3})$(acetone).

also brought about a spin gap (Anderson 1973). It is possible that the ground state of the material can be characterized in terms of the RVB state.

(c) *Heat capacity of* m-MPYNN $\cdot$ X

The temperature dependence of the heat capacities c_p was examined down to 0.12 K on m-MPYNN $\cdot$ BF$_4$ $\cdot$ $(\frac{1}{3})$(acetone) (Wada *et al.* 1997). The results below 3 K are shown in figure 6. The value of c_p gradually increases with a decrease in temperature. After making a broad maximum at 1.4 K, c_p show a decrease. Below 0.24 K, where χ_{AC} shows the spin gap ground state, the value of c_p increases again. The temperature of the maximum c_p, namely 1.4 K, almost agrees with $|J_2|/k_B$. Monte Carlo calculation indicates that the heat capacity of the spin-1/2 Kagome Heisenberg antiferromagnet exhibits a maximum of short-range magnetic ordering at 1.4$|J|/k_B$ (Nakamura & Miyashita 1995). The observed anomaly at 1.4 K is probably due to the short-range ordering made by J_2. The reason for the increase in c_p below 0.24 K is not clear, but it suggests another anomaly below 0.1 K. It is notable that there is no signal of long-range ordering in the temperature range down to 0.1 K, which corresponds to 8% of $|J_2|/k_B$. This supports the spin frustration in this magnetic system.

The plots of c_p/T show a gradual increase as the temperature is decreased down to 0.12 K (not shown). We calculated the entropy change accompanied with the anomaly at 1.4 K to be $\Delta S = 4.7$ J K^{-1} mol^{-1}, by using the data above 0.12 K and by subtracting the contributions of the lattice and the excited state due to J_1. Since

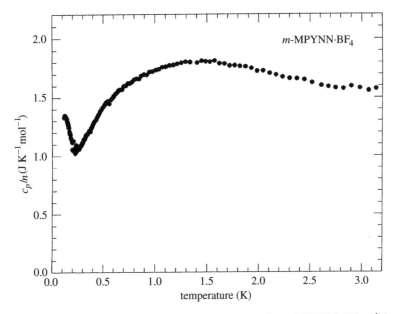

Figure 6. Temperature dependence of the heat capacities c_p for m-MPYNN $\cdot$ BF$_4$ $\cdot$ ($\frac{1}{3}$)(acetone).

the triplet spins on the m-MPYNN dimers lose the magnetic freedom in the short-range ordering, the magnetic entropy is theoretically calculated to be $S_{\mathrm{m}} = (\frac{1}{2}R)\ln 3 = 4.567$ J K^{-1} mol^{-1}. The observed value of ΔS is already larger than S_{m}, despite the fact that ΔS was obtained using the data above 0.12 K. This means that there is an unknown degree of freedom besides the magnetic degree of freedom, which cooperates with the short-range magnetic ordering in this ultralow-temperature range.

(d) μ^+SR of m-MPYNN $\cdot$ X

We carried out muon spin relaxation on m-MPYNN $\cdot$ BF$_4$ $\cdot$ ($\frac{1}{3}$)(acetone) in the temperature range down to 30 mK, to clarify whether magnetic transitions exist or not, and to confirm the non-magnetic ground state. Positive muon spin relaxation (μ^+SR) is a good microscopic probe to sense such a magnetic state of the system. A muon spin is completely polarized along a beam direction even in the zero-field (ZF) condition and depolarized after the stop at a potential-minimum position in the crystal of m-MPYNN $\cdot$ BF$_4$ interacting with a local field at a muon site (Uemura *et al.* 1985). A long-range or a short-range ordering of the dimer spins can be recognized as a change of the depolarization behaviour of the muon spin, because a static or a dynamically fluctuating component of the internal field which is accompanied by the magnetic transition affects strongly the muon spin polarization (Uemura *et al.* 1985; Mekata 1990).

Figure 7 shows ZF-μ^+SR time spectra obtained at 265 K, 100 K, 2.9 K and 30 mK. In this figure the asymmetry parameter of the muon spin at time t, $A(t)$, is defined as $[F(t) - B(t)]/[F(t) + B(t)]$, where $F(t)$ and $B(t)$ are muon events counted by the forward and backward counters, respectively. The asymmetry at each temperature is

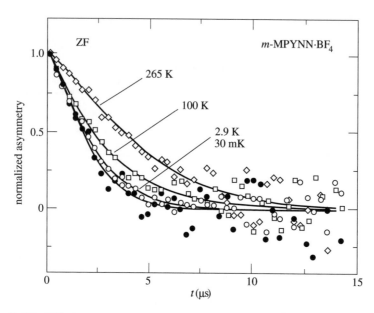

Figure 7. ZF-μ^+SR time spectra obtained for m-MPYNN $\cdot$ BF$_4$ $\cdot$ ($\frac{1}{3}$)(acetone) at 265 K, 100 K, 2.9 K and 30 mK.

normalized to be one at $t = 0$, to compare the difference of the depolarization behaviour. The depolarization behaviour cannot be described by either a simple Gaussian function or a Lorentzian function. For convenience sake, the obtained ZF-μ^+SR spectra were analysed by a power function, $A_0 \exp(-\lambda t)^\beta$, where A_0 is the initial asymmetry at $t = 0$ and λ is the depolarization rate. The solid curves in figure 7 are the best-fit results using this power function.

The temperature dependence of A_0, λ and β was obtained from the best-fit analysis. All parameters show the temperature independence below about 100 K, showing that static and dynamical properties of the local field at the muon site are temperature independent. This fact is different from other types of Kagome magnets in which strong enhancement of the depolarization rate by the critical slowing-down behaviour of magnetic moments is observed around a magnetic transition temperature (Keren et al. 1994, 1996; Uemura et al. 1994; Dunsiger et al. 1996). The parameter λ shows a slight decrease above 150 K, which is probably due to the motional narrowing effect indicating that the muon starts to diffuse through the crystal. From longitudinal-field dependence of the μ^+SR time spectra (not shown), a half-width of the distribution of the static internal field at the muon site ΔH was estimated to be 10 ± 2 G.

It is known that the muon implanted into a crystal which contains F$^-$ ions forms the strong FμF state through a hydrogen bonding (Brewer et al. 1986). In this case, the distance between the F$^-$ ion and the muon is similar to a nominal F$^-$ ionic radius of 1.16 Å. Assuming that the distance between the stopped muon and the ^{19}F nucleus in m-MPYNN $\cdot$ BF$_4$ is the nominal F$^-$ ionic radius, the dipole field of the ^{19}F nucleus at the muon site is estimated to be about 8.5 G. This value is comparable to the obtained ΔH. Although the reason for the missing of the muon spin precession which has been observed in other fluorides (Brewer et al. 1986) is still unclear, it can be concluded that the implanted muon is expected to stop near the F$^-$ ion forming the

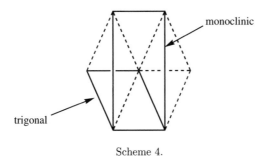

Scheme 4.

Table 1. *Cell and magnetic parameters for m-R-PYNN · I*

	m-MPYNN · I trigonal	*m*-EPYNN · I monoclinic	*m*-PPYNN · I monoclinic	*m*-BPYNN · I monoclinic
a (Å)	15.876(5)	16.14(6)	28.153(9)	9.794(2)
b (Å)		28.07(8)	16.407(14)	8.53(2)
c (Å)	23.583(6)	24.02(5)	24.705(11)	11.312(2)
β (deg)		91.1(2)	90.25(3)	104.19(2)
V (Å^3)	5147(3)	10 876(53)	11 411(11)	917(2)
Z	12	24	24	2
J_1/k_B	10.2	9.6	6.6	0.9
J_2/k_B	−1.6	−1.7	−1.0	−0.7

hydrogen bonding and that the static internal field at the muon site originates from the ^{19}F nuclear dipole field.

In conclusion, the temperature independent depolarization behaviour which is due to the distributed static internal field induced by the ^{19}F nuclear dipoles at the muon site was observed down to 30 mK. The width of the field distribution was 10 ± 2 G. No clear long-range magnetic ordering of the dimer spins was observed. Taking into account the results of the magnetic susceptibility measurements, the ground state of m-MPYNN · BF$_4$ is concluded to be non-magnetic.

(e) *Magnetic properties of distorted Kagome lattices*

We have already studied the crystal structures and magnetic properties of the *para*-R-PYNN · I series, changing the length of the N-alkyl chain (Awaga *et al.* 1994*b*). Depending on the length, various crystal structures and magnetism were found. The chemical modification was a powerful tool to investigate the magnetostructural correlation in the crystals of the R-PYNN series. In this work, we prepared the iodide salts of the *meta*-R-PYNN series, where R = ethyl (E), n-propyl (P) and n-butyl (B), and compared their crystal structures and magnetic properties with those of m-MPYNN · I.

The materials were obtained by the same procedure as for m-MPYNN · I. The recrystallizations of m-EPYNN · I and m-PPYNN · I from their acetone or acetone/benzene solutions resulted in hexagonal-shaped single crystals, while m-BPYNN · I gave the block-shape single crystals. The elemental analyses indicated the following chemical formulae: m-EPYNN · I · 1.5H$_2$O, m-PPYNN · I · 0.5H$_2$O and

(*a*)   (*b*)

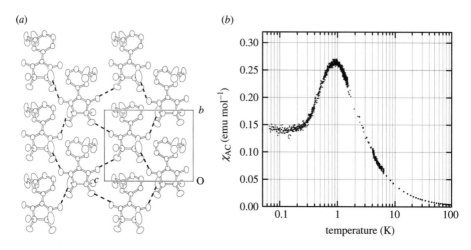

Figure 13. (*a*) A projection of the structure of *p*-BPYNN · I along the monoclinic *a*-axis. (*b*) Temperature dependence of the AC magnetic susceptibilities χ_{AC} for the *m*-BPYNN · I.

m-BPYNN · I. Table 1 shows the unit cell dimensions of the *m*-*R*-PYNN · I series, determined using X-ray diffraction data in the range $20° < 2\theta < 25°$. While the crystal of *m*-MPYNN · I belongs to the trigonal system, *m*-EPYNN · I, *m*-PPYNN · I and *m*-BPYNN · I crystallize into the monoclinic systems. However, it was found that lattice transformations for *m*-EPYNN · I and *m*-PPYNN · I lead to cell dimensions which were quite similar to those of *m*-MPYNN · I. Schematic comparison between the trigonal and monoclinic cells is shown in scheme 4. The transformed lattice constants are $a = 16.14(6)$ Å, $b = 16.19(4)$ Å, $c = 24.02(6)$ Å, $\alpha = 89.5(2)°$, $\beta = 90.1(2)°$, $\gamma = 119.9(1)°$, $V = 5440(53)$ Å^3 for *m*-EPYNN · I and $a = 16.30(1)$ Å, $b = 16.30(1)$ Å, $c = 24.71(1)$ Å, $\alpha = 89.7(4)°$, $\beta = 90.1(4)°$, $\gamma = 119.5(3)°$, $V = 5709(5)$ Å^3 for *m*-PPYNN · I. The obtained lattice parameters, *a*, *b* and *c*, and *V*, are slightly larger than the corresponding ones of *m*-MPYNN · I, but the differences between them are very small. Although we could not finish full structural analyses for *m*-EPYNN · I and *m*-PPYNN · I, probably because of positional disorders of the iodide ion and the crystal solvent, it is expected that they have a slightly distorted Kagome lattice. The decrease in crystal symmetry means distortion of the equilateral triangle to an isosceles triangle. This will significantly affect the low-temperature magnetic properties, as described later.

    m-BPYNN · I crystallizes into a completely different structure. Figure 8*a* shows a projection of the structure along the monoclinic *a*-axis. The *m*-BPYNN molecules form a two-dimensional network caused by the contacts between the NO group and the methyl group in the nitronylnitroxide moiety. The hydrogen atoms in the methyl group, namely β-H atoms, have negative spin densities. It is believed that such contacts between the NO group and β-H atoms brings about a ferromagnetic interaction (Togashi *et al.* 1996).

    Figure 9 shows the temperature dependence of $\chi_p T$ for the *m*-*R*-PYNN · I series. The plots for *m*-MPYNN · I, *m*-EPYNN · I and *m*-PPYNN · I show quite similar temperature dependence: the value of $\chi_p T$ increases, as temperature is decreased from room temperature down to *ca.* 10 K. After passing through a maximum, $\chi_p T$ shows a

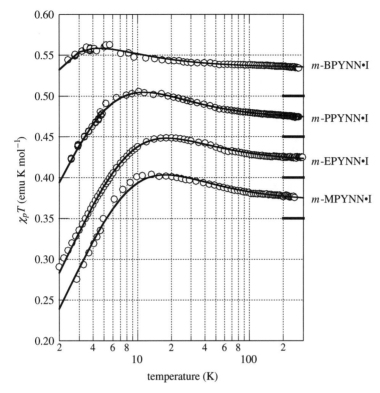

Figure 9. Temperature dependence of the paramagnetic susceptibilities χ_p for the m-R-PYNN $\cdot$ I series.

quick decrease. Their temperature dependences are well explained by equation (2.1). Exactly speaking, m-EPYNN $\cdot$ I and m-PPYNN $\cdot$ I include two kinds of J_2, because of the distortion of the Kagome lattice in them. However, the distortion is so small that we ignore the difference. The solid curves going through the plots for m-MPYNN $\cdot$ I, m-EPYNN $\cdot$ I and m-PPYNN $\cdot$ I in figure 9 are the theoretical best fits to the data, obtained with the parameters listed in table 1. The values of J_1 and J_2 systematically decrease with the extension of the N-alkyl chain. Presumably this is caused by the expansion of the two-dimensional lattice.

The value of $\chi_p T$ for m-BPYNN $\cdot$ I shows a slight increase with a decrease in temperature down to *ca.* 5 K, and it decreases after passing through a maximum. This behaviour indicates coexistence of a stronger ferromagnetic interaction and a weaker antiferromagnetic coupling between the ferromagnetic units, but their intensities seem smaller than those in the other three salts. The crystal of m-BPYNN $\cdot$ I consists of the two-dimensional network shown in figure 8a. Assuming a ferromagnetic intralayer interaction J_1 and an antiferromagnetic interlayer interaction J_2, the temperature dependence was fitted to the equation

$$\chi = \frac{\chi'}{1 - \dfrac{2zJ_2}{N_A g\mu_B}\chi'} \qquad (2.4)$$

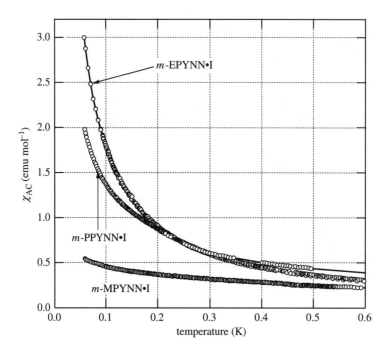

Figure 10. Temperature dependence of the AC susceptibilities χ_{AC} for m-MPYNN $\cdot$ I, m-EPYNN $\cdot$ I and m-PPYNN $\cdot$ I.

with

$$\chi' = \frac{C}{T}(1 + 2K + 2K^2 + \tfrac{4}{3}K^3 + \tfrac{13}{12}K^4) \quad \text{and} \quad K = \frac{J_1}{k_B T},$$

where z is the number of neighbouring layers and is equal to 2 in this case. The theoretical best fit was obtained with the parameters listed in table 1.

The temperature dependence of the AC susceptibilities χ_{AC} for m-MPYNN $\cdot$ I, m-EPYNN $\cdot$ I and m-PPYNN $\cdot$ I is shown in figure 10. Although m-MPYNN $\cdot$ I includes the regular antiferromagnetic Kagome lattice, as well as m-MPYNN $\cdot$ BF$_4$, the value of χ_{AC} continues to increase down to 0.05 K without showing the spin gap state. Probably this is caused by the fact that the salt is unstable because of evaporation of the crystal solvent. In fact the dependence can be explained by the equation

$$\chi = A\left(\frac{\Delta}{k_B T}\right)^f \exp\left(-\frac{\Delta}{k_B T}\right) + \frac{C_{\text{def}}}{T}, \tag{2.5}$$

where the first term is the same as equation (2.2) and the second term is for the Curie contribution of lattice defects. Fitting the data to equation (2.5), the parameters are obtained as follows: $A = 0.64$, $\Delta/k_B = 0.25$ K and $C_{\text{def}} = 0.025$ emu K mol^{-1} (6.6%). The values of A and Δ are very close to the corresponding parameters for m-MPYNN $\cdot$ BF$_4$. The distorted Kagome materials, m-EPYNN $\cdot$ I and m-PPYNN $\cdot$ I, show a similar temperature dependence, but their values of χ_{AC} below 0.2 K are approximately five times larger than that of m-MPYNN $\cdot$ I. It seems hard to attribute them to the contribution of lattice defects. It is considered that their behaviour is intrinsic and the spin gap state is easily collapsed by the small distortion of the Kagome lattice.

The temperature dependence of χ_{AC} for m-BPYNN $\cdot$ I is shown in figure 8b. The value increases with decreasing temperature down to 0.9 K and, after passing through a maximum, it becomes nearly two-thirds of the maximum value at the lowest temperature. The heat capacity makes a maximum at 0.56 K (not shown). The magnetic and thermal behaviour strongly indicate an antiferromagnetic transition at 0.56 K. The antiferromagnetic transition is rationalized, because of the intralayer ferromagnetic interaction and the interlayer antiferromagnetic interaction. It is interesting that the m-BPYNN $\cdot$ I exhibits the highest transition temperature in the series, in spite of the fact that the magnetic interactions in it are weaker than those in the other three salts. This suggests that the magnetic properties of the Kagome and Kagome-like materials are governed by spin-frustration and that the frustration on the Kagome lattice prevents usual antiferromagnetic ordering.

(*f*) *Summary*

The crystal structures and magnetic properties of the m-R-PYNN $\cdot$ X series were studied. In the crystal of m-MPYNN $\cdot$ X, the ferromagnetic dimers formed the triangular lattice with the weak interdimer antiferromagnetic coupling. The magnetic system can be regarded as a spin-1 Kagome antiferromagnet at low temperatures. The single-crystal EPR concluded the two-dimensional Heisenberg character of the spin system. The low-temperature magnetic behaviour indicated the spin-gap ground state, which was possibly identical with the RVB state. The temperature dependence of the heat capacity showed the short-range magnetic ordering caused by the interdimer antiferromagnetic interaction, but did no long-range ordering down to 0.1 K. In addition, it suggested an unknown degree of freedom which cooperated with the short-range magnetic ordering. The μ^+SR exhibited the temperature-independent depolarization behaviour down to 30 mK. In other words, this study strongly supports no long-range magnetic ordering of the dimer spins and the non-magnetic ground state. With an extension of the N-alkyl chain in the m-R-PYNN $\cdot$ I series, distortion of the Kagome lattice took place, which resulted in collapse of the spin-gap ground state.

3. Molecular spin ladder, p-EPYNN $\cdot$ [Ni(dmit)$_2$]

There is an increasing interest in preparation of molecule-based materials which show multifunctional properties (Day 1993). One of the promising routes to the multi-functional materials is to synthesize molecular complexes consisting of two components with different physical properties. N-alkylpyridinium nitronylnitroxides are useful as a component of ferromagnetic property because they often exhibit ferromagnetic intermolecular interactions (Awaga *et al.* 1994b, 1992b). Nickel bis(4,5-dithiolato-1,3-dithiole-2-thione), abbreviated as [Ni(dmit)$_2$], is an electron acceptor molecule. The centric Ni ion can take three kinds of valence states: $+2$, $+3$ and $+4$, and the molecule shows corresponding oxidation states, namely [Ni(dmit)$_2$]$^{2-}$, [Ni(dmit)$_2$]$^-$ and [Ni(dmit)$_2$]^0, respectively. In particular, the mixed-valence state between [Ni(dmit)$_2$]$^-$ and [Ni(dmit)$_2$]^0, which can be achieved by an electrochemical method, attracts much interest because it often shows metallic conductivity and

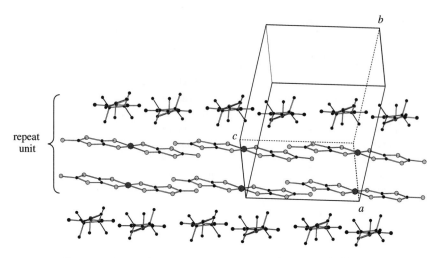

Figure 11. View of the crystal structure of p-EPYNN $\cdot$ [Ni(dmit)$_2$].

superconductivity (Brossard *et al.* 1986, 1988, 1989; Kobayashi *et al.* 1987, 1991, 1992; Tajima *et al.* 1993).

In this work, we studied the complexes of N-alkylpyridinium nitronylnitroxide cations and [M(dmit)$_2$] (M = Ni and Au) anions. While the aim of this research was to combine the magnetic property of nitronylnitroxide and the conductive property of [M(dmit)$_2$] and to obtain multifunctional molecule-based materials, we found a ladder structure of [Ni(dmit)$_2$] in the crystal of p-EPYNN $\cdot$ [Ni(dmit)$_2$], where p-EPYNN is p-N-ethylpyridinium nitronylnitroxide. After comparing the structures and magnetic properties of p-EPYNN $\cdot$ [Ni(dmit)$_2$] and p-EPYNN $\cdot$ [Au(dmit)$_2$], we will report the impurity effects on the molecular spin ladder.

(a) Crystal structure and magnetic properties of p-EPYNN $\cdot$ [Ni(dmit)$_2$]

The molecular complex, p-EPYNN $\cdot$ [Ni(dmit)$_2$], was obtained by a double-decomposition reaction between p-EPYNN $\cdot$ I and TBA $\cdot$ [Ni(dmit)$_2$] (Imai *et al.* 1996). Figure 11 shows a view of the crystal structure. Figure 12a represents the structure of p-EPYNN, in which slightly non-planar p-EPYNN cation radicals form an alternating chain along the c-axis with a side-by-side head-to-head arrangement. This chain is formed by short contacts between the nitronylnitroxide oxygen atom and the pyridinium ring, which are probably due to an electrostatic interaction. As explained in the previous section, a ferromagnetic coupling can be expected in the structure. Figure 12b shows the structure of [Ni(dmit)$_2$], in which [Ni(dmit)$_2$] exists as a face-to-face dimer. Further, the dimers are stacked along the c-axis with a translational relation, forming a ladder structure. There are unusually short S$\cdots$S contacts in the interdimer arrangement. Since each [Ni(dmit)$_2$] is a doublet spin species, this structure can be regarded as a two-leg spin ladder. To our knowledge, this is the first example of a spin ladder found in a molecule-based material. As shown in figure 11, there is an alternation of the chain of p-EPYNN and the ladder of [Ni(dmit)$_2$] along the b-axis. In the arrangement between p-EPYNN and [Ni(dmit)$_2$], one [Ni(dmit)$_2$] molecule interacts with two pyridinium rings of p-EPYNN. Since the unpaired electron is

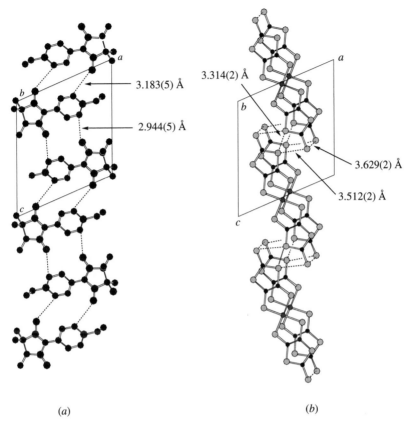

Figure 12. Projections of the structures of (*a*) *p*-EPYNN and (*b*) [Ni(dmit)$_2$] in *p*-EPYNN · [Ni(dmit)$_2$] along the *b*-axis.

localized on the NO groups in the nitronylnitroxide, the magnetic interaction between the cation and the anion seems weak.

The temperature dependence of $\chi_p T$ for *p*-EPYNN · [Ni(dmit)$_2$] is depicted in figure 13 (Imai *et al.* 1996). With an increase in temperature from 2 K, the value of $\chi_p T$ quickly decreases and, after passing through a plateau in the temperature range 40–150 K, it gradually increases. This behaviour can be well understood by taking into account two contributions: the dark grey part is from the chain of *p*-EPYNN, which shows a weak ferromagnetic property; the light grey part is from [Ni(dmit)$_2$], which exhibits an activation-type susceptibility. The susceptibility data are analysed using the following equation:

$$
\begin{aligned}
\chi(T) = {} & \frac{2C_1}{\sqrt{\pi(a/k_B)\,T}}\,\mathrm{e}^{-\Delta/k_B T} \\
& + \frac{C_2}{T}\left(\frac{1 + A_1 K + A_2 K^2 + A_3 K^3 + A_4 K^4 + A_5 K^5}{1 + A_6 K + A_7 K^2 + A_8 K^3}\right)^{2/3},
\end{aligned}
\tag{3.1}
$$

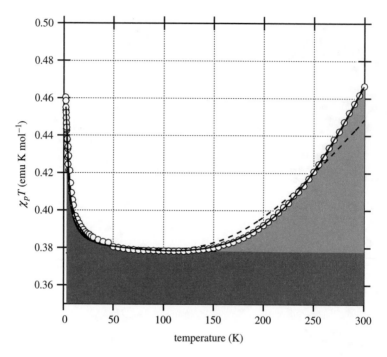

Figure 13. Temperature dependence of the paramagnetic susceptibilities χ_p for p-EPYNN · [Ni(dmit)$_2$].

with

$$K = \frac{J_{\mathrm{F}}}{k_{\mathrm{B}}\,T},$$

where the first term expresses the contribution of the spin ladder chain and the second term expresses the contribution of the one-dimensional ferromagnetic chain. The parameter Δ is the magnetic gap, $a = \frac{1}{2}[\partial^2\varepsilon(k)/\partial k^2]_{k=\pi}$, where $\varepsilon(k)$ is the energy dispersion of the spin wave excitation in the spin ladder of [Ni(dmit)$_2$], J_{F} is the ferromagnetic exchange coupling constant of p-EPYNN, and C_1 and C_2 correspond to the Curie constants for [Ni(dmit)$_2$] and p-EPYNN, respectively. The parameters A_1–A_9 are defined in Baker $et\ al.$ (1964). The first term deriving from the gap equation, namely equation (2.2), was obtained to fit the magnetic susceptibility of the spin ladders (Troyer $et\ al.$ 1994). The solid curve in figure 13 represents the best fit to the experimental data obtained with $\Delta/k_{\mathrm{B}} = 940$ K, $a/k_{\mathrm{B}} = 13.6$ K and $J_{\mathrm{F}}/k_{\mathrm{B}} = 0.32$ K. The parameters C_1 and C_2 are fixed to be 0.375 emu K mol^{-1} ($g = 2.00$) and 0.394 emu K mol^{-1} ($g = 2.05$), respectively. The broken curve in this figure is the theoretical trial fit of the dimer model with $\Delta/k_{\mathrm{B}} = 870$ K, which indicates significant deviation from the experimental plots.

Quantum spin ladders, which fill the gap between one- and two-dimensional quantum magnets, are attracting much interest. It is well known that one-dimensional antiferromagnetic chains of Heisenberg doublet spins exhibit a non-zero magnetic susceptibility at absolute zero. Their ground states are non-magnetic, but the excitation of a paramagnetic spin wave from the ground state costs no energy. This gap-less structure in energy results in the non-zero susceptibility. Recently, theorists have found that the crossover between the dimensions is not at all smooth, depending on whether the number of legs in the ladder is even or odd (Rice *et al.* 1994). They found a spin gap for the even-leg spin ladders, because of formation of the RVB states. The presence of spin gap in the two-leg spin ladders was already proved in several inorganic materials (Johnston *et al.* 1987; Azuma *et al.* 1994), and the above analysis for the molecular spin ladder also concluded the spin gap: the intrinsic χ_p for [Ni(dmit)$_2$] became zero at low temperatures.

We also carried out conductivity measurements on *p*-EPYNN · [Ni(dmit)$_2$]. The material was an anisotropic semiconductor whose conductivity along the *c*-axis, namely, the ladder direction, was about 10^{-4} S cm^{-1} and the activation energy was 0.27 eV. The important point was that the conductivity perpendicular to the *c*-axis was worse than this by three orders of magnitude.

To our knowledge, there have been two reports on new molecular spin ladders since we found *p*-EPYNN · [Ni(dmit)$_2$] (Rovira *et al.* 1997; Komatsu *et al.* 1997). It is highly possible that the molecular spin ladders will add valuable contributions to the studies of the spin ladders, the RVB states, etc.

(b) *Crystal structure and magnetic properties of p-EPYNN · [Au(dmit)$_2$]*

The complex *p*-EPYNN · [Au(dmit)$_2$] was prepared by the same procedure as for *p*-EPYNN · [Ni(dmit)$_2$]. *p*-EPYNN · [Au(dmit)$_2$] crystallizes into a structure that is different form that of *p*-EPYNN[Ni(dmit)$_2$], but there are interesting similarities between the two structures. Figure 14*a* shows the structure of *p*-EPYNN in *p*-EPYNN · [Au(dmit)$_2$], projected along the *b*-axis. There is a one-dimensional chain of *p*-EPYNN along the *c*-axis in which the molecules are arranged head-to-tail and side-by-side. This pattern is very similar to that in the crystal of *p*-EPYNN · [Ni(dmit)$_2$]. Figure 14*b* shows the structure of [Au(dmit)$_2$] projected along the *b*-axis. The [Au(dmit)$_2$] anions make two independent stacking chains, A and B, along the *c*-axis with short S···S contacts. However, there is no face-to-face overlapping between the [Au(dmit)$_2$] chains, in contrast to the [Ni(dmit)$_2$] anions which exhibit the face-to-face dimers in *p*-EPYNN · [Ni(dmit)$_2$]. This is probably caused by the fact that the diamagnetic [Au(dmit)$_2$] cannot gain exchange energy by dimerization. Along the *a*-axis there is an alternation of the *p*-EPYNN chain and the chain A of [Au(dmit)$_2$], in which the cation–anion arrangement is very similar to that in *p*-EPYNN · [Ni(dmit)$_2$].

The temperature dependence of $\chi_p T$ for *p*-EPYNN · [Au(dmit)$_2$] is shown in figure 15. The value is almost constant at high temperatures, and shows an increase below 50 K. The behaviour can be interpreted in terms of the ferromagnetic chain, using

$$\chi = \frac{C}{T}\left(\frac{1 + A_K + A_2 K^2 + A_3 K^3 + A_4 K^4 + A_5 K^5}{1 + A_6 K + A_7 K^2 + A_8 K^3}\right)^{2/3}, \tag{3.2}$$

with

$$K = \frac{J_{\mathrm{F}}}{k_{\mathrm{B}} T}.$$

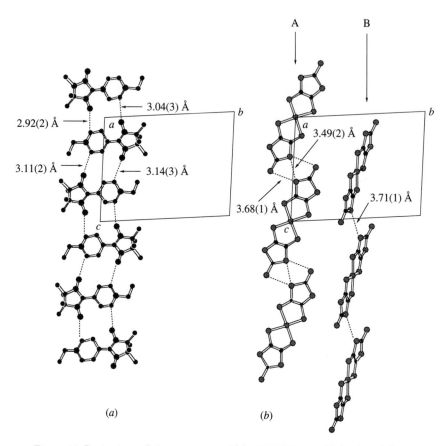

Figure 14. Projections of the structures of (a) p-EPYNN and (b) [Au(dmit)$_2$] in p-EPYNN · [Au(dmit)$_2$] along the b-axis.

The best fit, depicted as a solid curve in figure 15, is obtained with the parameters $C = 0.375$ emu K mol^{-1} (fixed) and $J_F/k_B = 0.38$ K. This justifies the analysis of the magnetic data of p-EPYNN · [Ni(dmit)$_2$] shown in figure 13.

(c) Magnetic properties of p-EPYNN · [Ni(dmit)$_2$]$_{1-x}$ · [Au(dmit)$_2$]$_x$

In the crystal of p-EPYNN · [Ni(dmit)$_2$], the [Ni(dmit)$_2$] anions form the spin ladder with the spin gap of 940 K, sandwiched by the ferromagnetic chains of p-EPYNN. In the crystal of p-EPYNN · [Au(dmit)$_2$], p-EPYNN forms the ferromagnetic chain which is very similar to that in p-EPYNN · [Ni(dmit)$_2$] and [Au(dmit)$_2$] forms the two independent chains. It is known that the monovalent anions, [Ni(dmit)$_2$]$^-$ and [Au(dmit)$_2$]$^-$, make a solid solution in the salt of n-Bu$_4$N$^+$ (Kirmse et al. 1980). In this work we prepared a solid solution system, p-EPYNN · [Ni(dmit)$_2$]$_{1-x}$ · [Au(dmit)$_2$]$_x$, in order to elucidate impurity effects on the magnetic properties of the molecular spin ladder.

The solid solutions were obtained by the reaction of the three components: p-EPYNN · I, TBA · [Ni(dmit)$_2$] and TBA · [Au(dmit)$_2$]. The cell parameters for the

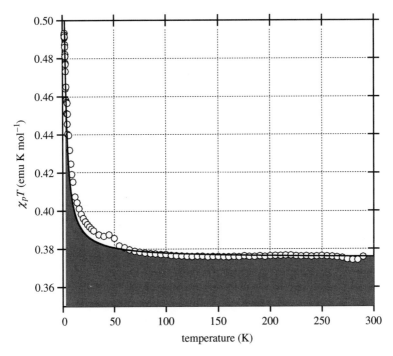

Figure 15. Temperature dependence of the paramagnetic susceptibilities χ_p for p-EPYNN $\cdot$ [Au(dmit)$_2$].

Table 2. *Cell parameters for p-EPYNN $\cdot$ [Ni(dmit)$_2$]$_{1-x}$ $\cdot$ [Au(dmit)$_2$]$_x$*

	$x = 0.0$ triclinic	$x = 0.5$ triclinic	$x = 1.0$ triclinic
a (Å)	11.647(4)	11.704(6)	14.399(4)
b (Å)	11.986(3)	12.062(9)	16.524(4)
c (Å)	12.047(6)	12.091(5)	12.743(3)
α (deg)	103.25(3)	102.92(6)	92.71(2)
β (deg)	106.01(3)	106.07(5)	100.08(2)
γ (deg)	109.91(2)	110.49(4)	103.65(2)
V (Å^3)	1419.4(9)	1435(2)	2888(1)
Z	2	2	4

three crystals, p-EPYNN $\cdot$ [Ni(dmit)$_2$]$_{1-x}$ $\cdot$ [Au(dmit)$_2$]$_x$ of $x = 0.0$, 0.5 and 1.0, are compared in table 2. It is found that the $x = 0.5$ crystal is isomorphous to the $x = 0$ crystal, namely p-EPYNN $\cdot$ [Ni(dmit)$_2$], which involves the spin ladder. It is reasonable to assume that the solid solutions of $0 \leqslant x \leqslant 0.5$ are all isomorphous to p-EPYNN $\cdot$ [Ni(dmit)$_2$].

The temperature dependence of the magnetic susceptibilities for the solid solutions of $x = 0$, 0.1, 0.25 and 0.5 were examined. The results are shown in figure 16, where the product $\chi_p T$ is plotted as a function of temperature. The closed circles show the

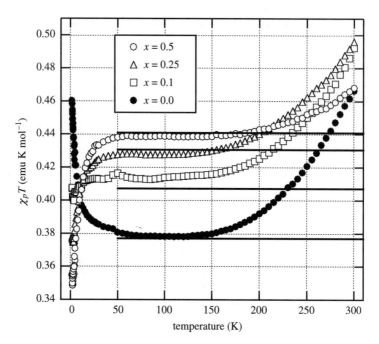

Figure 16. Temperature dependence of the paramagnetic susceptibilities χ_p for
p-EPYNN $\cdot$ [Ni(dmit)$_2$]$_{1-x}$ $\cdot$ [Au(dmit)$_2$]$_x$ of $x = 0$, 0.1, 0.25 and 0.5.

results on p-EPYNN $\cdot$ [Ni(dmit)$_2$], which were already explained in the previous
section. As the value of x is increased, the $\chi_p T$ value at the plateau shows a significant
increase. Probably this is caused by an increase of the paramagnetic [Ni(dmit)$_2$]
anions which pair with the doped [Au(dmit)$_2$] in the ladder. Assuming that such
paramagnetic lattice defects were randomly distributed in the ladder and followed the
Curie law, the $\chi_p T$ values at the plateaus for the four crystals were calculated. The
solid lines in figure 16 show the theoretical values. They can explain the rise of the
plateaus quantitatively. The probable Curie behaviour of [Ni(dmit)$_2$] on the defects
indicates that the exchange coupling constant perpendicular to the ladder direction
$J_\perp$ is much larger than that parallel to the ladder $J_\parallel$ (see scheme 5).

The most significant and unexpected effect of the impurity doping can be seen at
low temperatures. While the $\chi_p T$ value for the $x = 0$ crystal increases below 40 K
because of the ferromagnetic interaction in the p-EPYNN chain, those for the solid
solutions show an opposite tendency: $\chi_p T$ decreases below 40 K. This tendency is
enhanced with an increase in x. The doping of non-magnetic [Au(dmit)$_2$] results in the
antiferromagnetic behaviour at low temperatures. In order to clarify the origin of this
behaviour, the AC susceptibilities χ_{AC} for the solid solutions were recorded in the
temperature range down to 50 mK. The results are shown in figure 17, where $\chi_{AC} T$
are plotted as a function of temperature. The $\chi_{AC} T$ value for the $x = 0$ material
increases with a decrease in temperature, and, after making a maximum at $ca.$ 1 K, it
shows a decrease, which is probably due to an antiferromagnetic coupling between the
ferromagnetic p-EPYNN chains through the diamagnetic [Ni(dmit)$_2$] spin ladder. On
the other hand, the $\chi_{AC} T$ values for the solid solutions of $x = 0.1$, 0.25 and 0.5

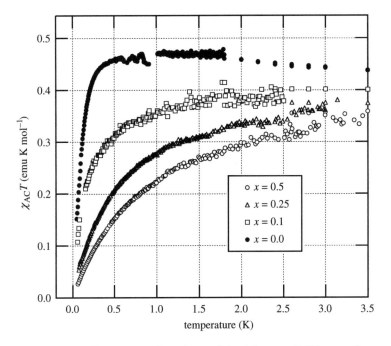

Figure 17. Temperature dependence of the AC susceptibilities χ_{AC} for
p-EPYNN $\cdot$ [Ni(dmit)$_2$]$_{1-x}$ $\cdot$ [Au(dmit)$_2$]$_x$.

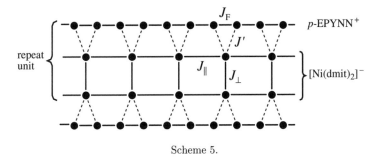

Scheme 5.

monotonically decrease and approach zero at absolute zero. It is reasonably concluded that the p-EPYNN radicals are involved in the antiferromagnetic properties. There should be an enhanced interchain interaction between the ferromagnetic chains of p-EPYNN, and/or an antiferromagnetic intrachain interaction in the p-EPYNN chain.

Scheme 5 represents the magnetic system in the material. The spin ladder of [Ni(dmit)$_2$] is sandwiched by the chains of p-EPYNN. One [Ni(dmit)$_2$] molecule interacts with two p-EPYNNs. The parameter J_F is a ferromagnetic coupling constant in the p-EPYNN chain, while $J_\parallel$ and $J_\perp$ are antiferromagnetic couplings in the [Ni(dmit)$_2$] ladder. The parameter J' represents the interaction between p-EPYNN and [Ni(dmit)$_2$], and could be antiferromagnetic. When [Ni(dmit)$_2$] forms a strong singlet pair at the regular sites, J' is negligibly small. However, the antiferromagnetic

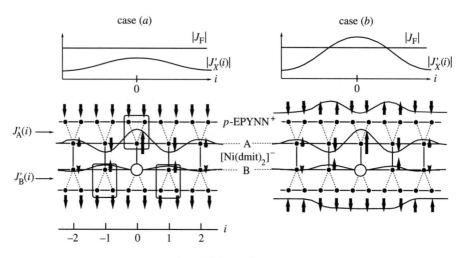

Scheme 6.

property induced by the impurity indicates an enhancement of the antiferromagnetic interaction at the impurity site. The intermolecular arrangements between p-EPYNN and $[\mathrm{Ni(dmit)_2}]$ are almost the same at the regular and the impurity site, but there should be different J' to understand the observation. Fukuyama et $al.$ (1996) studied the impurity effects to the spin ladders theoretically and predicted a spin polarization around the impurity site. Since this idea is widely accepted, it is reasonable to assume a spin density distribution on the molecular spin ladder, as shown in scheme 6. In this case, we can define the pairwise coupling constants between the spin on p-EPYNN and the polarized spin on $[\mathrm{Ni(dmit)_2}]$, $J'_X(i)$, where i represents the site number and X represents the leg, A or B. It is expected that $J'_X(i)$ is site dependent: it exhibits a maximum at the impurity site and decreases when going far from $i = 0$. Now we think about the relative intensity between $|J_F|$ and $|J'_X(i)|$. If $|J_F|$ is always larger than $|J'_X(i)|$ (case (a) in scheme 6), the paramagnetic lattice defect operates as a magnetic coupler between the ferromagnetic p-EPYNN chains. The strongest coupling on the leg-A side appears at $i = 0$, while the strongest coupling on the B side appears at $i = -1$ and 1. By assuming local antiferromagnetic coupling between p-EPYNN and the polarized spin on $[\mathrm{Ni(dmit)_2}]$, the magnetic coupling between the p-EPYNN chains is concluded to be ferromagnetic. Since the paramagnetic lattice defects operate as a ferromagnetic coupler, this model cannot explain the observed behaviour. On the other hand, if $|J_F| < |J'_X(i)|$ around the impurity site (case (b) in scheme 6), the spin polarization in the spin ladder is transferred into the organic radical chain through the pairwise magnetic interactions, and an antiferromagnetic domain is induced in the p-EPYNN chain. This can be regarded as a 'template' effect, and reasonably explains the experimental results.

(d) Summary

The crystal structures and the magnetic properties of p-EPYNN · $[\mathrm{Ni(dmit)_2}]$ and p-EPYNN · $[\mathrm{Au(dmit)_2}]$ were studied. The crystal of p-EPYNN · $[\mathrm{Ni(dmit)_2}]$ included the ladder structure of $[\mathrm{Ni(dmit)_2}]$ sandwiched by the ferromagnetic chain of p-EPYNN. The magnetic measurements concluded the spin gap for the spin ladder. p-EPYNN $\cdot$ [Au(dmit)$_2$] was not isomorphous to p-EPYNN $\cdot$ [Ni(dmit)$_2$], but their structures had interesting similarity. We prepared the solid solutions p-EPYNN $\cdot$ [Ni(dmit)$_2$]$_{1-x}$ $\cdot$ [Au(dmit)$_2$]$_x$ with $0 \leqslant x \leqslant 0.5$, in which the spin ladder structure was maintained. The doping of the non-magnetic impurity resulted in Curie defects, which suggested $|J_\perp| \gg |J_\parallel|$ in the ladder. The antiferromagnetic behaviour observed at low temperatures was interpreted in terms of the template effect: the spin polarization around the impurity in the ladder was transferred into the p-EPYNN chain and resulted in the antiferromagnetic domains.

4. Concluding remarks

We reported the two novel low-dimensional molecule-based magnets, m-MPYNN $\cdot$ X and p-EPYNN $\cdot$ [Ni(dmit)$_2$]. It was found that they exhibited spin-gap ground states, though there was a big difference between their gap intensities. We speculate that their spin gap states originate in their common characters.

(i) Their structures allow appearance of the RVB states.

(ii) Three-dimensional antiferromagnetic ordering is unstable in them, because of the spin frustration in m-MPYNN $\cdot$ X and because of the low dimensionality of the magnetic system in p-EPYNN $\cdot$ [Ni(dmit)$_2$].

In general, the molecule-based magnetic materials exhibit low-dimensional structures that are flexible enough to allow lattice distortions, such as dimerization. They are advantageous to formation of spin-gap ground states, rather than three-dimensional antiferromagnetic states. Low-dimensional molecule-based magnetic materials are quite useful in elucidating the spin-gap states and the RVB states.

We thank our co-workers (Masao Ogata, Tsunehisa Okuno, Akira Yamaguchi, Morikuni Hasegawa, Masahiro Yoshimaru, Wataru Fujita, Takeo Otsuka, Masao Ogata, Hideo Yano, Tatsuya Kobayashi, Seiko Ohira, Hiroyuki Imai) for their important contributions to the works reported herein.

References

Akita, T., Mazaki, Y., Kobayashi, K., Koga, N. & Iwamura, H. 1995 *J. Org. Chem.* **60**, 2092.

Anderson, P. W. 1963 *Magnetism* (ed. G. T. Rado & H. Suhl), vol. 1, p. 25. New York: Academic.

Anderson, P. W. 1973 *Mater. Res. Bull.* **8**, 153.

Awaga, K. & Maruyama, Y. 1989 *Chem. Phys. Lett.* **158**, 556.

Awaga, K., Sugano, T. & Kinoshita, M. 1987 *Chem. Phys. Lett.* **141**, 540.

Awaga, K., Inabe, T., Nagashima, U. & Maruyama, Y. 1989 *J. Chem. Soc. Chem. Commun.*, p. 1617.

Awaga, K., Inabe, T. & Maruyama, Y. 1992a *Chem. Phys. Lett.* **190**, 349.

Awaga, K., Inabe, T., Nakamura, T., Matsumoto, M. & Maruyama, Y. 1992b *Chem. Phys. Lett.* **195**, 21.

Awaga, K., Yokoyama, T., Fukuda, T., Masuda, S., Harada, Y., Maruyama, Y. & Sato, N. 1993 *Mol. Cryst. Liq. Cryst.* **232**, 27.

Awaga, K., Okuno, T., Yamaguchi, A., Hasegawa, M., Inabe, T., Maruyama, Y. & Wada, N. 1994a *Phys. Rev.* B **49**, 3975.

Awaga, K., Yamaguchi, A., Okuno, T., Inabe, T., Nakamura, T., Matsumoto, M. & Maruyama, Y. 1994b J. Mater. Chem. **4**, 1377.

Azuma, M., Hiroi, Z., Takano, M., Ishida, K. & Kitaoka, Y. 1994 Phys. Rev. Lett. **73**, 3463.

Baker Jr, G. A., Rushbrooke, G. S. & Gilbert, H. E. 1964 Phys. Rev. A **135**, 1272.

Benelli, C., Caneschi, A., Gatteschi, D. & Sessoli, R. 1993 Inorg. Chem. **32**, 4797.

Brewer, J. H., Kreitzman, S. R., Noakes, D. R., Ansaldo, E. J., Harshman, D. R. & Keitel, R. 1986 Phys. Rev. B **33**, 7813.

Brossard, L., Ribault, M., Bousseau, M., Valade, L. & Cassoux, P. 1986 C. R. Acad. Sci. Paris **302**, 205.

Brossard, L., Hurdequint, H., Ribault, M., Valade, L., Legros, J.-P. & Cassoux, P. 1988 Synth. Met. B **27**, 157.

Brossard, L., Ribault, M., Valade, L. & Cassoux, P. 1989 J. Physique **50**, 1521.

Bulaevskii, L. N. 1969 Sov. Phys. Solid State **11**, 921.

Caneschi, A., Gatteschi, D., Rey, P. & Sessoli, R. 1988 Inorg. Chem. **27**, 1756.

Caneschi, A., Gatteschi, D., Renard, J. P., Rey, P. & Sessoli, R. 1989 Inorg. Chem. **28**, 2940.

Chandra, P. & Coleman, P. 1991 Phys. Rev. Lett. **66**, 100.

Cirujeda, J., Ochando, L. E., Amigó, J. M., Rovira, C., Rius, J. & Veciana, J. 1995 Angew. Chem. Int. Ed. Engl. **34**, 55.

D'Anna, J. A. & Wharton, J. H. 1970 J. Chem. Phys. **53**, 4047.

Davis, M. S., Morukuma, K. & Kreilick, R. W. 1972 J. Am. Chem. Soc. **94**, 5588.

Day, P. 1993 Science **261**, 431.

de Panthou, F. L., Luneau, D., Laugier, J. & Rey, P. 1993 J. Am. Chem. Soc. **115**, 9095.

Dunsiger, S. R. (and 13 others) 1996 Phys. Rev. B **54**, 9091.

Estner, N., Singh, R. & Young, A. P. 1993 Phys. Rev. Lett. **71**, 1629.

Fazekas, P. & Anderson, P. W. 1974 Phil. Mag. **30**, 423.

Fukuyama, H., Nagaosa, N., Saito, M. & Tanimoto, T. 1996 J. Phys. Soc. Japan **65**, 2377.

Gatteschi, D., Laugier, J., Rey, P. & Zanchini, C. 1987 Inorg. Chem. **26**, 938.

Haldene, F. D. M. 1983 Phys. Lett. A **93**, 464.

Hasegawa, M., Yamaguchi, A., Okuno, T. & Awaga, K. 1995 Synth. Met. **71**, 1797.

Hernàndez, E., Mas, M., Molins, E., Rovira, C. & Veciana, J. 1993 Angew. Chem. Int. Ed. Engl. **32**, 882.

Imai, H., Inabe, T., Otsuka, T., Okuno, T. & Awaga, K. 1996 Phys. Rev. B **54**, 6838.

Inoue, K. & Iwamura, H. 1993 Chem. Phys. Lett. **207**, 551.

Jacobs, I. S., Bray, J. W., Hart Jr, H. R., Interrante, L. V., Kasper, J. S. & Watkins, G. D. 1976 Phys. Rev. B **14**, 3036.

Johnston, D. C., Johnson, J. W., Goshorn, D. P. & Jacobson, A. J. 1987 Phys. Rev. B **35**, 219.

Keren, A., Le, L. P., Luke, G. M., Wu, W. D., Uemura, Y. J., Ajiro, Y., Asano, T., Huriyama, H., Mekata, M. & Kikuchi, H. 1994 Hyperfine Interactions **85**, 181.

Keren, A., Kojima, K., Le, L. P., Luke, G. M., Wu, W. D., Uemura, Y. J., Takano, M., Dabkowska, H. & Gingras, N. J. P. 1996 Phys. Rev. B **53**, 6451.

Kinoshita, M., Turek, P., Tamura, M., Nozawa, K., Shiomi, D., Nakazawa, Y., Ishikawa, M., Takahashi, M., Awaga, K., Inabe, T. & Maruyama, Y. 1991 Chem. Lett., p. 1225.

Kirmse, R., Stach, J., Dietzsch, W., Steimecke, G. & Hoyer, E. 1980 Inorg. Chem. **19**, 2679.

Kittel, C. 1947 Phys. Rev. **71**, 270.

Kittel, C. 1948 Phys. Rev. **73**, 155.

Kobayashi, A., Kim, H., Sasaki, Y., Kato, R., Kobayashi, H., Moriyama, S., Nishio, Y., Kajita, K. & Sasaki, W. 1987 Chem. Lett., p. 1819.

Kobayashi, A., Kobayashi, H., Miyamoto, A., Kato, R., Clark, R. A. & Underhill, A. E. 1991 Chem. Lett., p. 2163.

Kobayashi, A., Bun, K., Naito, T., Kato, R. & Kobayashi, H. 1992 *Chem. Lett.*, p. 1909.

Komatsu, T., Kojima, N. & Saito, G. 1997 *Solid State Commun.* **103**, 519.

McConnell, H. M. 1967 *Proc. R. A. Welch Found. Chem. Res.* **11**, 144.

Maruta, G., Takeda, S., Yamaguchi, A., Okuno, T., Awaga, K. & Yamaguchi, K. 1999 *Mol. Cryst. Liq. Cryst.* (In the press.)

Matsushita, M. M., Izuoka, A., Sugawara, T., Kobayashi, T., Wada, N., Takeda, N. & Ishikawa, M. 1997 *J. Am. Chem. Soc.* **119**, 4369.

Mekata, M. 1990 *J. Magn. Magn. Mater.* **90/91**, 247.

Nakamura, T. & Miyashita, S. 1995 *Phys. Rev.* B **52**, 9174.

Okuno, T., Otsuka, T. & Awaga, K. 1995 *J. Chem. Soc. Chem. Commun.*, p. 827.

Otsuka, T., Okuno, T., Ohkawa, M., Inabe, T. & Awaga, K. 1997 *Mol. Cryst. Liq. Cryst.* **306**, 285.

Otsuka, T., Okuno, T., Awaga, K. & Inabe, T. 1998 *J. Mater. Chem.* **8**, 1157.

Ramirez, A. P. J. 1991 *Appl. Phys.* **70**, 5952.

Ressouche, E., Boucherle, J., Gillon, B., Rey, P. & Schweizer, J. 1993 *J. Am. Chem. Soc.* **115**, 3610.

Rice, T. M., Gopalan, S. & Sigrist, M. 1994 *Europhys. Lett.* **23**, 445.

Richards, P. M. & Salamon, M. B. 1974 *Phys. Rev.* B **9**, 32.

Rovira, C. (and 12 others) 1997 *Angew. Chem. Int. Ed. Engl.* **36**, 2324.

Siqueira, M., Nyki, J., Cowan, B. & Saunders, J. 1997 *Phys. Rev. Lett.* **76**, 1884.

Stump, H. O., Ouahab, L., Pei, Y., Grandjean, D. & Kahn, O. 1993 *Science* **261**, 447.

Sugano, T., Tamura, M., Kinoshita, M., Sakai, Y. & Ohashi, Y. 1992 *Chem. Phys. Lett.* **200**, 235.

Sugawara, T., Matsushita, M. M., Izuoka, A., Wada, N., Takeda, N. & Ishikawa, M. 1994 *J. Chem. Soc. Chem. Commun.*, p. 1723.

Syozi, I. 1951 *Prog. Theor. Phys.* **6**, 306.

Tajima, H., Inokuchi, M., Kobayashi, A., Ohta, T., Kato, R., Kobayashi, H. & Kuroda, H. 1993 *Chem. Lett.*, p. 1235.

Takeda, K. & Awaga, K. 1997 *Phys. Rev.* B **56**, 14560.

Takui, T., Miura, Y., Inui, K., Teki, Y., Inoue, M. & Itoh, K. 1995 *Mol. Cryst. Liq. Cryst.* **271**, 55.

Tamura, M., Shiomi, D., Hosokoshi, Y., Iwasawa, N., Nozawa, K., Kinoshita, M., Sawa, H. & Kato, R. 1993 *Mol. Cryst. Liq. Cryst.* **232**, 45.

Togashi, K., Imachi, R., Tomioka, K., Tsuboi, H., Ishida, T., Nogami, T., Takeda, N. & Ishikawa, M. 1996 *Bull. Chem. Soc. Japan* **69**, 2821.

Troyer, M., Tsunetsugu, H. & Würtz, D. 1994 *Phys. Rev.* B **50**, 13515.

Turek, P., Nozawa, K., Shiomi, D., Awaga, K., Inabe, T., Maruyama, Y. & Kinoshita, M. 1991 *Chem. Phys. Lett.* **180**, 327.

Uemura, Y. J., Yamazaki, T., Harshman, D. R., Senba, M. & Ansaldo, E. J. 1985 *Phys. Rev.* B **31**, 546.

Uemura, Y. J. (and 11 others) 1994 *Phys. Rev. Lett.* **73**, 3306.

Wada, N., Kobayashi, T., Yano, H., Okuno, T., Yamaguchi, A. & Awaga, K. 1997 *J. Phys. Soc. Japan* **66**, 961.

Wang, H., Zhang, D., Wan, M. & Zhu, D. 1993 *Solid State Commun.* **85**, 685.

Wen, X. G., Wilczek, F. & Zee, A. 1989 *Phys. Rev.* B **39**, 11413.

Wills, A. S. & Harrison, A. 1996 *J. Chem. Soc. Faraday Trans.* **92**, 2161.

Yamaguchi, K., Fueno, T., Nakasuji, K. & Iwamura, H. 1986 *Chem. Lett.*, p. 629.

Yamaguchi, A., Okuno, T. & Awaga, K. 1996 *Bull. Chem. Soc. Japan* **69**, 875.

Zheludev, A., Barone, V., Bonnet, M., Delley, B., Grand, A., Ressouche, E., Rey, P., Subra, R. & Schweizer, J. 1994 *J. Am. Chem. Soc.* **116**, 2019.

Discussion

M. VERDAGUER (*Chimie des Metaux de Transition, Université Pierre et Marie Curie, Paris, France*). In Professor Awaga's spin-ladder systems, the ferromagnetic interactions are very weak. He proposes a sophisticated model to explain the magnetic behaviour of the doped system. How can he be sure that under doping, the structure is not changed and hence that the ferromagnetic interactions are maintained?

K. AWAGA. (p-EPYNN)[Ni(dmit)$_2$] and (p-EPYNN)[Au(dmit)$_2$] crystallize into different structures. However, the structures of p-EPYNN in them are very similar: p-EPYNN exhibits a side-by-side, head-to-tail one-dimensional stacking. It is, therefore, natural to assume that the structure of p-EPYNN is maintained in the solid solutions, (p-EPYNN)[Ni(dmit)$_2$]$_{(1-x)}$[Au(dmit)$_2$]$_x$.

We studied the magnetic properties of

$$(p\text{-EPYNN})[\text{Ni(dmit)}_2],$$
$$(p\text{-EPYNN})[\text{Au(dmit)}_2],$$
$$(p\text{-EPYNN})[\text{Ni(dmit)}_2]_{(1-x)}[\text{Au(dmit)}_2]_x$$

($x > 0.1$). We found that p-EPYNN exhibited ferromagnetic behaviour in the former two but exhibited antiferromagnetic behaviour in the last one. We proposed that paramagnetic lattice defects in the [Ni(dmit)$_2$]$_{(1-x)}$[Au(dmit)$_2$]$_x$ change the magnetic coupling in the p-EPYNN chain from ferromagnetic to antiferromagnetic.

Muon-spin-rotation studies of organic magnets

By S. J. Blundell

*Department of Physics, University of Oxford, The Clarendon Laboratory,
Parks Road, Oxford OX1 3PU, UK*

A muon is an unstable spin-$\frac{1}{2}$ particle with a lifetime of 2.2 µs. Beams of spin-polarized positive muons can be prepared at accelerator facilities and then subsequently implanted in various types of condensed matter. Both the time and direction dependence of the subsequent positron emission can be monitored. This allows the precession and relaxation of the average muon-spin polarization to be measured and the local magnetic field in the sample to be directly inferred. The muon thus behaves essentially as a 'microscopic magnetometer' and is used to follow the magnetic order at a local level and to investigate both static and dynamic effects. This article outlines the principles of various experimental techniques that involve implanted muons, and reviews some recent experimental data on organic and molecular magnets.

Keywords: muon-spin rotation; organic magnets;
molecular magnets; nitronyl nitroxides

1. Introduction

The technique of muon-spin rotation (µSR) is extremely useful for studying various magnetic and superconducting systems (for reviews of the technique, see Schenck (1985), Cox (1987), Dalmas de Réotier & Yaouanc (1997)). This is because the frequency of the spin precession of the implanted muon (as measured by the time dependence of the spatial asymmetry in the decay positron emission) is directly related to the magnetic field at the muon site; hence, the muon can be used as a 'microscopic magnetometer'. Muons have been found to be effective probes of various types of condensed-matter physics phenomena and their use has been aided by the development of a number of accelerator facilities, most notably TRIUMF (Vancouver), PSI (Villigen, near Zürich), ISIS (Rutherford Appleton Laboratory, Oxfordshire) and KEK (Tsukuba). The technique requires the use of bulk samples because the incident muons are formed with an energy of 4 MeV and penetrate a few hundred micrometres into any sample. Surface studies may be possible in the future with the development of 'slow muon' beams, in which the energy of the muon beam is reduced down to ~1–10 eV. This is achieved either by moderation in thin layers of rare-gas solid (Morenzoni *et al.* 1994) or by resonant ionization of thermal muonium (μ^+e^- produced from the surface of a hot tungsten foil placed in a pulsed proton beam) by a pulsed laser source (Nagamine *et al.* 1995), although the efficiency of both of these processes is currently rather low.

µSR has been used extensively in the study of various organic materials (Blundell 1997), including conducting polymers (Hayes 1995; Pratt *et al.* 1997) and organic superconductors (Lee *et al.* 1997). The technique has its most obvious application in the study of magnetic systems (Denison *et al.* 1979). As a general probe of magnetic materials it can be particularly useful because

(a) it is a *local* probe of internal fields;

(b) it can be used to follow an order parameter as a function of temperature;

(c) it works very well at millikelvin temperatures (the incident muons easily pass through the dilution refrigerator windows);

(d) it provides information on antiferromagnets, spin-gap systems and spin glasses as well as on ferromagnets;

(e) if there is a range of muon sites, it can provide information about internal magnetic-field distributions; and

(f) it provides information about magnetic fluctuations and spin dynamics, even above the magnetic transition temperature.

In this article, I will describe the technique and how it has been used to study various organic and molecular magnetic systems (for a review of organic and molecular magnetism, see Kahn (1993)).

2. Experimental techniques

At a number of locations in the world (see above), intense beams of muons are prepared artificially for research in condensed-matter physics. These are made by colliding a high-energy proton beam with a suitable target, which produces pions. The pions decay very quickly into muons; if one selects the muons arising from pions that have stopped in the target, the muon beam emerges completely spin polarized.

These muons can then be implanted into a sample but their energy is large, at least 4 MeV. Following implantation they lose energy very quickly (in 0.1–1 ns) to a few keV by ionization of atoms and scattering with electrons. A muon then begins to undergo a series of successive electron capture and loss reactions, which reduces the energy to a few hundred eV in *ca.* 1 ps. If muonium is ultimately formed, then electron capture ultimately wins and the last few eV are lost by inelastic collisions between the muonium atom and the host atoms. All of these effects are very fast so that the muon (or muonium) is thermalized very rapidly. Moreover, the effects are all Coulombic in origin and do not interact with the muon spin, so that the muon is thermalized in matter without appreciable depolarization. This is a crucial feature for muon-spin-rotation experiments. One may be concerned that the muon may only measure a region of sample that has been subjected to radiation damage by the energetic incoming muon. This does not appear to be a problem since there is a threshold energy for vacancy production, which means that only the initial part of the muon path suffers much damage. Beyond this point of damage, the muon still has sufficient energy to propagate through the sample a further distance, thought to be *ca.* 1 μm, leaving it well away from any induced vacancies (Chappert 1984).

The muon decays with a mean lifetime of 2.2 μs as follows:

$$\mu^+ \rightarrow e^+ + \nu_e + \bar{\nu}_\mu.$$

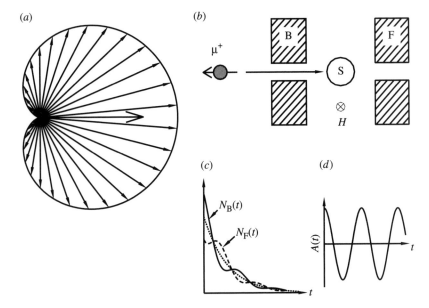

Figure 1. (*a*) The angular distribution of emitted positrons with respect to the initial muon-spin direction. The expected distribution for the most energetically emitted positrons is shown. (*b*) Schematic illustration of a μSR experiment. A spin-polarized beam of muons is implanted in a sample S. Following decay, positrons are detected in either a forward detector F or a backward detector B. If a transverse magnetic field H is applied to the sample as shown, then the muons will precess. (*c*) The number of positrons detected in the forward (dashed line) and backward (solid line) detectors. The dotted line shows the average of the two signals. (*d*) The asymmetry function.

The decay involves the weak interaction, which does not conserve parity, and this leads to a propensity for the emitted positron to emerge predominantly along the direction of the muon-spin when it decayed. The angular distribution of emitted positrons is shown in figure 1*a* for the case of the most energetically emitted positrons. In fact, positrons over a range of energies are emitted, so that the net effect is something not quite as pronounced, but the effect nevertheless allows one to follow the polarization of an ensemble of precessing muons with arbitrary accuracy, providing one is willing to take data for long enough.

A schematic diagram of the experiment is shown in figure 1*b*. A muon, with its polarization aligned antiparallel to its momentum, is implanted in a sample. (It is antiparallel because of the way that it was formed, see above, so the muon enters the sample with its spin pointing along the direction from which it came.) If the muon is unlucky enough to decay immediately, then it will not have time to precess, and a positron will be emitted preferentially into the backward detector. If it lives a little longer, it will have time to precess so that if it lives for half a revolution, the resultant positron will be preferentially emitted into the forward detector. Thus, the positron beam from an ensemble of precessing muons can be likened to the beam of light from a lighthouse.

The time evolution of the number of positrons detected in the forward and backward detector is described by the functions $N_F(t)$ and $N_B(t)$, respectively, and these are shown in figure 1c. Because the muon decay is a radioactive process, these two terms sum to an exponential decay. Thus the time evolution of the muon polarization can be obtained by examining the normalized difference of these two functions via the asymmetry function $A(t)$, given by

$$A(t) = \frac{N_B(t) - N_F(t)}{N_B(t) + N_F(t)},\qquad(2.1)$$

and is shown in figure 1d.

This experimentally obtained asymmetry function has a calculable maximum value, A_{max}, for a particular experimental configuration that depends on the initial beam polarization (usually very close to 1), the intrinsic asymmetry of the weak decay, and the efficiency of the detectors for positrons of different energies, and usually turns out to be $A_{max} \sim 0.25$. The function can be normalized to 1, in which case it expresses the spin autocorrelation function of the muon, $G(t) = A(t)/A_{max}$, which represents the time-dependent spin polarization of the ensemble of muons.

3. Experimental results

(a) Nitronyl nitroxides

Many organic radicals exist that have unpaired spins, but few are stable enough to be assembled into crystalline structures. Moreover, even when that is possible, *aligning* these spins ferromagnetically is usually impossible. Ferromagnets are rather rare, even among the elements, and are found exclusively in the d-block or f-block. Thus the discovery of ferromagnetism, albeit at rather low temperatures, in certain nitronyl nitroxide molecular crystals was particularly remarkable. The first material of this sort to be found was *p*-nitrophenyl nitronyl nitroxide (*p*-NPNN), which showed ferromagnetism up to $T_C \sim 0.65$ K only in one of its crystal phases (Tamura *et al.* 1991; Kinoshita 1994). Nitronyl nitroxides contain only the elements C, H, N and O and are, therefore, fully organic. On each nitronyl nitroxide molecule there is an unpaired spin associated with the two N—O groups. Small chemical changes to the rest of the molecule lead to significant changes in crystal structure, thereby altering the intermolecular overlaps and, thus, the magnetic interactions between unpaired spins on neighbouring molecules. Thus, different compounds have greatly different magnetic ground states. Muon studies of this and related materials began soon after the initial discovery (Le *et al.* 1993*a*; Pratt *et al.* 1993; Blundell *et al.* 1994, 1995, 1996).

Following muon implantation, it is thought that muonium (Mu = μ^+e^-), with a single electronic spin, attaches to a particular nitronyl nitroxide and combines with the unpaired spin on the nitronyl nitroxide. As shown in figure 2, the resulting electronic spin state of the muonated radical may be a singlet ($S = 0$, leading to a diamagnetic state) or a triplet ($S = 1$, leading to a paramagnetic state; see Blundell *et al.* (1995)). Both states are found in experiments. In a diamagnetic state, the muon-spin precesses at a frequency $\nu_\mu = \gamma_\mu B = 0.1355$ MHz mT^{-1} × B, so the frequency of the precession signal directly yields the local field B at the muon site. (γ_μ is the muon gyromagnetic ratio.) In a paramagnetic state, the muon-spin precesses at very high

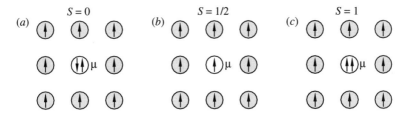

Figure 2. Electronic spin states following muon implantation in a nitronyl nitroxide system. In each case, the (*a*) singlet, (*b*) doublet and (*c*) triplet states are surrounded by nearest neighbours each with an unpaired spin. (*a*) and (*c*) are formed by Mu = μ^+e^- addition. (*b*) is formed by μ^+ addition. The muon-spin is not shown in each case (after Blundell *et al.* 1995).

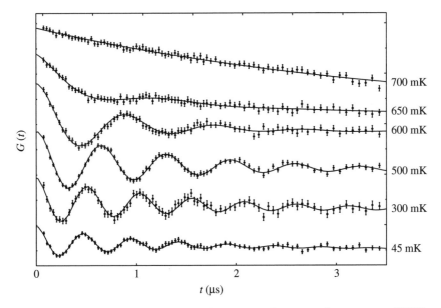

Figure 3. Zero-field muon-spin-rotation frequency in the organic ferromagnet *p*-NPNN (Blundell *et al.* 1995). The data for different temperatures are offset vertically for clarity.

frequency in the hyperfine field, and the presence of this state can be detected via a loss of muon polarization at $t = 0$.

In figure 3, the time evolution of the muon polarization in polycrystalline *p*-NPNN is plotted as a function of temperature (Blundell *et al.* 1995). This is an example of an experiment with zero applied magnetic field (in fact, a small magnetic field was applied to compensate the effect of the Earth's field). At *ca.* 0.67 K there is a clear change between the high-temperature paramagnetic state, in which there are no oscillations, and the low-temperature ordered state, in which clear oscillations can be seen. As the sample is warmed, the frequency of oscillations decreases as the internal field decreases until it is above the Curie temperature and no oscillations can be observed, only a weak spin relaxation arising from spin fluctuations. The temperature dependence of the precession frequency ($\nu_\mu(T)$) of these oscillations is shown in figure 4 together with the calculated local internal field. This is fitted to a

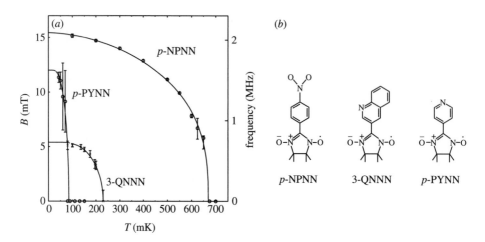

Figure 4. Temperature dependence of the zero-field muon-spin-rotation frequency in p-NPNN, 3-QNNN and p-PYNN (Blundell *et al.* 1994, 1995, 1997b). The fitted curve for p-NPNN is as described in the text. For 3-QNNN and p-PYNN, the fitted curve assumes a mean-field dependence. The molecular structure of each compound is also shown.

functional form $\nu_\mu(T) = (1 - (T/T_C)^\alpha)^\beta$, yielding $\alpha = 1.7 \pm 0.4$ and $\beta = 0.36 \pm 0.05$. This is consistent with three-dimensional long-range magnetic order. Near T_C, the critical exponent is as expected for a three-dimensional Heisenberg model (one expects $\beta \approx 0.36$ in this case). At low temperatures, the reduction in local field is consistent with a Bloch-$T^{3/2}$ law, indicative of three-dimensional spin waves (Blundell *et al.* 1995).

For p-NPNN, the magnetic properties were first determined using conventional magnetic measurements (Tamura *et al.* 1991). However, the ease of combining μSR with very low temperatures has led to ordered states being discovered using muons. In 3-quinolyl nitronyl nitroxide (3-QNNN), μSR oscillations have been observed below 210 mK and a lower local field at the muon site. The reduced local field is consistent with a canted magnetic structure (Pratt *et al.* 1993; Blundell *et al.* 1997b). The crystal structure of p-pyridyl nitronyl nitroxide (p-PYNN) is very different and consists of one-dimensional chains in which molecules are arranged side-by-side and head-to-tail (Blundell *et al.* 1994). This favours ferromagnetic interactions along the chain, but the inter-chain interactions are thought to be antiferromagnetic. Weak μSR oscillations below ~90 mK are observed superimposed on a large background (Blundell *et al.* 1994), reflecting this more complicated magnetic structure. The temperature dependence of the precession frequencies of these three compounds, together with their molecular structures, is shown in figure 4.

The dependence of the magnetic ground state on the molecular shape is well illustrated by the chemical isomers 1-NAPNN and 2-NAPNN (NAPNN is naphthyl nitronyl nitroxide), whose different molecular shapes lead to different crystal packing and, consequently, different magnetic properties, and, hence, μSR behaviour (Blundell *et al.* 1996) (1-NAPNN shows a magnetic transition, 2-NAPNN does not). For the isomers 2-HOPNN and 4-HOPNN (HOPNN is hydroxy-phenyl nitronyl nitroxide), the former gives oscillatory μSR data which are characteristic of three-dimensional ordering below *ca.* 0.5 K, but the latter shows only low-frequency oscillations below *ca.* 0.7 K with evidence of a dimensional magnetic crossover at 0.1 K (Garçia-Muñoz *et al.* 1998). A magnetic transition has also been found in *p*-CNPNN (CNPNN is cyanophenyl nitronyl nitroxide, see Blundell *et al.* (1997*a*)). The overlaps in all these materials that favour ferromagnetism appear to agree with the McConnell mechanism: as a result of spin polarization effects, positive and negative spin density may exist on different parts of each molecule; intermolecular exchange interactions tend to be antiferromagnetic, so, if the dominant overlaps are between positive (majority) spin density on one molecule and negative (minority) spin density on another molecule, the overall intermolecular interaction may be ferromagnetic. Though the mechanism for ferromagnetism is electronic, the low values of T_C imply that the dipolar interactions will play an additional role in contributing to the precise value of T_C and determining the easy magnetization axis. This too depends on the crystal structure, which in turn depends on the molecular shape (Sugano *et al.* 1997).

(b) Other organic magnetic materials

An early candidate for organic ferromagnetism was the galvinoxyl radical, in which an unpaired electron is shared between the two symmetrically related halves of the molecule. Magnetic measurements revealed rather large intermolecular ferromagnetic interactions in the temperature range 85–300 K (Mukai *et al.* 1967). The magnetic susceptibility increased much faster than expected for isolated radicals as the temperature was decreased. The data give a Curie–Weiss constant of +11 K, which corresponds to $J \sim 10$ cm^{-1}, assuming that galvinoxyl behaves magnetically as a linear chain of spins. However, at 85 K an abrupt phase transition occurs and most of the paramagnetism vanishes. It is thought that this phase transition is accompanied by a dimerization of the galvinoxyl units, affording antiferromagnetically coupled pairs. This interpretation is supported by the fact that at 77 K the EPR spectrum shows a fine structure characteristic of a triplet state (Mukai *et al.* 1982). The intermolecular interaction between adjacent units is, therefore, ferromagnetic above 85 K and antiferromagnetic below 85 K. μSR data show Gaussian relaxation across the entire temperature range and no precession signal. This is indicative of the fact that there is no transition to long-range magnetic order.

Tanol suberate is a biradical with formula $(C_{13}H_{23}O_2NO)_2$. The susceptibility follows a Curie–Weiss law with a positive Curie temperature (+0.7 K). The specific heat exhibits a λ anomaly (Saint Paul & Veyret 1973) at 0.38 K and is found to be an antiferromagnet, but, in a field of 6 mT, it undergoes a metamagnetic transition (Benoit *et al.* 1985; Chouteau & Veyret-Jeandey 1981). μSR experiments yield clear spin precession oscillations (Sugano *et al.* 1999). The temperature dependence of the precession frequency follows the equation $\nu_\mu(T) = \nu_\mu(0)(1 - T/T_C)^\beta$, where $\beta = 0.22$. This critical exponent is consistent with a two-dimensional XY magnet (Bramwell & Holdsworth 1993) and also with the temperature dependence of the magnetic susceptibility (Sugano *et al.* 1999). The relaxation rate of the oscillations rises to a maximum at the transition temperature and then falls dramatically, as shown in figure 5.

A variety of other organic magnetic materials exist and have been studied using μSR, but the nitronyl nitroxides remain the main source of purely organic bulk ferromagnets to date (recently, magnets based on TEMPO radicals have been

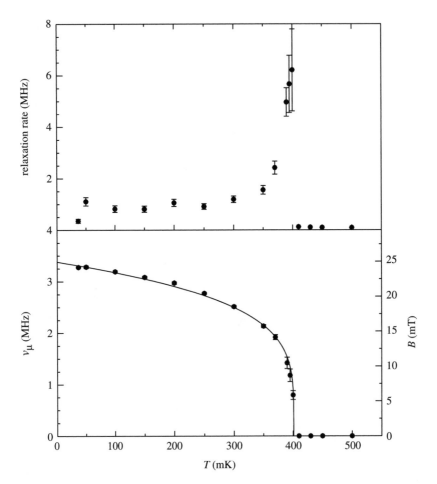

Figure 5. μSR data measured on tanol suberate showing the temperature dependence of the relaxation rate and frequency of the muon-spin oscillations.

discovered and a μSR study has been reported (Imachi *et al.* 1997)). Another initially promising candidate was prepared by using the organic donor TDAE to make a charge-transfer salt with C_{60} (Allemand *et al.* 1991). The resulting material, TDAE-C_{60}, is electrically conducting and shows a ferromagnetic-like transition at 16.1 K. However, this is not metallic and a μSR experiment (Lappas *et al.* 1995) shows a broad distribution of internal fields at the muon site, which could be consistent with incommensurate magnetic order resulting from density-wave formation.

Some of the most technologically promising materials are molecular magnets in which a transition metal ion provides the localized moment and organic bridges act as exchange pathways. Progress has been achieved using materials with unpaired electrons on both the metal ions and on the organic molecules, as in (DMeFc)TCNE (Chittipeddi *et al.* 1987) (where DMeFc=[Fe((CH$_3$)$_5$C$_5$)$_2$] is a donor and TCNE is an acceptor). This material has a chain structure but shows bulk ferromagnetic order below 4.8 K. Muons have been used to measure the development of short-range spin correlations above this temperature, which are very slow in this quasi-one-dimensional material (Uemura *et al.* 1994). Molecular magnets can also be made using the dicyanamide ligand; in particular, $Ni(N(CN)_2)_2$ and $Co(N(CN)_2)_2$ have been prepared and show long-range ferromagnetic order below 21 K and 9 K, respectively (Kurmoo & Kepert 1998). μSR experiments on these materials detect the magnetic order by a loss of muon polarization below T_C but have not been able to resolve a precession signal, suggesting a broad internal field distribution. Significant spin relaxation is observed in the paramagnetic state, which persists well above T_C.

μSR can also be useful for studying gapped systems. One such organic example is $MEM(TCNQ)_2$, which undergoes a spin-Peierls (SP) transition at low temperature. This is an intrinsic lattice instability in spin-$\frac{1}{2}$ antiferromagnetic Heisenberg chains. Above the transition temperature T_{SP}, there is a uniform antiferromagnetic next-neighbour exchange in each chain; below T_{SP}, there is an elastic distortion resulting in dimerization, and, hence, two unequal alternating exchange constants. The alternating chain possesses an energy gap between the singlet ground state and the lowest lying band of triplet excited states that closes up above T_{SP}. μSR studies indicate a slowing down of the electronic fluctuations resulting from the opening of a gap in the magnetic excitation spectrum as the temperature is lowered below T_{SP} (Blundell *et al.* 1997*c*; Lovett *et al.* 1999).

(c) *Charge-transfer salts*

Various charge-transfer salts of the organic molecules TMTSF and BEDT-TTF are found to be very good metals. The properties of these salts can be tuned by small variations of the anion (for example, $(BEDT-TTF)_2NH_4Hg(SCN)_4$ is a super-conductor with $T_C = 1$ K, while the isostructural $(BEDT-TTF)_2KHg(SCN)_4$ is not). In a number of these metallic charge-transfer salts it is found that there is a competition between a spin-density wave (SDW) ground state and a superconducting ground state. Thus the SDW state is of great interest to study, since its presence precludes superconductivity. If the muon occupies one site per unit cell and the SDW is commensurate with the crystal lattice, a number of distinct muon-spin precession frequencies would be expected to be measured. If the SDW is incommensurate, a Bessel function relaxation (Le *et al.* 1993*b*) is predicted if the field at the muon site varies sinusoidally, easily recognized since the maxima and minima appear shifted by a $\pi/4$ phase. The SDW state in $(TMTSF)_2X$, where $X=PF_6$, NO_3 or ClO_4, has been detected using μSR with similar amplitude for all three compounds (Le *et al.* 1993*b*). The observed oscillations are consistent with an incommensurate SDW. In $(BEDT-TTF)_2KHg(SCN)_4$ a very weak SDW (of estimated amplitude $3 \times 10^{-3}\mu_B$) has been detected (Pratt *et al.* 1995) below 12 K. Too small to be seen by NMR, this SDW state was suspected on the basis of susceptibility and Fermi surface experiments, but appeared in μSR data as a small change in zero-field spin relaxation.

4. Dipole fields

In many magnetic systems the implanted muon is located at a *unique* crystallographic site, so that the muon-spin precession frequency can be used to deduce directly the magnetic field at that site (which is similarly unique, neglecting any time-dependent fluctuating fields). In more complicated cases, muons may occupy more than one crystallographically independent site, leading to more than one muon-spin precession frequency. In both cases the temperature dependence of these frequencies measured in zero applied field can be used to follow the order parameter (Denison *et al.* 1979).

This situation can be contrasted with that of type II superconducting systems in an applied field. Here it is immaterial whether the muon stops in a unique crystallographic site. The magnetic field at the muon site depends mainly on its position with respect to the (much larger) vortex lattice. The vortex lattice is usually incommensurate with the crystallographic lattice. In these experiments an ensemble of muons measures the *field distribution* $p(B)$, which is, in effect, the probability of measuring a field B for a muon placed 'at random'. The detailed form of $p(B)$ can yield the penetration depth (Brandt 1988; Seeger & Brandt 1986) and also information concerning vortex lattice melting (Lee *et al.* 1997).

In very complex organic systems one may have something of an intermediate situation. One does not expect to observe a single muon precession frequency, but, rather, a large (perhaps very large) number of different frequencies, which reflects the large distribution of muon sites in the unit cell. Therefore, it may be fruitful to approximate this situation by the convenient fiction that the muon occupies *all* possible positions in the unit cell. Thus one is concerned with a magnetic-field distribution (as in the case of experiments on type II superconductors) that arises from the dipolar fields from a lattice of magnetic moments.

The magnetic field $B(r)$ at position r in the unit cell is given by

$$B(r) = \frac{\mu_0}{4\pi} \sum_i \frac{3(m_i \cdot \hat{R}_i)\hat{R}_i - m_i}{R_i^3}, \tag{4.1}$$

where $R_i = r - r_i$, $R_i = |R_i|$, $\hat{R}_i = R_i/R_i$, and the sum runs over all magnetic moments m_i at sites r_i. The sum is taken over the infinite lattice, but it is well known that this sum converges in such a way that it is necessary only to sum over points inside a sphere centred on r with sufficiently large radius. Calculations of dipole fields using this method have been performed for some time (McKeehan 1933). (An alternative method of calculation is provided by Ewald; for details see Bowden & Clark (1981)). For a spherical sample this calculation would yield the correct local field, but for other shapes the Lorentz field, $\mu_0 M/3$, and the demagnetizing field, must be considered, though for spherical samples these two terms cancel. For an antiferromagnetic lattice of spins, these two terms do not contribute. I will ignore these corrections below since they change only the details, rather than the main features, of the field distribution in each case. In what follows, we will be concerned with the distribution of the magnitude of the field $|B(r)|$, since in polycrystalline samples this is the quantity that is directly relevant to μSR.

As an example, consider a cubic unit cell of side 10 Å with $1\mu_B$ moments on each corner (this is typical for a real organic system, although many real examples may be more magnetically dilute). All moments in the lattice are aligned as shown. Figure 6*a* shows the unit cell of this example, and contours of the field distribution $B = |B|$ are drawn on each face of the cube. Close to each corner (say a distance d away) the local field is dominated by the corner moment and grows as d^{-3}. Near each corner the contours are, therefore, not shown. The corresponding field distribution $p(B)$ is plotted in figure 6*b*. The high-field tail of this distribution falls as B^{-2} and corresponds

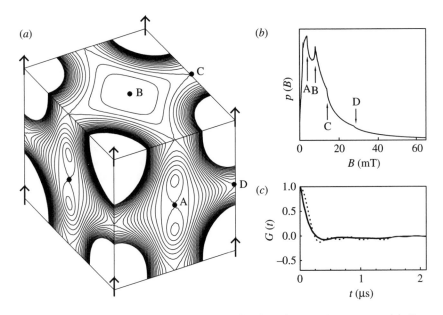

Figure 6. A simple cubic lattice of ferromagnetically aligned magnetic moments. (*a*) Contours of $|B|$ are shown on the faces of the cubic unit cell. (Moments shown as vertical arrows.) (*b*) Field distribution $p(B)$. The singularities marked correspond to the points marked in (*a*). (*c*) Corresponding muon relaxation function $G(t)$ (solid line). The dotted line shows the result when only considering sites with $B < 20$ mT.

to the regions near each corner. The asymmetric lineshape is typical of these types of dipolar lattice. In addition, various discontinuities can be seen (four of them are labelled A, B, C and D), and these correspond to van Hove singularities in the field distribution. (They correspond to the four points marked A, B, C and D in figure 6*a*.) If we assume that muons find static sites in this lattice completely at random, then they will show a precession frequency distribution that exactly follows the field distribution. The corresponding muon relaxation function,

$$G(t) = \int_0^\infty dB\, p(B) \cos(\gamma_\mu B t) \bigg/ \int_0^\infty dB\, p(B),$$

is shown by the solid line in figure 6*c*. It is characterized by a fast initial decay followed by a very weakly oscillating plateau.

In a real material, the ensemble of muons would not, of course, occupy *every* position in the unit cell, but would select a small subset of points, chosen roughly 'at random' from the $p(B)$ probability distribution. (The choice is, of course, governed by electrostatic considerations (Valladares *et al.* 1998), but is, in general, uncorrelated with the magnetic-field distribution.) However, we should ignore muons that stop very close to the electronic moments at the corners since this leads to the formation of paramagnetic states (Blundell *et al.* 1995). In order to gain some qualitative understanding of this effect, one should examine the dotted line in figure 6*c*, which shows $G(t)$ calculated from $p(B)$ but excludes any muon-site with a local field larger than 20 mT (i.e. cutting off $p(B)$ above 20 mT). This truncation produces weak oscillations in the relaxation function. The asymmetric $p(B)$ lineshape produces a fast initial relaxation followed by very weak oscillations in the calculated muon relaxation function $G(t)$. This feature is likely to be robust even if the muon samples only a few points ('at random') from the $p(B)$ distribution, except, of course, that the oscillations in $G(t)$ will be much stronger.

This fast initial relaxation behaviour is rather reminiscent of that observed in a number of experiments on organic magnetic materials, in particular the organic nitronyl nitroxides p-PYNN and 1-NAPNN (Sugano et al. 1995, 1997; Blundell et al. 1994, 1996). In the former case, weak oscillations were observed below the magnetic transition temperature T_C superimposed upon a fast relaxation (Blundell et al. 1994); in the latter case, no oscillations were observed, but the magnetic transition temperature was inferred by the development of a fast relaxation (Blundell et al. 1996). It was not clear from these experiments whether this fast relaxation could be ascribed to a static or dynamic effect; the present work provides some rationale for interpreting the fast relaxation as a purely static effect. Moreover, μSR experiments are currently being attempted on more chemically complex magnetic materials, in which there is a likelihood of many muon stopping sites in the unit cell; it may therefore be possible to use methods akin to those outlined above to interpret the results, because in such a regime it is appropriate to consider field *distributions*. It has been assumed throughout that the magnetic moments are point dipoles; if, instead, they have some small spatial distribution, this will have most effect on the high-field tail of the $p(B)$ distribution, rather than on the shape of the lower-field part. In addition, the extension to non-cubic lattices is straightforward.

Of course, in many organic systems a single precession frequency can be observed (e.g. in p-NPNN and tanol suberate described in §3), but the analysis presented above is likely to be relevant in systems of increasing complexity.

5. Conclusions

Muon-spin rotation has recently been attempted on complex organic systems that show a variety of types of magnetic order. It is clear from these studies that μSR has a lot to offer by providing magnetic information from a purely local level. Experiments can be performed on powder or single-crystal samples over a large range of temperatures. The precession frequencies are typically low in comparison with inorganic systems because the magnetic moments are less dense. This makes the experiments easy to perform at muon facilities with low precession frequency resolution, such as are found at synchrotron sources. A significant issue to be resolved for each material studied is the nature of the implanted muon site. In the case of a single muon site the analysis is straightforward, but for more complex materials, the magnetic-field distributions discussed above may be of relevance. For materials that show gaps in the spin excitation spectrum due to dimerization and singlet formation, μSR is particularly useful in measuring the temperature dependence of the spin fluctuations via the muon-spin relaxation. In conclusion, it is expected that μSR will continue to play an important role in characterizing and studying organic and molecular magnets.

This work has been supported by the EPSRC (UK). I thank F. L. Pratt, T. Sugano and M. Kurmoo for many fruitful discussions and for their contribution to this work. I am also very grateful to my other collaborators, including W. Hayes, A. Husmann, Th. Jestädt, B. W. Lovett and I. M. Marshall.

References

Allemand, P.-M., Khemani, K. C., Koch, A., Wudl, F., Holczer, K., Donovan, S., Grüner, G. & Thompson, J. D. 1991 *Science* **253**, 301.

Benoit, A., Flouquet, J., Gillon, B. & Schweizer, J. 1985 *J. Magn. Magn. Mater.* **31–34**, 1155.

Blundell, S. J. 1997 *Appl. Magn. Res.* **13**, 155.

Blundell, S. J., Pattenden, P. A., Valladares, R. M., Pratt, F. L., Sugano, T. & Hayes, W. 1994 *Solid St. Commun.* **92**, 569.

Blundell, S. J., Pattenden, P. A., Pratt, F. L., Valladares, R. M., Sugano, T. & Hayes, W. 1995 *Europhys. Lett.* **31**, 573.

Blundell, S. J., Sugano, T., Pattenden, P. A., Pratt, F. L., Valladares, R. M., Chow, K. H., Uekusa, H., Ohashi, Y. & Hayes, W. 1996 *J. Phys.: Condens. Matter* **8**, L1.

Blundell, S. J., Pattenden, P. A., Chow, K. H., Pratt, F. L., Sugano, T. & Hayes, W. 1997*a Synth. Met.* **85**, 1745.

Blundell, S. J., Pattenden, P. A., Pratt, F. L., Chow, K. H., Hayes, W. & Sugano, T. 1997*b Hyp. Int.* **104**, 251.

Blundell, S. J., Pratt, F. L., Pattenden, P. A., Kurmoo, M., Chow, K. H., Jestädt, T. & Hayes, W. 1997*c J. Phys.: Condens. Matter* **9**, L119.

Bowden, G. J. & Clark, R. G. 1981 *J. Phys.* C **14**, L827.

Brandt, E. H. 1988 *Phys. Rev.* B **37**, 2349.

Bramwell, S. T. & Holdsworth, P. C. W. 1993 *J. Phys.: Condens. Matter* **5**, L53.

Chappert, J. 1984 In *Muons and pions in materials research* (ed. J. Chappert & R. I. Grynszpan). Elsevier.

Chittipeddi, S., Cromack, K. R., Miller, J. S. & Epstein, A. J. 1987 *Phys. Rev. Lett.* **58**, 2695.

Chouteau, G. & Veyret-Jeandey, Cl. 1981 *J. Physique* **42**, 1441.

Cox, S. F. J. 1987 *J. Phys.* C **20**, 3187.

Dalmas de Réotier, P. & Yaouanc, A. 1997 *J. Phys.: Condens. Matter* **9**, 9113.

Denison, A. B., Graf, H., Kündig, W. & Meier, P. F. 1979 *Helv. Phys. Acta* **52**, 460.

Garçia-Muñoz, J. L., Cirujeda, J., Veciana, J. & Cox, S. F. J. 1998 *Chem. Phys. Lett.* **293**, 160.

Hayes, W. 1995 *Phil. Trans. R. Soc. Lond.* A **350**, 249.

Imachi, R., Ishida, T., Nogami, T., Ohira, S., Nishiyama, K. & Nagamine, K. 1997 *Chem. Lett.*, p. 233.

Kahn, O. 1993 *Molecular magnetism*. New York: VCH.

Kinoshita, M. 1994 *Japan. J. Appl. Phys.* **33**, 5718.

Kurmoo, M. & Kepert, C. J. 1998 *New J. Chem.* **22**, 1515.

Lappas, A., Prassides, K., Vavekis, K., Arcon, D., Blinc, R., Cevc, P., Amato, A., Feyerherm, R., Gygax, F. N. & Schenck, A. 1995 *Science* **267**, 1799.

Le, L. P., Keren, A., Luke, G. M., Wu, W. D., Uemura, Y. J., Tamura, M., Ishikawa, M. & Kinoshita, M. 1993*a Chem. Phys. Lett.* **206**, 405.

Le, L. P. (and 13 others) 1993*b Phys. Rev.* B **48**, 7284.

Lee, S. L., Pratt, F. L., Blundell, S. J., Aegerter, C. M., Pattenden, P. A., Chow, K. H., Forgan, E. M., Sasaki, T., Hayes, W. & Keller, H. 1997 *Phys. Rev. Lett.* **79**, 1563.

Lovett, B. W., Blundell, S. J., Jestädt, Th., Pratt, F. L., Hayes, W., Kurmoo, M. & Tagaki, S. 1999 *Synth. Met.* **103**, 2034.

McKeehan, L. W. 1933 *Phys. Rev.* **43**, 913.

Morenzoni, E., Kottmann, F., Maden, D., Matthias, B., Meyberg, M., Prokscha, T., Wutzke, T. & Zimmermann, U. 1994 *Phys. Rev. Lett.* **72**, 2793.

Mukai, K., Nishiguchi, H. & Deguchi, Y. 1967 *J. Phys. Soc. Japan* **23**, 125.

Mukai, K., Ueda, K., Ishizu, K. & Deguchi, Y. 1982 *J. Chem. Phys.* **77**, 1606.

Nagamine, K., Miyake, Y., Shimomura, K., Birrer, P., Iwasaki, M., Strasser, P. & Kuga, T. 1995 *Phys. Rev. Lett.* **74**, 4811.

Pratt, F. L., Valladares, R., Caulfield, J., Deckers, I., Singleton, J., Fisher, A. J., Hayes, W., Kurmoo, M., Day, P. & Sugano, T. 1993 *Synth. Met.* **61**, 171.

Pratt, F. L., Sasaki, T., Toyota, N. & Nagamine, K. 1995 *Phys. Rev. Lett.* **74**, 3892.

Pratt, F. L., Blundell, S. J., Hayes, W., Ishida, K., Nagamine, K. & Monkman, A. P. 1997 *Phys. Rev. Lett.* **79**, 2855.

Saint Paul, M. & Veyret, C. 1973 *Phys. Lett.* **45A**, 362.

Schenck, A. 1985 *Muon spin rotation spectroscopy.* Bristol: Adam Hilger.

Seeger, A. & Brandt, E. H. 1986 *Adv. Phys.* **35**, 189.

Sugano, T., Kurmoo, M., Day, P., Pratt, F. L., Blundell, S. J., Hayes, W., Ishikawa, M., Kinoshita, M. & Ohashi, Y. 1995 *Mol. Cryst. Liq. Cryst.* **271**, 107.

Sugano, T., Blundell, S. J., Pratt, F. L., Hayes, W., Uekusa, H., Ohashi, Y., Kurmoo, M. & Day, P. 1997 *Mol. Cryst. Liq. Cryst.* **305**, 435.

Sugano, T., Blundell, S. J., Pratt, F. L., Jestadt, T., Lovett, B. W., Hayes, W. & Day, P. 1999 *Mol. Liq. Cryst.* (In the press.)

Tamura, M., Nakazawa, Y., Shiomi, D., Nozawa, K., Hosokoshi, Y., Ishikawa, M., Takahashi, M. & Kinoshita, M. 1991 *Chem. Phys. Lett.* **186**, 401.

Uemura, Y. J., Keren, A., Le, L. P., Luke, G. M., Sternlieb, B. J. & Wu, W. D. 1994 *Hyp. Int.* **85**, 133.

Valladares, R. M., Fisher, A. J., Blundell, S. J. & Hayes, W. 1998 *J. Phys.: Condens. Matter* **10**, 10 701.

Discussion

M. VERDAGUER (*Chimie des Metaux de Transition, Université Pierre et Marie Curie, Paris, France*). Some data related to high spin molecules can be added (or will be added soon) to the impressive list of muon studies on molecular magnets.

S. J. BLUNDELL. Constraints of time did not allow me to mention these, but work on $Mn_{12}O_{12}$ acetate has been published by Lascialfari *et al.* (1998) that shows that muons are sensitive to the fluctuations in thermal equilibrium of the orientation of the magnetization. Related work has been recently performed on an Fe_8 system (Salman *et al.* 1999). Muons can therefore be used to study effects such as quantum tunnelling of magnetization in systems for which the dynamics falls in the appropriate time window.

M. VERDAGUER. One of the problems with muon studies is the location of implantation of the muon. If we think that the muon is a small proton, are there evidences of 'chemical' interactions (acid–base interactions) between the muon and some specific sites of molecular crystals (perchlorates, amines), or is Dr Blundell engaged in such studies?

S. J. BLUNDELL. In molecular systems, such interactions are indeed important and have begun to be studied. Muonium can form various muonated radicals that can be studied using level-crossing resonances. In ionic crystals, μ^+ tends to form a hydrogen bond with species such as F^- or O^{2-}. To be sensitive to local magnetism, one needs a significant fraction of muons implanted in diamagnetic states. In *p*-NPNN we observed a large diamagnetic fraction giving rise to spin precession, but we also observed a paramagnetic fraction, indicating that more than one state may be formed in a single experiment. The Oxford group are currently involved in experiments on organic molecules such as TCNQ, TTF and BEDT-TTF, which are building blocks for various organic magnets and superconductors, in order to study the nature of the implanted muon states in some detail. We are also performing similar experiments on conducting polymers and liquid crystals. In any case, for organic magnets, any uncertainty in the muon site does not prevent us from following the temperature dependence of the order parameter and studying any associated fluctuations.

Additional references

Lascialfari, A., Jang, Z. H., Borsa, F., Carretta, P. & Gatteschi, D. 1998 *Phys. Rev. Lett.* **81**, 3773.

Salman, Z., Keren, A., Mendels, P., Scuiller, A. & Verdaguer, M. 1999 *Physica.* (In the press.)

High-spin polymeric arylamines

By Richard J. Bushby, Daniel Gooding and Matthew E. Vale

School of Chemistry, University of Leeds, Leeds LS2 9JT, UK

High-spin p-doped polyarylamines have been created containing small clusters (tens) of ferromagnetically coupled unpaired electrons. In these doped polymers neighbouring spins couple through a pathway analogous to that found in *meta*-quinodimethane and the spin-carriers are triarylamine radical cations. The use of polymers in which the triarylamine radical cation centres are stabilized by Ar_2N rather than RO substituents gives improved stability but much poorer spin-coupling properties. Furthermore, the much greater ease of dication (bipolaron) formation makes it difficult to dope these amine-stabilized polymers to the requisite level. Work by other groups exploiting phenoxy and triarlymethyl spin-carriers and the prospects for producing polymers with much larger clusters of ferromagnetically coupled spins are discussed.

Keywords: magnet; polymer; synthesis; radical cation

1. Polymer design

The long-term aim in creating 'polymer magnets' and 'molecular magnets' is not to replace applications of established magnetic materials but to create new materials with novel combinations of properties. 'Polymer magnets' should share many of the advantages of 'molecular magnets' in being insulating transparent light materials that are soluble and easy to process at low temperatures. However, unlike molecular magnets, polymer magnets should make it possible to exploit strong through-bond exchange interactions, and higher Curie temperatures should result. Whether we are dealing with atomic, molecular or macromolecular systems, the conditions for Hund's Rule to apply are the same (Bushby *et al.* 1996). The singly occupied orbitals need to be more-or-less degenerate (approximately of the same energy), coextensive (overlap in their spatial distributions) (Hughbanks & Yee 1990; Dougherty 1991) and orthogonal (have a zero overlap integral). These conditions are always fulfilled in those atomic systems with which we are most familiar. Hence, as shown in figure 1*a*, the degenerate singly occupied 2p atomic orbitals of atomic carbon are coextensive (they share space in common) but orthogonal. One orbital is symmetric and the other antisymmetric with respect to a vertical mirror plane so regions of positive and negative overlap exactly cancel. As is shown in figure 1*b*, exactly the same applies to the singly occupied π molecular orbitals ϕ_4 and ϕ_5 of *meta*-quinodimethane (MQDM) **1**. The singly occupied non-bonding molecular orbitals are degenerate, coextensive and orthogonal. In the case of atomic carbon the triplet state is favoured over the singlet state by 1.25 eV (Condon & Odataoi 1980) and in the case of MQDM **1** the triplet state is favoured over the singlet state by 0.43 eV. There are many other molecular π diradicals which have triplet ground states (Allinson *et al.* 1994), but this paper is almost wholly concerned with MQDM derivatives.

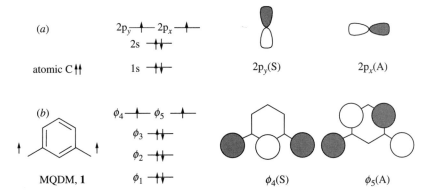

Figure 1. Basis of 'Hund's rule' ferromagnetic coupling of spins (*a*) in atomic carbon; and (*b*) in metaquinodimethane.

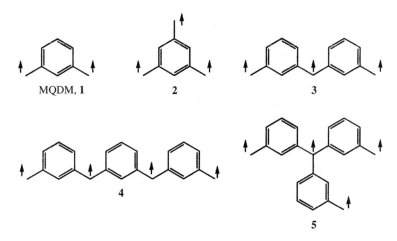

Figure 2. Oligomeric polyradicals which are extended versions of the MQDM motif.

In both MQDM and atomic carbon there are two ferromagnetically coupled spins but this number can easily be extended. Hence, in atomic nitrogen there are three degenerate singly occupied 2p orbitals and the quartet state is favoured over the doublet state by 2.37 eV (Condon & Odataoi 1980). It is predicted that the triradicals **2** and **3** (figure 2) have quartet ground states and the tetraradicals **4** and **5** have quintet ground states.

The maximum number of ferromagnetically coupled spins which can be created using cocentred atomic orbitals is small (up to three if we use three p orbitals and up to five if we use five d orbitals), but in molecular systems theory predicts that there is no limit. At least in principle, it is possible to extend the motifs shown in figure 2 to create macromolecular polyradicals containing tens, hundreds, even thousands of ferromagnetically coupled spins and perhaps even systems with bulk superparamagnetic or ferromagnetic properties (Mataga 1968; Ovchinnikov 1978). It is important to note that macromolecular polyradicals involving an extended MQDM motif are not

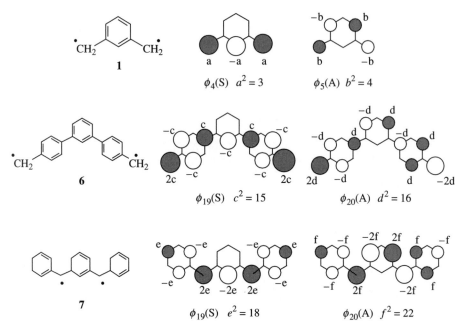

Figure 3. The effect of extending the conjugated system on the Hueckel NBMOs of MQDM.

necessarily linear (like **3** and **4**). As shown in figure 2, it is possible to create derivatives branched either through the benzene ring (as in triradical **2**) or through the α-carbon (as in tetraradical **5**). This is significant since ferromagnetic coupling always breaks down in one-dimensional systems if the chains are sufficiently long. If the ultimate aim is to make a bulk ferromagnet, the incorporation of these branched 'building blocks' is crucial since it allows the creation of two- and three-dimensional networked macromolecular structures which solve the 'dimensionality problem' (Bushby *et al.* 1996). The low thermal stability of carbon-based radicals and their poor air stability are much more challenging problems. Solutions have been found through extension of the conjugated system, by the introduction of steric hindrance (substituents like *tert*-butyl), and by the introduction of heteroatoms into the π system. Figure 3 shows two ways of extending the conjugated system. As shown, inserting a *para*-phenylene between the CH_2 and the benzene ring as in compound **6** or by introducing peripheral C_6H_5 substituents as in compound **7** does not affect the essential symmetry or the coextensive nature of the singly occupied non-bonded molecular orbitals. However, the greatest improvement in stability is seen in π-heteroatomic systems in which the atom carrying the largest part of the spin density is a nitrogen or oxygen rather than carbon.

In MQDM both of the α-carbons can be (formally) replaced by $N^{+\cdot}$. This gives an isoelectronic structure which is more stable. It leaves the symmetry of the putative singly occupied orbitals essentially unchanged but it also slightly lifts their degeneracy (Bushby 1999). Nevertheless, the spin states of amminium cation diradicals and triradicals seem to follow those of the corresponding hydrocarbons. Hence, the dication diradicals **8** (Sato *et al.* 1997), **9** (Stickley & Blackstock 1994) and **12** (Wienk

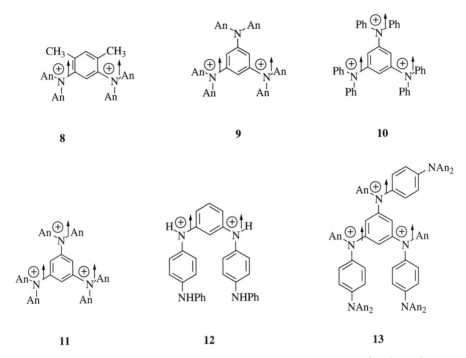

Figure 4. High-spin oligomers based on amminium radical cation spin-bearing units: Ph, phenyl (C_6H_5); An, anisyl (p-$CH_3OC_6H_4$).

& Janssen 1996 *a*, *b*) (figure 4), which are analogous to the all-carbon diradical **1** (figure 2), have triplet ground states. The trication triradicals **10** (Yoshizawa *et al.* 1992 *a*, *b*, 1996; Yoshizawa 1993), **11** (Stickley & Blackstock 1994; Yoshizawa *et al.* 1996) and **13** (Stickley *et al.* 1997) (figure 4), which are analogous to the all-carbon triradical **2** (Figure 2), give quartet spectra. The stability of aryl substituted amminium radical cations is highly structure dependent (Stickley & Blackstock 1995), but some are stable enough to be considered for use as polymer building blocks. Hence tris(o-bromophenyl)amminium pentachloroantimonate is a commercially available stable solid which is used by synthetic organic chemists as a one-electron oxidant; most of the polyradicals shown in figure 4 are stable in solution, and the quartet trication **13** is an isolable air-stable solid (Stickley *et al.* 1997). In general, triarylamminium radical cations are expected to be stabilized by extending the conjugated system or by the incorporation of *ortho-* or *para*-halogen, alkoxy, or amino substituents. The last should provide the greatest stabilization. However, for the trication **13**, the presence of the *para*-amino substituent also has an undesired effect. It results in a marked reduction in the splitting between the doublet and quartet states so that they become almost equal in energy.

2. Synthesis and characterization of the polymers

In this paper we describe the synthesis and characterization of a high-spin polymer **15**$^{\cdot+}$ (figure 5) based on the simple monomer **8** and of a high-spin polymer **22**$^{\cdot+}$

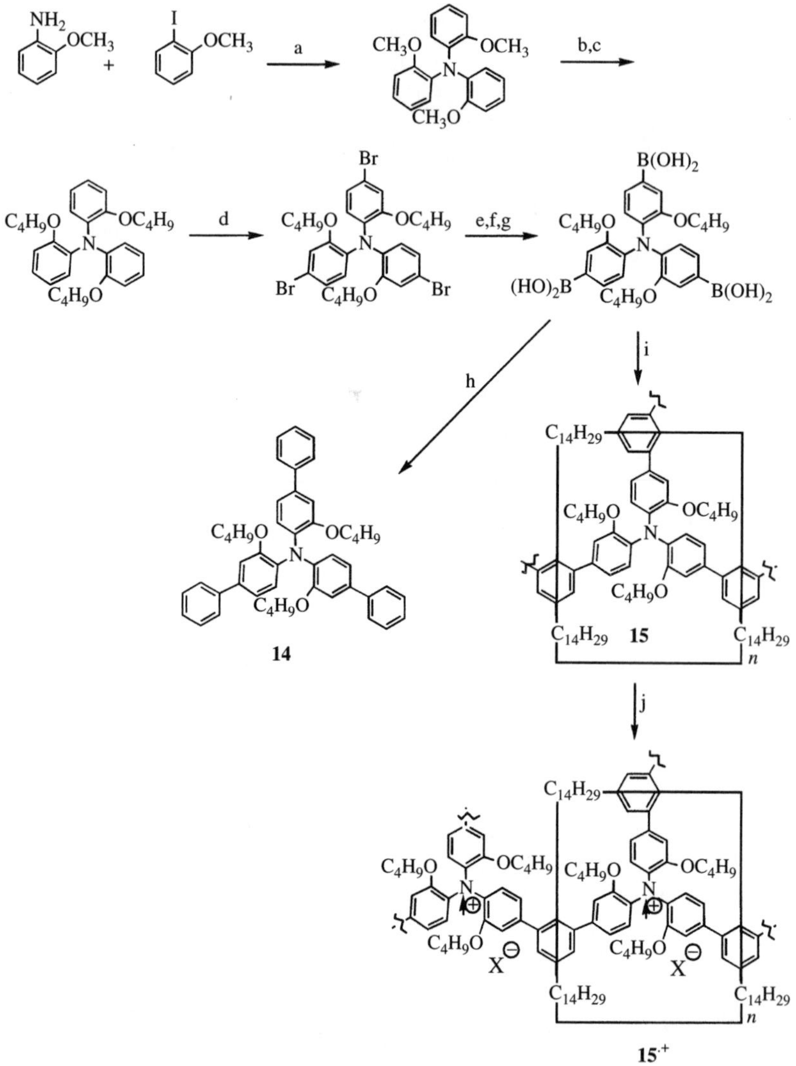

Figure 5. Synthesis of the doped polymer 15^+. Reagents: (a) $K_2CO_3/Cu/reflux$; (b) $BBr_3/-78\,°C$; (c) $C_4H_9Br/K_2CO_3/reflux$; (d) $Br_2/CHCl_3/0\,°C$; (e) $BuLi/THF/-78\,°C$; (f) $(^iPrO)_3B/-78\,°C$; (g) HCl/H_2O; (h) $Pd(PPh_3)_4/toluene/aq.\ Na_2CO_3/C_6H_5Br$; (i) $Pd(PPh_3)_4/toluene/aq.\ Na_2CO_3/3,5\text{-}dibromotetradecylbenzene}$; (j) $NOBF_4/CH_2Cl_2$.

(figure 6) based on the Blackstock monomer **12**. The synthetic routes employed are summarized in figures 5 and 6 together with those of model compounds **14** and **23** which contain the essential chromophores/electrophores of the undoped polymers. The synthesis of polymer **15** has been described previously (Bushby & Gooding 1998) and details for the synthesis of polymer **22** are given in the experimental section. The choice of systems **18–22** with *two* methoxy substituents was dictated by the fact that the synthesis of 'parent' systems (without the methoxy groups) failed because of the

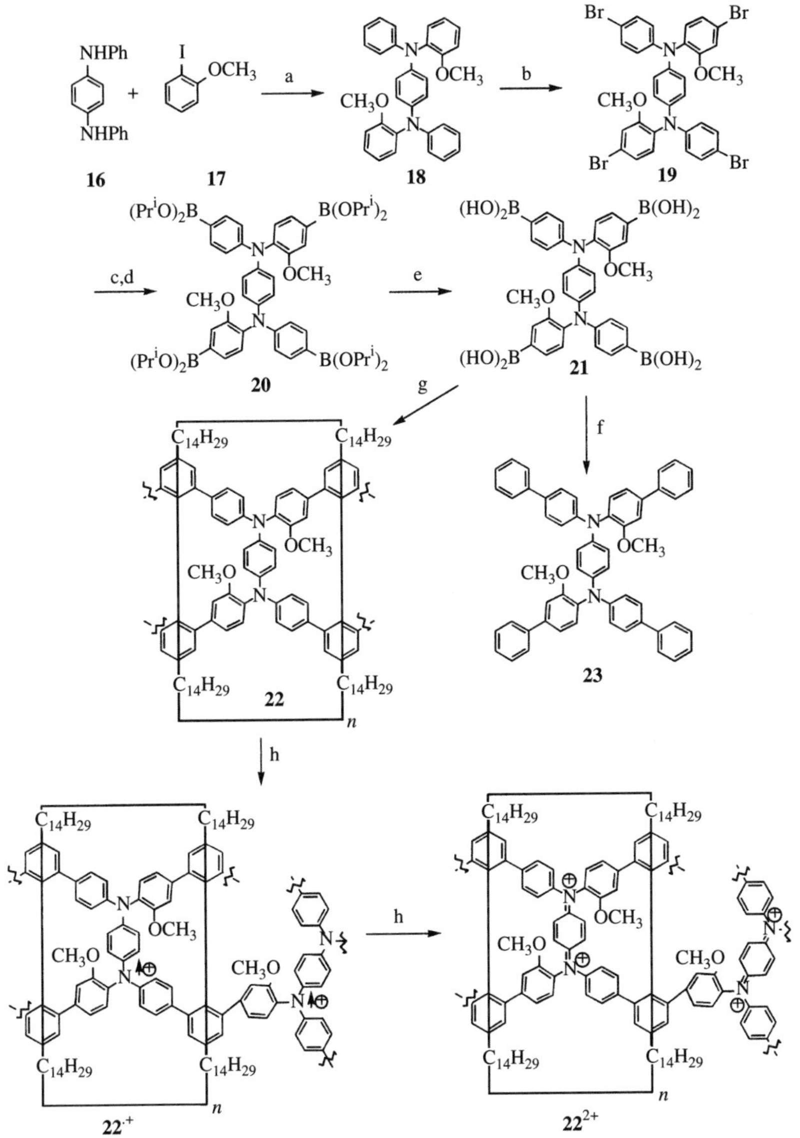

Figure 6. Synthesis of the doped polymers $22^{.+}$ and 22^{2+}. Reagents: (a) $K_2CO_3/Cu/reflux$; (b) $Br_2/CHCl_3/0\,°C$; (c) $BuLi/THF/-78\,°C$; (d) $(^iPrO)_3B/-78\,°C$: (e) HCl/H_2O; (f) $Pd(PPh_3)_4/$ toluene/aq. Na_2CO_3/C_6H_5Br; (g) $Pd(PPh_3)_4/toluene/aq.$ $Na_2CO_3/3,5$-dibromotetradecylbenzene; (h) $NOBF_4/CH_2Cl_2$. $Pr^i = iso$-propyl.

poor solubility of the corresponding tetrabromide and intermediates in the synthesis of systems with four methoxy substituents were found to be highly photolabile. Polymerization reactions were based on a Suzuki coupling reaction (Hoshino *et al.* 1988) and were stopped before the 'gel' point. The chloroform-soluble fraction was purified by repeated reprecipitation from chloroform/methanol. Gel permiation chromatography against a polystyrene standard for two fractions of the polymer **15** gave a 'low M_w fraction' with $M_w = 8500$, $M_n = 1600$, $M_w/M_n = 5.2$ and a 'high M_w fraction' with $M_w = 86\,000$, $M_n = 10\,000$, $M_w/M_n = 8.5$. Polymer **22** had $M_w = 36\,000$, $M_n = 19\,000$, $M_w/M_n = 1.9$. It should be noted, however, that molecular weights obtained in this way are usually underestimated for networked polymers.

A single crystal X-ray diffraction study of the model for the repeat unit **14** gave a molecular geometry with a propeller-like twist of the aryl residues about the central nitrogen (Bushby et al. 1999) and this is very similar to that of other triaryl amines. This is expected to change little on oxidation to the radical cation. The dihedral angles between the aryl rings in the biphenyl groups were 44, 45 and 25°. Some degree of twisting about the aryl–aryl bonds is essential to give a three-dimensional network and calculations show that, within limits, it does not affect the ferromagnetic coupling. However, near-orthogonality of the rings does reverse the ordering of the singlet and triplet states and complete orthogonality of the rings is fatal since it totally destroys the conjugated pathway on which the exchange interaction depends.

3. Doping of the polymers

Initially we experienced difficulty in obtaining quantitative or even nearly quantitative doping of these arylamine polymers (Bushby et al. 1996, 1997a; Bushby & Gooding 1998) but we have subsequently shown that these problems can be overcome (Bushby et al. 1999). The polymer needs to be doped with an excess of NOBF$_4$ in very dilute solution otherwise it precipitates in a partly doped state. Hence, when a very dilute solution of the amine **14** in dry chloroform was titrated with a solution of NOBF$_4$ in dry chloroform, the absorption λ_{max} at 347 nm was progressively and quantitatively (1 mole reacting with 1 mole) replaced by that for the radical cation **14**$^{+}$, λ_{max} at 418 nm. When the same experiment was repeated using a ca. 0.006 g l^{-1} solution of the polymer **15**, except that a roughly four-fold excess of NOBF$_4$ had to be used and that the reaction was slower, the results were almost identical and the resultant spectra are shown in figure 7. Critically it was found that when a very large excess of NOBF$_4$ was used the UV spectrum was unchanged. This system shows no tendency to overoxidize to the spinless bipolaron level.

When a ca. 0.006 g l^{-1} solution of the polymer **22** was doped in the same manner there was a two-stage oxidation. In the first stage the chromophore of the neutral phenylenediamine (λ_{max} at 349 nm) was replaced by that of the radical cation (λ_{max} at 422 nm, figure 8 left). In the second stage the chromophore of the radical cation was replaced by that of the dication (λ_{max} at 812 nm, figure 8 right). The solution of the radical cation **22**$^{+}$ was green and that of the dication **22**$^{2+}$ was blue.

The progress of the oxidation reactions was also studied by EPR spectroscopy. Integrating the signals obtained by NOBF$_4$ oxidation of a ca. 0.006 g l^{-1} solution of compound **14** and comparison with equivalent solution of the polymer **15** treated in exactly the same gave an average doping level of 99%. In the case of polymer **22**, the spin concentration was followed in parallel to the UV experiment shown in figure 8. The integrated intensity rose to a maximum which corresponded to $92 \pm 4\%$ of the theoretical maximum at the stage we have identified as that of the radical cation/polaron (λ_{max} at 422 nm, figure 8 left) and fell to almost zero again once the dication/bipolaron (λ_{max} at 812 nm, figure 8 right) was formed.

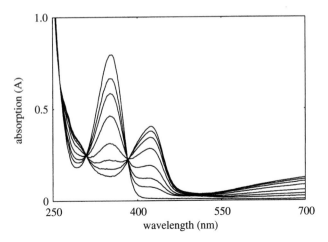

Figure 7. UV spectra showing the doping of polymer **15** in chloroform using aliquots of a saturated NOBF$_4$ solution in chloroform.

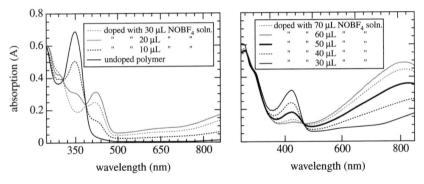

Figure 8. The UV spectra showing the doping of polymer **22**, firstly to the monoradical cation level (left) **22**$^{\cdot+}$ and subsequently to the dication level (right) **22**$^{2+}$ using a solution of NOBF$_4$.

Although, in general, all operations with the doped polymers **15**$^{\cdot+}$ and **22**$^{\cdot+}$ were carried out under an inert atmosphere and using a glove-box, the doped polymer **22**$^{\cdot+}$ was found to be air stable.

Cyclic voltammetry studies of a solution of the polymer **15** showed a first oxidation potential of 0.22 V versus a silver–silver chloride standard electrode and a second oxidation potential 0.75 V higher, at 0.97 V (Bushby *et al.* 1997*b*). Whereas, the first and second oxidation potentials of polymer **15** are well separated, Blackstock has shown that systems containing a similar electrophore to polymer **22** not only have a lower first oxidation potential but also the first and second oxidation potentials are closer together, typically separated by only 0.4 V (Stickley *et al.* 1997). This is consistent with our observation that, although polymer **14** could be doped with a large excess of reagent without bipolaron formation, the doping of polymer **22** is much more difficult to control.

The solutions used for the UV experiments were too dilute (*ca.* 0.006 g l^{-1}) for doping on a preparative scale. For the preparative doping experiments a concentration

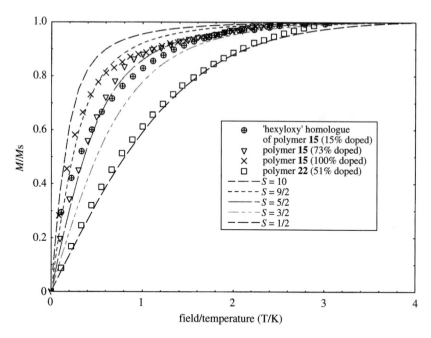

Figure 9. Field dependence of the relative magnetization of doped arylamine polymers **15** and **22** at 2 K plotted against theoretical Brillouin functions.

of 0.6 g l^{-1} was employed. Under these conditions the polymer **15** with $M_w = 8500$ was essentially quantitatively doped but the polymer **15** with $M_w = 86000$ was only *ca.* 73% doped. A sample of polymer **22** with $M_w = 36\,000$ gave 51% radical cations.

4. Magnetic properties of the doped polymers

We have previously shown that when 15% of the available sites in the hexyloxy 'homologue' of polymer **15** were oxidized to the N$^{\cdot+}$ level, the field dependence of the bulk magnetization at 2 K fitted a Brillouin function, $S = \frac{5}{2}$ (Bushby *et al.* 1996, 1997*b*), and these data are shown in figure 9. Nutation resonance studies of this 15% doped material showed the presence of high-spin species from $S = \frac{1}{2}$ up to at least $S = 3$ (Shiomi *et al.* 1997).

Using the dilute-solution doping method for the 'high-molar-mass' ($M_w = 86\,000$) fraction of polymer **15** gave *ca.* 73% doping and, as shown in figure 9, 'improved' the spin-coupling behaviour. However, using the 'low-molar-mass' ($M_w = 8500$) fraction of polymer **15** it is possible to achieve almost 100% doping. Hence, a *ca.* 0.6 g l^{-1} solution of the polymer **15** in chloroform was oxidized with a 20-fold excess of powdered NOBF$_4$. The UV spectrum of the green solid obtained when this solution was evaporated, both before and after the SQUID magnetometer measurements, showed almost 100% oxidation to the N$^{\cdot+}$ level and that there had been no detectable deterioration of the sample. The absolute value of its saturation magnetization was 6.19 emu g^{-1}, which actually corresponds to 96% doping, although, like UV, EPR spectroscopy gave a value of 100% doped. The susceptibility of the doped polymer was measured as a function of temperature (2 K to room temperature) at constant field

(0.5 T) and as a function of field (0–5 T) at constant temperature (2 K). The susceptibility measurements were corrected to allow for the diamagnetic contribution of the polymer and of the sample holder. Curie-law behaviour was observed between *ca.* 100 K and room temperature. Figure 9 shows relative magnetization as a function of field. It is clear that, as we progress from 15% to 73% to almost 100% doping, the size of the ferromagnetically coupled spin clusters increases. The behaviour is typical of a polydisperse spin system. In each case the relative magnetization rises too rapidly at low H/T (when the behaviour is dominated by the larger moments) and too slowly at high H/T (when the behaviour is dominated by the $S = \frac{1}{2}$, 1, etc., sites) as compared to Brillouin functions for monodisperse spin systems (Murray *et al.* 1994). For a disperse spin system we can write

$$M = Ng\mu_B \Sigma_S p_S S B(S, H/T),$$

where $B(S, H/T)$ is the Brillouin function and p_S is the fractional contribution to the magnetization from species of spin S. The difficulty in interpreting the data shown in figure 9 is that the starting polymer is highly polydisperse and the distribution of spin states (the p_S terms) are unknown. Attempts to use a variety of distribution functions shows that the data cannot be fitted without assuming a wide distribution involving, at one extreme, a considerable contribution from low spin ($S = \frac{1}{2}$, 1, etc.) species and, at the other extreme, from species $S > 15$. Figure 10 shows a fit to the linear distribution function $p_S = S$ of $\frac{1}{2}$–22 shown in the inset. The data used in figures 9 and 10 were all obtained at 2 K. Figure 11 shows typical data for the temperature dependence of the moment. The moment at room temperature for a 73% doped sample is a little greater than that expected for an $S = 1$ system and this rises to that expected for an $S = \frac{5}{2}$ system just above 2 K. The Brillouin function fit for this same sample at 2 K (shown in figure 9) perhaps looks closest to $S = \frac{5}{2}$ or 3.

Characterization of the magnetic behaviour of the doped polymer $\mathbf{22}^{\cdot +}$ caused some difficulty. When a dilute solution of the polymer *ca.* 0.6 g l^{-1} was 'titrated' with NOBF$_4$ and the reaction monitored by UV spectroscopy it could be stopped at the radical cation stage $\mathbf{22}^{\cdot +}$ but this required more than one equivalent of reagent. When the resultant solution was evaporated and the solid obtained was redissolved it was found to be overoxidized and to be a mixture of $\mathbf{22}^{\cdot +}$ and $\mathbf{22}^{2+}$. On the other hand, use of just one equivalent of NOBF$_4$ gave a product which was a mixture of $\mathbf{22}$ and $\mathbf{22}^{\cdot +}$. This made it difficult to dope the polymer exactly to the radical cation level. Figure 9 shows relative magnetization data for a sample which, on the basis of its saturation magnetization and UV spectrum, was a 51:49 mixture of $\mathbf{22}^{\cdot +}$:$\mathbf{22}^{2+}$. The behaviour is essentially that of an $S = \frac{1}{2}$ material. In view of the observation that even imperfectly doped samples of polymer $\mathbf{15}$ showed some evidence of '$S > \frac{1}{2}$' behaviour, this implies that there is no significant ferromagnetic coupling of the spins in $\mathbf{22}^{\cdot +}$.

In conclusion, doped forms of the polymer $\mathbf{22}$ (based on the paraphenylenediamine motif of the Blackstock monomer) are much more stable than doped forms of polymer $\mathbf{15}$ but, because the oxidation potential is lower and the first and second oxidation potentials are closer together, the doping process is much more difficult to control. Furthermore, we have not obtained high-spin products from this polymer. In the related monomer $\mathbf{13}$, the high- and low-spin states are almost degenerate and in this less-symmetrical polymer system it appears that the low-spin state wholly dominates. However, the p-doped polymer $\mathbf{15}^{\cdot +}$ does show the desired ferromagnetic

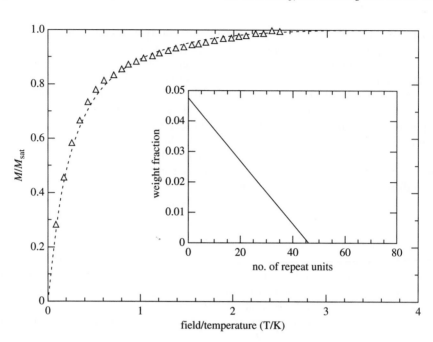

Figure 10. Field dependence at 2 K of the magnetization of polymer **15**, $M_w = 8500$, 100% doped fitted to a theoretical line calculated by summation of Brillouin functions and based on the distribution of spin states shown in the insert.

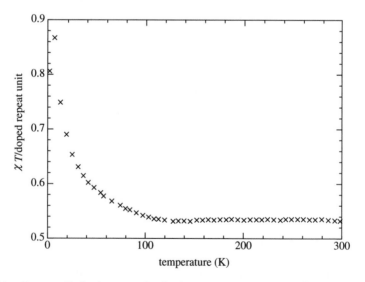

Figure 11. χT versus T plot for a sample of polymer **15**, $M_w = 86\,000$, 73% doped and measured at a field of 0.5. χ is expressed in emu per repeat unit. On this basis (for monodisperse spin systems) we expect $\chi T = 0.37$ (for $S = \frac{1}{2}$), 0.87 (for $S = \frac{5}{2}$) and 1.12 (for $S = \frac{7}{2}$). The Brillouin function 'fit' for this same sample at 2 K shown in figure 9 suggests $S \approx \frac{7}{2}$.

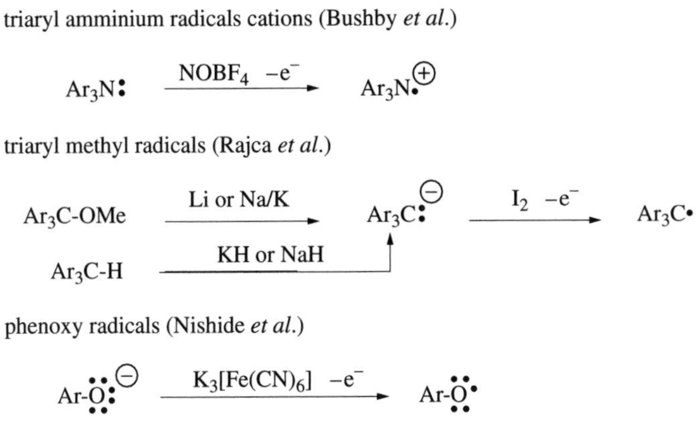

triaryl amminium radicals cations (Bushby *et al.*)

$$Ar_3N\colon \xrightarrow{\text{NOBF}_4 \ -e^-} Ar_3N\overset{\oplus}{\cdot}$$

triaryl methyl radicals (Rajca *et al.*)

$$Ar_3C\text{-OMe} \xrightarrow{\text{Li or Na/K}} Ar_3C\overset{\ominus}{\colon} \xrightarrow{\text{I}_2 \ -e^-} Ar_3C\cdot$$

$$Ar_3C\text{-H} \xrightarrow{\text{KH or NaH}}$$

phenoxy radicals (Nishide *et al.*)

$$Ar\text{-}\overset{..}{\underset{..}{O}}\colon^{\ominus} \xrightarrow{\text{K}_3[\text{Fe(CN)}_6] \ -e^-} Ar\text{-}\overset{..}{\underset{..}{O}}\colon^{\cdot}$$

Figure 12. Methods of generating spin-carriers (p-doping methods) for the systems discussed in this paper.

coupling. Although the spin states achieved are lower than expected on the basis of the molecular weight distribution (Bushby *et al.* 1999) and the moments are temperature dependent, even when imperfectly doped and at room temperature, it behaves like a system for which $S > 1$. In this polymer there is likely to be a range of dihedral angles between the component elements of the π system and this will lead to a range of different couplings between adjacent centres. This is the most likely explanation of the observed behaviour but it is a problem that can be overcome by modifying the structure of the polymer.

5. Related work by other groups

The idea of producing polymer magnets based on repeating non-Kekule π biradical units is simple but one which has proved difficult to translate into reality. The area of 'polymer magnets' has been poorly served by a number of overoptimistic spurious claims. However, a systematic understanding of the problem is beginning to emerge. Among systems being investigated by other groups around the world, the triaryl-methyl radical systems developed by the group of A. Rajca and phenoxy radical systems by the groups of P. M. Lathi and N. Nishide seem particularly promising.

(a) Triarylmethyl polyradical systems (Rajca and co-workers)

Rajca's triarylmethyl polyradical systems use spin carriers based on the Schlenk hydrocarbon and MQDM to mediate the ferromagnetic spin coupling (Rajca 1994). The method of generating the radical sites is analogous to ours (figure 12). In Rajca's systems a one-electron oxidation of a carbanion gives a neutral carbon radical and in ours a one-electron oxidation of an amine gives the amminium radical cation. The chemistry involved in Rajca's systems appears to be more difficult since the carbanion intermediates are sensitive to moisture. One very impressive aspect of their work has been the way in which this group has produced specific (monodisperse) oligomers of well-defined structure such as the 'star-branched' decaradical **23** (Rajca *et al.* 1992) and the dendritic system **24** (Rajca & Utamapanya 1993) (figure 13). In systems of this type, spin coupling has been shown to be very sensitive to defects, particularly

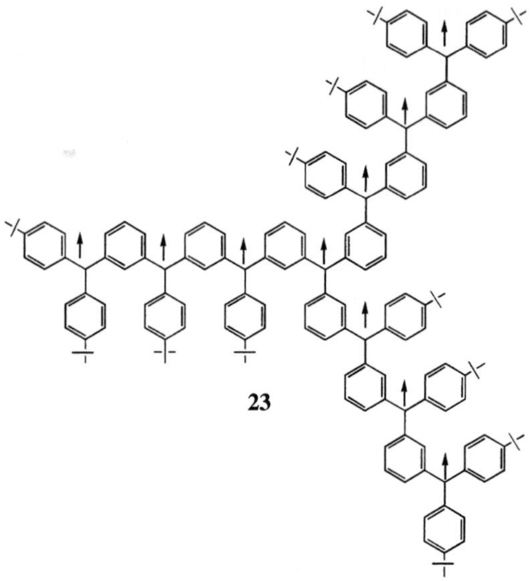

Figure 13. The 'star-branched' decaradical **23** and dendritic system **24**
of Rajca and co-workers.

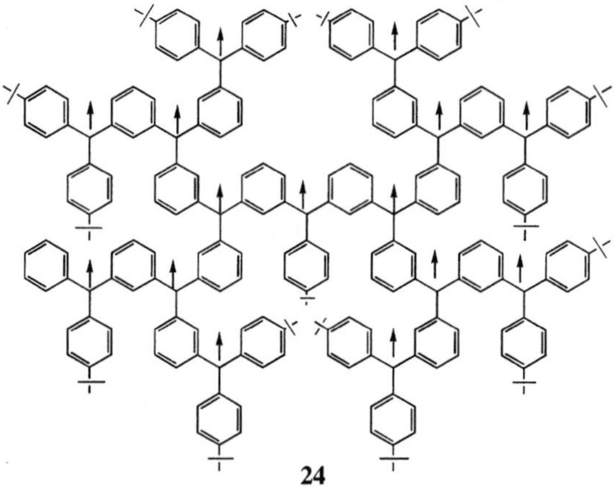

defects close to the centre of the molecule. Recently, this group attempted to tackle
this problem and have produced several 'closed-loop' structures, the most interesting
of which is polyradical **25** (Rajca *et al.* 1998) (figure 14). This should have a ground
state $S = 12$. Magnetic susceptibility data give a value of $S = 10$. The authors ascribe
this discrepancy to the presence of a small density of defects, which result either from
less than 100% generation of unpaired electrons, or from synthetic impurities present
in the polyether precursor, but, as in our systems, a dispersity in the site-to-site spin
interactions may also be a significant problem and it is interesting to note that **25** also
shows evidence for a temperature-dependent moment.

The group of Rajca has also successfully synthesized and characterized a polymeric

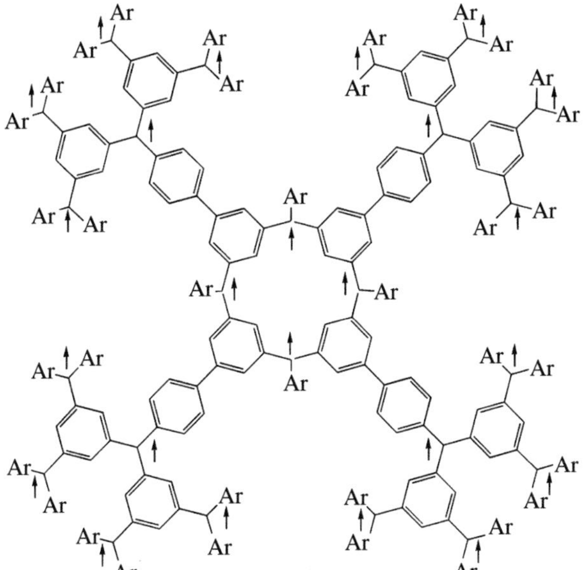

25

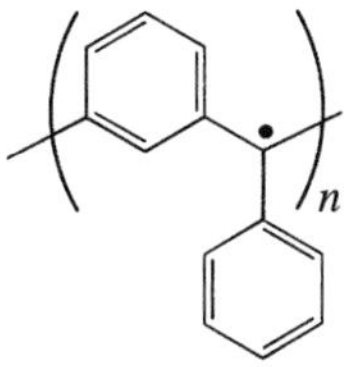

26

Figure 14. The 'closed-loop' polyradical **25** (should be $S = 12$, actually $S = 10$), Ar = $^{t}BuC_6H_4-$ and the 1,3-connected polyarylmethane radical 26 of Rajca and co-workers.

1,3 connected arylmethane **26** (Utamapanya *et al.* 1993) (figure 14). However, incomplete generation of spin sites proved to be the limiting factor. Indeed, defects are always particularly damaging in linear/one-dimensional systems. A fit of the normalized plot of the magnetization to those of theoretical Brillouin functions gave best agreement with a system $S = 2$.

(b) Phenoxy radical-based systems (Lathi, Nishide and co-workers)

The groups of Lathi, and particularly Nishide, have developed a variety of polymers and oligomers in which the spin carriers are of the phenoxy or galvinoxyl type (figure 14). The polyradical **27** gives an average spin quantum number, $S = \frac{5}{2}$, with a spin concentration of 0.7 spins per unit (Nishide *et al.* 1996). Star-branched oligomers **28** and **29** (figure 15) were also studied. The polyradical **28** displays an average $S = 4$ (Nishide *et al.* 1998). In comparison to our systems and those of Rajca, high doping levels seem more difficult to achieve and, unlike Rajca's systems, the materials are polydisperse. However, these phenoxy radicals have the advantage over

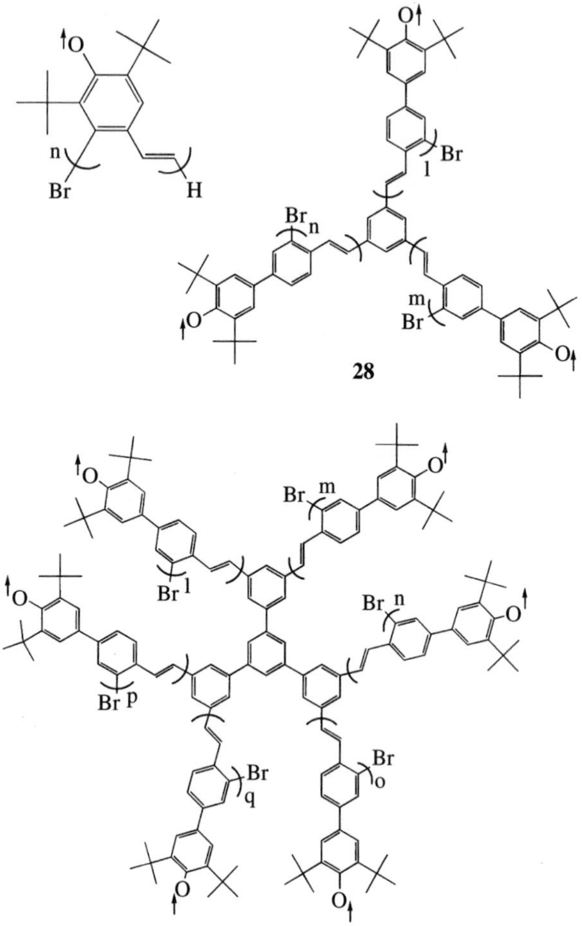

Figure 15. Phenoxy-radical based systems of Lathi and Nishide.

others in that they are chemically stable and easily handled even at room temperature.

6. Conclusions and future prospects

The dream of polymer magnets has proved difficult to translate into reality and progress has been slow. The best systems made so far have involved ferromagnetic coupling of a few tens of spins at temperatures close to absolute zero. However, real progress has been made in the last few years. The feasibility of creating high-spin clusters has been established, as has the feasibility of exploiting at least three types of spin carrier. These leads will doubtless be developed and higher spin species will result. The great virtue of organic molecular materials is that the range of synthetic methods now at our disposal allows us to engineer structures almost at will at a molecular level and to continuously refine our designs. It will become clear whether or not this will lead to materials with useful or interesting bulk properties.

7. Experimental

(a) General

The general experimental procedures and instrumentation used and those for the magnetometer studies (Bushby *et al.* 1997*a*; Bushby & Gooding 1998), for the UV studies and EPR studies (Bushby *et al.* 1999) and for the cyclic voltammetry studies (Bushby *et al.* 1997*b*) have all been described in previous publications, as have the syntheses of the polymer **14** (Bushby & Gooding 1998) and the synthesis and X-ray crystallography of compound **15** (Bushby *et al.* 1999).

(b) Preparation of N, N'-bis-(2-methoxyphenyl)-N, N'-diphenyl-paraphenylenediamine **18**

A stirred mixture of N, N'-diphenylphenylenediamine (9.44 g, 0.04 mol), 2-iodoanisole (25.46 g, 0.11 mol), powdered anhydrous potassium carbonate (59.62 g, 0.43 mol), copper powder (13.73 g, 0.22 mol), and degassed 1,2-dichlorobenzene (120 cm^3) was refluxed for 2 h under an argon atmosphere in the absence of light. After cooling to *ca.* 70 °C, the insoluble inorganic material was filtered off and the dark brown filtrate collected. The insoluble material was washed with dichloromethane (2 × 100 cm^3). The combined filtrate and organic phase was washed with dilute aqueous ammonia and thoroughly with water. After drying with magnesium sulphate the mixture was cooled in ice causing precipitation of a light grey solid, which was recrystallized from hexane: CHCl$_3$ (2:1) to give the product **18** (8.3 g, 72%) as light brown crystals with m.p. (melting point) 163–164 °C. (Found: C, 79.90%; H, 5.60%; N, 5.45%. C$_{32}$H$_{28}$N$_2$O$_2$ requires: C, 80.37%; H, 5.93%; N, 5.93%.) δ_H 3.63 (6H, s, -OMe), 6.81–6.94 (10H, m, ArH), 6.93 (4H, s,p-disubstituted ArH) and 7.12–7.22 (8H, m, ArH); *m/z*, 472 (M$^+$, 100%), 259 (33%) and 2 (14%).

(c) Preparation of N, N'-bis-(4-bromo-2-methoxyphenyl)-N, N'-bis-(4-bromophenyl)-paraphenylenediamine **19**

A solution of bromine (9.47 g, 59.28 mmol) in dry chloroform (70 cm^3) was added to a stirred solution of amine **18** (7.00 g, 14.82 mmol) in dry chloroform (70 cm^3) at 0 °C over 60 min. After the addition, a blue–green reaction mixture formed which was allowed to warm to room temperature with stirring overnight. The reaction mixture then was washed with water (2 × 150 cm^3), dilute sodium *meta*-bisulphite solution (150 cm^3) and brine (150 cm^3), and dried with magnesium sulphate. The solvent was removed under reduced pressure to give a white solid. The product was purified by column chromatography on silica gel, eluting with 30% dichloromethane in hexane to give the product **19** as small white needles (8.31 g, 72%), m.p. 136–137 °C. (Found: C, 48.30%; H, 3.20%; N, 3.25%; Br, 40.35%. C$_{32}$H$_{24}$Br$_4$N$_2$O$_2$ requires: C, 48.53%; H, 3.03%; N, 3.54%; Br, 40.36%.) δ_H (CDCl$_3$) 3.64 (6H, s, -OMe), 6.74 (4H, d, $J = 12.0$ Hz, ArH), 6.89 (4H, s, ArH), 7.0–7.15 (6H, m, ArH) and 7.24 (4H, d, $J = 12.0$ Hz, ArH); *m/z*, 788 (M$^+$, 100%) and 391 (13%).

(d) *Preparation of* N, N'-*bis-(2-methoxyphenyl-4-boronic acid)*-N, N'-*bis-*
(phenyl-4-boronic acid)-paraphenylenediamine **21**

Under rigorously dry conditions, *n*-butyl lithium in hexane (24.5 cm³, 1.6 M, 37.9 mmol) was added to a stirred solution of the brominated amine **20** (7.46 g, 9.39 mmol) in dry tetrahydrofuran (300 cm³) under an atmosphere of argon at −78 °C over a period of 30 min. The initial development of a pale yellow solution was followed by the formation of a thick white suspension 1 h after addition. After stirring for a further 3 h at −78 °C, a cold solution of tri-*iso* -propylborate (28.24 g, 150.0 mmol) in dry tetrahydrofuran (70 cm³) at −78 °C was cannulated into the reaction mixture under argon, and the entire content stirred for another 2 h before being warmed to room temperature overnight (*ca.* 12 h). The mixture was again cooled to −78 °C and the intermediate ester **20** was hydrolysed by adding 2 M hydrochloric acid (42 cm³), and slowly warming to room temperature with stirring for a further 1 h before workup. Ether (100 cm³) was added and the organic layer separated. The aqueous layer was extracted with further ether (2 × 20 cm³). The combined organic layers were washed with water and dried with magnesium sulphate. Solvents were removed under reduced pressure without heating and the green residue redissolved in tetrahydrofuran (100 cm³). The product was precipitated by adding hexane (100 cm³) and the pale green precipitate was collected by filtration. The solid was washed thoroughly with acetone to give the product **21** (4.08 g, 67%) as a pale green solid. The product is insoluble in ether, acetone, dichloromethane, chloroform and hexane, but is soluble in tetrahydrofuran and ethanol. (Found: C, 60.60%; H, 5.30%; N, 3.70%. $C_{32}H_{32}B_4N_2O_{10}$ requires: C, 59.95%, H, 4.94%; N, 4.33%.) In order to obtain spectroscopic data and to confirm the structure, an ester derivative was made by heating the product in the presence of ethane diol. (Found: C, 62.40%; H, 5.30%; N, 3.35%. $C_{40}H_{40}B_4N_2O_{10}$ requires: C, 62.7%; H, 5.50%; N, 3.65%.) δ_H (CDCl₃) 3.65 (6H, s, -OMe), 4.33 (8H, s, O—CH₂—CH₂—O), 4.38 (8H, s, O—CH₂—CH₂—O), 7.01 (4H, s, ArH of symmetrical AB system), 7.10–7.24 (2H, m, ArH), 7.25 (8H, dd $J = 12.0$ and 200 Hz, ArH of unsymmetrical AB system), 7.30–7.47 (4H, m, ArH); m/z, 752 (M⁺, 100%).

(e) *Preparation of polymer* **22**

The boronic acid **21** (2.03 g, 3.13 mmol) was stirred in ethanol (20 cm³) for 1 h and the solution degassed using a stream of argon bubbles. A mixture of 3,5-dibromo-tetradecylbenzene (Murray *et al.* 1994) (2.71 g, 6.26 mmol) and palladium tetrakis-triphenylphosphine (0.12 g, 0.11 mmol) was stirred for 30 min in dry degassed toluene (70 cm³) under argon. The ethanol solution was added followed by dilute sodium carbonate (2 M, 25 cm³), via a cannula to maintain inert conditions. The entire mixture was heated under reflux under an argon atmosphere while stirring for 4 days. When cooled, the aqueous layer was removed and extracted with chloroform (2 × 50 cm³). A quantity of black 'gel' was found in the remaining organic layer, which was also extracted in the same manner. The combined organic layers were washed with dilute HCl (50 cm³), water (2 × 100 cm³) and dried with magnesium sulphate. The volume of solvent was reduced to *ca.* 15 cm³ and the solution dropped into cold methanol, causing precipitation of a fine light-green powder which was collected at the pump and washed with acetone. The process was repeated several times to remove impurities, eventually yielding the polymer product as a green powder (2.30 g, 67%). (Found: C, 84.05%; H, 9.05%; N, 2.55%; Br, 0%. Repeat unit $C_{60}H_{84}O_2N_2$ requires: C, 85.10%; H, 8.75%; N, 2.76%; Br, 0.00%.) δ_H (CDCl$_3$) 0.83 (m, b, -CH$_3$ of C$_{14}$ alkyl chain), 1.03–1.42 (m, b, methylenes of C$_{14}$ alkyl chain), 1.70 (m, b, Ph—CH$_2$—CH$_2$), 2.64 (m, b, Ph—CH$_2$—), 3.76 (m, b, -OMe), 6.81–7.28 (m, b, ArH of phenylenediamine groups) and 7.26–7.32 (m, b, ArH of 1,3,5-trisubstituded benzene groups). $M_w = 36\,000$ ($M_n/M_w = 1.9$) from GPC against a polystyrene standard.

We thank the EPSRC for funding this work, Dr H. Blythe (Physics, Sheffield University) and Dr K. U. Neumann (Physics, Loughborough University) for the SQUID magnetometry and N. D. Tyers (IRC in Polymer Science and Technology, School of Chemistry, Leeds University) for the GPC measurements.

References

Allinson, G., Bushby, R. J. & Paillaud, J.-L. 1994 Organic molecular magnets: the search for stable building blocks. *J. Mater. Sci.: Mater. Electronics* **5**, 67–74.

Bushby, R. J. 1999 High-spin building blocks. In *Magnetic properties of organic materials* (ed. P. M. Lathi), pp. 179–196. New York: Dekker.

Bushby, R. J. & Gooding, D. 1998 Higher-spin π multiradical sites in doped polyarylamine polymers. *J. Chem. Soc. Perkin Trans.* **2**, 1069–1075.

Bushby, R. J., McGill, D. R. & Ng, K. M. 1996 Organic magnetic polymers. In *Magnetism: a supramolecular function* (ed. O. Kahn), pp. 181–204. Dordrecht: Kluwer.

Bushby, R. J., McGill, D. R., Ng, K. M. & Taylor, N. 1997a p-doped high spin polymers. *J. Mater. Chem.* **7**, 2343–2354.

Bushby, R. J., McGill, D. R., Ng, K. M. & Taylor, N. 1997b Disjoint and coextensive diradical diions. *J. Chem. Soc. Perkin Trans.* **2**, 1405–1414.

Bushby, R. J., Gooding, D., Thornton-Pett, M. & Vale, M. E. 1999 High-spin p-doped arylamine polymers. *Mol. Cryst. Liquid. Cryst.* (In the press.)

Condon, E. U. & Odataoi, H. 1980 *Atomic structure.* Cambridge University Press.

Dougherty, D. A. 1991 Spin control in organic molecules. *Acc. Chem. Res.* **24**, 88–94.

Hoshino, Y., Miyaura, N. & Suzuki, A. 1988 Novel synthesis of isoflavones by the palladium-catalysed cross-coupling reactions of 3-bromochromones with arylboronic acids. *Bull. Chem. Jap.* **61**, 3008–3010.

Hughbanks, T. & Yee, K. A. 1990 Superdegeneracies and orbital delocalization in extended organic systems. In *Magnetic molecular materials* (ed. D. Gatteschi, O. Kahn, J. S. Miller & F. Palacio), pp. 133–144. Dordrecht: Kluwer.

Mataga, A. 1968 Possible ferromagnetic states of some hypothetical hydrocarbons. *Theor. Chim. Acta* **10**, 372–376.

Murray, M. A., Kaszynski, P., Kasisaki, D. A., Chang, W.-H. & Dougherty, D. A. 1994 Prototypes for the polaronic ferromagnet. Synthesis and characterisation of high-spin organic polymers. *J. Am. Chem. Soc.* **116**, 8152–8161.

Nishide, H., Kaneko, T., Nii, T., Katoh, K., Tsuchda, E. & Lathi, P. M. 1996 Poly(phenylenevinylene)-attached phenoxy radicals: ferromagnetic interactions through planarized and π-conjugated skeletons. *J. Am. Chem. Soc.* **118**, 9695–9704.

Nishide, H., Miyasaka, M. & Tsuchda, E. 1998 High-spin polyphenoxyls attached to star-shaped poly(phenylenevinylene)s. *J. Org. Chem.* **63**, 7399–7407.

Ovchinnikov, A. A. 1978 Multiplicity of the ground states of large alternant organic molecules with conjugated bonds. *Theor. Chim. Acta* **47**, 297–304.

Rajca, A. 1994 Organic diradicals and polyradicals. From spin coupling to magnetism? *Chem. Rev.* **94**, 871–893.

Rajca, A. & Uttamapanya, S. 1993 Towards organic synthesis of a magnetic particle: dendritic polyradicals with 15 and 31 centres for unpaired electrons. *J. Am. Chem. Soc.* **115**, 10 688–10 694.

Rajca, A., Utamapanya, S. & Thayumanavan, S. 1992 Polyarylmethyl octet ($S = \frac{7}{2}$) heptaradical and undecet ($S = 5$) decaradical. *J. Am. Chem. Soc.* **114**, 1884–1885.

Rajca, A., Wongsrirat-Anakul, J., Rajca, S. & Cerny, R. 1998 A dendritic macrocyclic organic polyradical with a very high spin of $S = 10$. *Angew. Chem. Int. Edn. Engl.* **37**, 1229–1232.

Sato, K., Yano, M., Furuichi, M., Shiomi, D., Takui, T., Abe, K., Itoh, K., Higuchi, A., Katsuhiko, K. & Shirota, Y. 1997 Polycation high-spin states of one- and two-dimensional (diarylamino)benzenes, prototypical model units for purely organic ferromagnets as studied by pulsed ESR/electron spin transient nutation spectroscopy. *J. Am. Chem. Soc.* **119**, 6607–6613.

Shiomi, D., Sato, K., Takui, T., Itoh, K., McGill, D. R., Ng, K. M. & Bushby, R. J. 1997 High-spin states of hyperbranced polycationic organic polymers as studied by FT pulse-ESR/electron spin transient nutation. *Mol. Cryst. Liq. Cryst.* **306**, 513–520.

Stickley, K. R. & Blackstock, S. C. 1994 Triplet dication and quartet trication of a triaminobenzene. *J. Am. Chem. Soc.* **116**, 11 576–11 577.

Stickley, K. R. & Blackstock, S. C. 1995 Cation radicals of 1,3,5-tris(diarylamino)benzene. *Tet. Lett.* **36**, 1585–1588.

Stickley, K. R., Selby, T. D. & Blackstock, S. C. 1997 Isolable polyradical cations of polyphenylenediamines with populated high-spin states. *J. Org. Chem.* **62**, 448–449.

Utamapanya, S., Kakegawa, H., Bryant, L. & Rajca, A. 1993 High-spin polymers. Synthesis of 1,3-connected polyarylmethane and its carbopolyanion and polyradical. *Chem. Mater.* **5**, 1053–1055.

Wienk, M. M. & Janssen, R. A. J. 1996*a* Stable triplet state di(cation radicals) of a N-phenylaniline oligomer by acid doping. *Chem. Commun.*, pp. 267–268.

Wienk, M. M. & Janssen, R. A. J. 1996*b* Stable triplet state di(cation radicals) of a *meta-para* aniline oligomer by acid doping. *J. Am. Chem. Soc.* **118**, 10 626–10 628.

Yoshizawa, K. 1993 Molecular orbital study of quartet molecules with trigonal axis of symmetry. *Mol. Cryst. Liq. Cryst.* **233**, 323–332.

Yoshizawa, K., Chano, A., Ito, A., Tanaka, K., Yamabe, T., Fujita, H. & Yamauchi, J. 1992*a* Electron spin resonance of the quartet state of 1,3,5-tris(diphenylamino)benzene. *Chem. Lett.* 369–372.

Yoshizawa, K., Chano, A., Ito, A., Tanaka, K., Yamabe, T., Fujita, H., Yamauchi, J. & Shiro, M. 1992*b* ESR of the cationic triradical of 1,2,3-tris(diphenylamino)benzene. *J. Am. Chem. Soc.* **114**, 5994–5998.

Yoshizawa, K., Hatanaka, M., Ago, H., Tanaka, K. & Yamabe, T. 1996 Magnetic properties of 1,3,5-tris[bis(p-methoxyphenyl)amino]benzene cation radicals. *Bull. Chem. Soc. Jap.* **69**, 1417–1422.

Room-temperature molecule-based magnets

BY MICHEL VERDAGUER, A. BLEUZEN, C. TRAIN, R. GARDE,
F. FABRIZI DE BIANI AND C. DESPLANCHES

Laboratoire de Chimie Inorganique et Matériaux Moléculaires,
Unité Associée au CNRS 7071, Case 42, Université Pierre et Marie Curie,
75252 Paris Cedex 05, France

Room-temperature magnets belonging to the Prussian blue family were obtained
recently through mild chemistry methods, i.e. molecular solution chemistry at room
temperature and pressure. The paper describes the rational way followed to reach this
goal and the prospects opened. First, the structure of Prussian blues and how it allows
variation of the electronic structure and exchange interaction through the cyanide
bridge is recalled. Then it is shown how the systematic use of orbital models and
simple semiempirical calculations, combined with the Néel molecular field approach,
helps in increasing the Curie temperature up to room temperature in vanadium–
chromium derivatives. Some methods are then proposed to improve the magnetic
properties and some examples of applications in demonstrators, devices, photomag-
netism, etc, are given. Finally, we mention some exciting challenges in molecular
magnetism, including the preparation of single molecule magnets at room temperature.

Keywords: Prussian blue analogues; molecule-based magnets; hexacyanometallates;
vanadium–chromium system; thin layers; devices

1. Introduction

Ten years ago, writing a paper with such a title would have been impossible: no room-
temperature molecule-based magnet was available on the molecular market. Many
efforts in this direction had been made with some remarkable achievements, both
theoretical and experimental, which have been reported in conference proceedings (see
Güdel 1985; Miller *et al.* 1989; Gatteschi *et al.* 1991), but it was necessary to wait until
1991 to read the first report, by Miller and coworkers (see Manriquez *et al.* 1991), of a
room-temperature molecule-based magnet. Since then, a rapid development of the
field has taken place, thanks to the endeavours of inorganic chemists engaged in
molecular magnetism and particularly in the synthesis of systems with a tunable
three-dimensional magnetic ordering temperature T_C. This allows the presentation of
some examples and prospects.

One of the driving forces of such fundamental research is the need for high-density
storage materials: the molecular level is indeed the ultimate step for storage capacity.
Moreover, the accessibility of polyfunctional materials may answer some unsolved
technological problems. Truly, in the field of materials, molecular chemistry is
expected to present several drawbacks: the solids are diluted, their mechanical
properties are often weak and their thermal stability is low. However, this is
compensated by many other advantages: solubility in various solvents and the ability

to crystallize beautiful complex structures; the uniform character of the structures and properties of the molecules; the frequent biocompatibility; the low density; the optical properties (transparency, etc.); and the ability to present cofunctions (magnetism and optics, conductivity and magnetism, etc.). Some of the reasons that convinced chemists of the possibility of working in this direction are also related to the feasibility of working in 'mild chemistry' conditions; that is, to easily realize experiments in solution at room temperature and pressure.

A strong theoretical corpus relies on the ligand field theory and on the mechanisms of the interactions between neighbouring unpaired electrons in a molecule or in a solid. The existence of various correlations between structures and properties facilitates the prevision of the physical properties and of the related structures and is another strong argument to attract scientists to this field. It is therefore not surprising to encounter an active community trying different steps to cover the domain from the isolated complex to the three-dimensional ordered magnet.

In the present paper we describe our approach to room-temperature magnets using old-fashioned Prussian blue analogues, but looking at them with new eyes. The nature of the exchange interaction in these materials is discussed to explain the strategy developed to obtain room-temperature magnets. The experimental endeavours to improve the properties of room-temperature magnets are then presented. The use of these compounds in devices, their synthesis as thin films and, finally, a prospective concerning single-molecule magnets concludes the paper.

2. How to get room-temperature magnets?

Everyday life is full of examples of useful room-temperature metallic or oxide magnets. Our purpose is not to reproduce or to expand such results by high-temperature preparations of metals, alloys or oxides. Our challenge is to start from molecules, existing as such in solution, suitably designed to build a magnetic solid where intermolecular interactions are strong enough in the three directions of space for the magnetic ordering to survive at room temperature.

Three situations may arise in a molecule-based solid:

(A) the molecules are neutral, independent, with weak van der Waals interactions between them—the solid is truly molecular, the cooperativity is weak;

(B) the molecules are charged and well insulated from each other by counterions— the solid is made of molecular ions, the cooperativity is weak;

(C) the molecules have strong interactions between each other (bonding and exchange)—the cooperativity is strong.

Figure 1 shows the Lewis acid–base interaction between two molecules in case (C). An important feature is that most frequently the unpaired 'magnetic' electrons are non-bonding or antibonding electrons.

In this way it is possible, on the one hand, to design the molecular precursor by choosing the metallic ions (electronic structure, charge, spin), the nature of the ligands (charge, bonding, spin), the nature of the metal–ligand interaction and therefore the ligand field. This flexibility is the one of molecular chemistry. On the other hand, it is possible to combine the molecular precursors with different

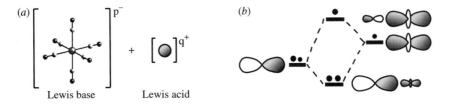

Figure 1. Schematic representation of Lewis acid–base interactions using hexacyanometallate and paramagnetic cations: (*a*) structural model; (*b*) orbital interaction.

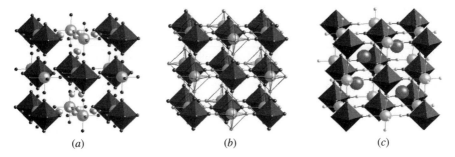

Figure 2. Typical structures of cubic Prussian blue analogues: (*a*) $M_3'^{II}[M^{III}(CN)_6]_2$, nH_2O, $M_3'M_2$; (*b*) $M'^{III}[M^{III}(CN)_6]$, $M_1'M_1$; (*c*) $CsM'^{II}[M^{III}(CN)_6]$, $Cs_1M_1'M_1$.

assembling units: the resulting periodical crystal structure allows the chemist to play with the vacancies and the doping. This flexibility is the one of solid-state chemistry. We chose the polycyanometallates $[M(CN)_n]^{p-}$ as the molecular precursors ten years ago since we found several of their properties promising. Indeed, they have a large structural flexibility, from one to eight coordination positions, and they have a wide coordination chemistry, being good Lewis bases. They are soluble in water, generally very stable and, some of them, especially with d^3 and d^6 ions, are quite inert; all these conditions are required to avoid secondary reactions of the cyanide ligand. The cyanide ligand is amphiphilic and allows the binding of different metallic ions on the two sites. Polycyanometallates allow the preparation of three-dimensional networks through the easy precipitation of neutral solids from solution. The overall symmetry of the bimetallic systems obtained from polycyanometallates is easy to control: the very symmetric isotropic cubic Prussian blues derived from the octahedral $[M(CN)_6]^{p-}$ analogues are well known (see Ludi & Güdel 1973); nevertheless, anisotropic and lower symmetry precursors like $[M(CN)_7]^{p-}$ or $[M(CN)_8]^{p-}$ lead to more intricate three-dimensional structures (see Larionova *et al.* 1998).

Our first attempts focused mainly on *hexacyanometallates* since the octahedral $[M(CN)_6]^{p-}$ precursors allow one to develop interactions with six neighbours in the three directions of space. They give rise to highly symmetrical systems, allowing the nature of the interactions in the linear arrangement M—CN—M′ to be understood. They span a large range of M/M′ stoichiometries, thanks to the choice of the charges on M and M′, and this is further increased by the possible insertion of monovalent cations in the tetrahedral sites of the structure (figure 2).

The T_C of the Prussian blue $Fe_4^{III}[Fe^{II}(CN)_6]_3 \cdot 14H_2O$ was measured as $T_C = 5.6$ K (see Ito *et al.* 1968). Replacing Fe^{III} and Fe^{II} by other transition metal ions, in a systematic study of three-dimensional cyanides, Babel, at Marburg university, demonstrated that an ordering temperature above liquid nitrogen could be reached: the highest T_C was obtained with the ferrimagnet $CsMn^{II}[Cr^{III}(CN)_6]$, $T_C = 90$ K (see Babel 1986). One of the key points in Babel's study is the use of two neighbouring paramagnetic ions M and M′ in the lattice, otherwise a paramagnetic system is obtained.

A short-range ferromagnetic coupling leads to a total spin ground state which is the sum of the spins: $S_T = S_A + S_B$; a short-range ferromagnetic coupling leads to a total spin ground state which is the difference of the spins: $S_T = |S_A - S_B|$. The key point is that antiferromagnetism between two neighbours bearing different spins leads to a magnetic ground spin state, known as ferrimagnetism after Néel (see Néel 1948). The ground-state spin value is less than when the coupling is ferromagnetic, but it is amazing to observe that orbital overlap can transform a phenomenon (antiferromagnetism) into the opposite phenomenon (magnetism) in a dialectic process.

To demonstrate the feasibility of the molecular approach to high-T_C systems, we used two theoretical tools: (*a*) the molecular field, proposed by Néel in three-dimensional systems and well known among physicists and solid-state chemists (see Néel 1948; Herpin 1968; Goodenough 1963); and (*b*) the orbital model of exchange interaction designed by Kahn for molecular chemistry (see Girerd *et al.* 1981; Kahn 1993).

In Néel (1948) an expression is given of the susceptibility of an AB ferrimagnet near the ordering temperature from which we can deduce

$$kT_C = \frac{Z|J|}{N_A g^2 \beta^2} \sqrt{C_A C_B}, \qquad (2.1)$$

where Z is the number of magnetic neighbours, $|J|$ is the absolute value of the exchange interaction, C_A and C_B are the Curie constants of A and B, N_A is the Avogadro constant, g is a mean g Landé factor and β is the Bohr magneton. Everything being equal, to increase T_C, it is therefore necessary to enhance Z, $|J|$ and the Curie constants C_A and C_B ($C_i = S_i(S_i + 1)$, if S_i is the spin of site i).

Z is controlled by the stoichiometry; with hexacyanometallates its maximum is 6 when the ratio M′/M is 1; its average becomes 4 when M′/M$_{2/3}$. More generally, for M′/M$_z$, $Z = 6z$ and one easily finds that $T_C(M'/M_z) = zT_C(M'/|MM)$.

To foresee the $|J|$ value, Hoffmann's orthogonalized magnetic orbitals model (see Hay *et al.* 1975) or Kahn's magnetic orbitals model can be used. Both models predict that orthogonal orbitals give rise to ferromagnetism and that orbital overlap gives rise to antiferromagnetism. Due to its better versatility with asymmetric systems we prefer to use Kahn's model. The following expression summarizes the model in the case of two electrons on two sites, described by two identical orbitals: the singlet–triplet energy gap, equal to J ($J = E_S - E_T$), is given by

$$J = 2k + 4\beta S. \qquad (2.2)$$

Here k is the bielectronic exchange integral (positive) between the two non-orthogonalized magnetic orbitals a and b, β is the monoelectronic resonance integral (negative) and S is the monoelectronic overlap integral (positive) between a and b. The positive term represents the ferromagnetic contribution J_F, favouring parallel alignment of the spins, while the negative term is the antiferromagnetic contribution J_{AF}, favouring antiparallel alignment of the spins. When the two a and b orbitals are different, in the

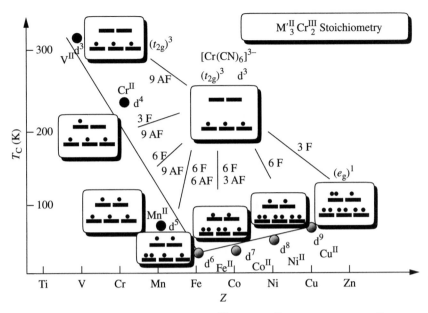

Figure 3. Exchange pathways in a binuclear [Cr^{III}—CN—M$'^{II}$] linear unit when M$'^{II}$ is a divalent ion of the first period of the transition metals and $T_C = f(Z_{M'})$ in related M$_3'^{II}$Cr$_2^{III}$ systems.

absence of a rigorous analytical treatment, the semiempirical relation proposed by Kahn can be used (see Kahn 1993):

$$J = 2k + 2S(\Delta^2 - \delta^2)^{1/2}, \tag{2.3}$$

where δ is the initial energy gap between the magnetic orbitals and Δ is the energy gap between the molecular orbitals derived from them.

When several electrons are present on each centre, n_A on one side, n_B on the other, J can be described by the sum of the different 'orbital pathways', $J_{\mu\nu}$, weighted by the number of electrons (see Kahn 1993):

$$J = (\Sigma_{\mu\nu} J_{\mu\nu})/n_A n_B. \tag{2.4}$$

The kind and the number of orbital interactions between the magnetic orbitals of M (t_{2g} local symmetry) and M$'$ (t_{2g} and e_g local symmetries) in the linear arrangement M—C≡N—M$'$ are described in figure 3, where M is Cr^{III} and $(t_{2g})^3$ and M$'$ are the divalent cations of the first period of transition metal ions.

It is clear from this simple scheme that ferromagnetic interactions can be expected with M$' = $ Cu^{II} and Ni^{II}, weak interactions with Co^{II} and Fe^{II} and stronger and stronger antiferromagnetic interactions from Mn^{II} to V^{II}. The experimental results in the M$_3'^{II}$Cr$_2^{III}$ stoichiometry series confirm the predictions of the model: the Cu^{II} and Ni^{II} derivatives are indeed ferromagnetic (see Gadet *et al.* 1992); the Co^{II} and Fe^{II} derivatives display the lowest T_C. The Mn^{II} analogue is ferrimagnetic (see Babel 1986), a Siberian ambient temperature is reached with Cr^{II} (see Mallah *et al.* 1993) and the first magnetic Prussian blue above European room temperature is obtained with V^{II}, $T_C = 315$ K or 42 °C, ferrimagnetic (see Ferlay *et al.* 1995). A very important point is the magnitude of the ferromagnetic contributions: all the Cr^{III}(CN)$_6$

antiferromagnetic interactions   no interaction

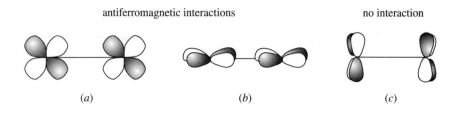

(a)   (b)   (c)

ferromagnetic interactions

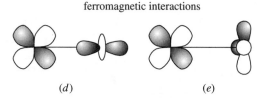

(d)   (e)

Figure 4. Orbital interactions in a binuclear unit $[(CN)_5M'^{II}(\mu\text{-}NC)M^{III}(CN)_5]^{6-}$.

derivatives with M' going from Fe^{II} to Ni^{II} are ferromagnetic. When the ferromagnetic contributions are progressively suppressed from Mn^{II} to V^{II}, large antiferromagnetic J values and high T_C are obtained.

The validity of the above simple orbital approach can be confirmed by semiempirical extended Hückel calculations (see Landrum 1992) on a series of heterobinuclear complexes with formula $[(CN)_5M'^{II}(\mu\text{-}NC)M^{III}(CN)_5]^{6-}$ and new insights on the systems are obtained.

Interactions were computed in the binuclear unit $[(CN)_5M^{II}(\mu\text{-}NC)M'^{III}(CN)_5]^{6-}$, where M^{II} = Ti, V, Cr, Mn, Fe, Co and M^{III} = Ti, V, Cr, Mn, Fe, so that there was always at least one antiferromagnetic pathway between M' and M. We therefore chose M ions possessing no more than five electrons, which occupy the t_{2g} orbitals, and M' ions possessing no more than seven electrons, so that in a high spin state they also have t_{2g} orbitals that are at least partly filled. Bond distances were kept fixed (M—$C = 2.07$ Å, C—$N = 1.13$ Å, M'—$N = 2.10$ Å) for all the combinations to monitor separately the influence of the electronic configuration.

As shown in figure 4, in the binuclear systems two couples (a and b) of the six t_{2g} orbitals strongly interact, the strength of the coupling being the same. In a three-dimensional network the missing interaction becomes possible (c), as strong as the other ones, so that it can be summed with the others. Due to the symmetry of the system it is sufficient to analyse the interaction for a single couple and to sum up on all the combinations present in a three-dimensional network. Two of the possible interactions between t_{2g} and e_g orbitals (d and e) are also shown in figure 4.

For each pair of metals the value of Δ and δ have been calculated (see equation (2.3)). Some features may be revealed by decomposing the antiferromagnetic term as $(\Delta^2 - \delta^2) = (\Delta - \delta)(\Delta + \delta)$. Two effects contribute to the tendency of the electrons to pair: (i) a strong interaction between the magnetic orbitals is represented by the term $(\Delta - \delta)$; and (ii) a possible charge transfer between the two magnetic orbitals, represented by the term $(\Delta + \delta)$. The dependence of Δ on δ is non-trivial. Δ contains δ itself, but it is enhanced when δ is small, since the orbitals better fit in

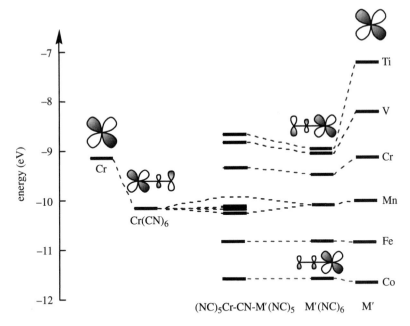

Figure 5. Composition of the magnetic orbitals and variation
of π and π^* interactions when M′ is varied.

energy. Therefore, to have a strong antiferromagnetic interaction it is necessary to have the best combination between these two terms, which have opposite constraints: $(\Delta - \delta)$ is higher when δ is smaller, while $(\Delta + \delta)$ is higher when δ is higher. In cases when the interelectronic repulsion on one centre is strong (localized electrons), charge transfer is unlikely and the $(\Delta - \delta)$ becomes the leading term.

Figure 5 shows the interaction between the magnetic orbitals of the $Cr(CN)_6$ and $M'(NC)_6$ moieties. In both halves the t_{2g} orbitals are more or less involved in a back-donation interaction with the empty CN π^* orbitals and in a repulsion interaction with the full CN π orbitals. This double interaction severely affects the energy of the magnetic orbitals, which depends on the diffuseness of the d orbitals of the metallic ions (lower values of the ζ Slater coefficient in our calculation) and on the relative position of the d orbitals with respect to the energy of the π and π^* CN orbitals. For example, in the case presented in figure 5, all the M′ sites from Ti to Cr are mainly involved in a π^* back-donation interaction, while for M′ going from Mn to Co the π repulsion interaction almost compensates the first interaction. Two important consequences follow: (i) it is not obvious which magnetic orbitals will have the best fit in energy; and (ii) the composition of the magnetic orbitals is affected as shown in figure 5, as well as the overlap between them. The stronger interaction generally involves moieties possessing different metals and corresponds to cases where Δ and δ possess their *lower* values.

The corresponding $(\Delta^2 - \delta^2)$ values are shown in figure 6. The higher $(\Delta^2 - \delta^2)$ values are reached for the $[(CN)_5 Ti^{II}(\mu\text{-}NC)M^{III}(CN)_5]^{6-}$ binuclear units and, among them, the highest is found for the $[Ti^{II}Fe^{III}]$ pair. Such a result appears counter-intuitive at first sight. It is understood by noting that the better combination for

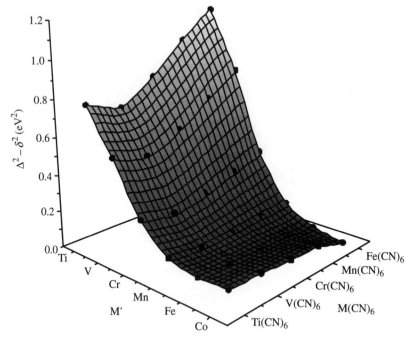

Figure 6. Values of $\Delta^2 - \delta^2$ for the different couples of $[M(CN)_6]$ and M'.

$(\Delta + \delta)$ and $(\Delta - \delta)$ is obtained due to both the high values of Δ and δ and to the diffuseness of the d orbital of Ti, which allows an efficient overlap between the magnetic orbitals (high $\Delta - \delta$). The calculated surface also suggests that the compounds containing an M cation chosen in the first members of the series should always possess the stronger antiferromagnetic interaction, whatever the M' metal ion chosen.

As foreseen from the Néel expression, the strength of J_{AF} is not the only parameter to determine the T_C value since the number of possible interactions and the spin of both sites affect the critical temperature. Indeed, the product $Z|J|/\sqrt{C_{M'}C_M}$ is much more reliable in predicting T_C for the different combinations of M and M' ions. If we consider $z = 1$ ($Z = 6$) we obtain, from equation (2.5),

$$T_C \propto \frac{|J|}{N_A g^2 \beta^2} \sqrt{C_M C_{M'}} \propto \frac{N(\Delta^2 - \delta^2)}{n_M n_{M'}} \sqrt{C_{M'} C_M}. \tag{2.5}$$

All the symbols maintain their usual meaning and N is the number of interactions. Since $C_i \propto n_i(n_i + 2)$, we multiplied the data plotted in figure 6 by the factor

$$N\sqrt{\frac{(n_M + 2)(n_{M'} + 2)}{n_M n_{M'}}},$$

so obtaining the surface shown in figure 7.

The result is in good agreement with the experimental values of T_C (see Dunbar et al. 1997) and confirms that in these series the best choice is indeed the vanadium–chromium pair.

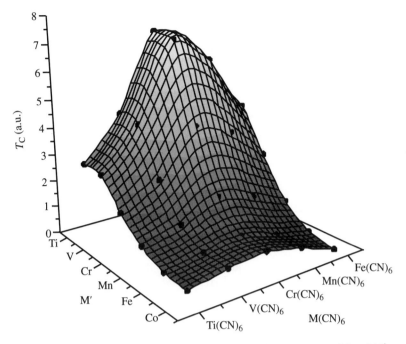

Figure 7. Computed values of T_C (in arbitrary units) when varying M and M'.

The ferromagnetic contributions cannot be explicitly calculated, being bielectronic in origin. Nevertheless, some hints can be attained even from these simple calculations. The exchange integral k_{ab} between two magnetic orbitals $|a\rangle$ and $|b\rangle$ is directly proportional to the squared overlap density $\rho_{ab} = |a\rangle|b\rangle$ (see Charlot & Kahn 1980). This contribution to the exchange interaction is always present and it becomes the only contribution when the overlap S_{ab} is zero, i.e. for the t_{2g}–e_g interactions. For example, in a binuclear complex $[(CN)_5Ni^{II}(\mu\text{-}NC)Cr^{III}(CN)_5]^{6-}$, the $\rho_{t_{2g}-e_g}$ overlap density can be computed and illustrated by the projection on the xz-plane of the product $|(Cr)t_{2g}\text{-}xz\rangle|(Ni)e_g\text{-}z^2\rangle$ of the orbitals of the fragments $Ni^{II}(NC)_6$ and $Cr^{III}(CN)_6$. The orbital composition is given by the extended Hückel output. Figure 8 shows the overlap density between the t_{2g}–xz magnetic orbital centred on chromium and the e_g–z^2 magnetic orbital centred on the nickel and the presence of high overlap density zones. The peak value of ρ is found in the surrounding of the nitrogen atom of the cyanide, which is a peculiar feature of the electronic structure of the cyanide bridge giving rise to strong ferromagnetic interactions, strong positive J values and high T_C.

To close this section we focus on the vanadium–chromium (VCr) system and study the role of the number of interactions when varying the stoichiometry. All the experimentalists working on this system are faced with the problem of the oxidation of vanadium (II) during the reaction. In some cases, a mixed-valence V^{II}–V^{III} system arises; in some other cases vanadyle $V^{IV}O$ is present. The general formula of a V^{II}–V^{III}/Cr^{III} system is written as $(C_y V_\alpha^{II} V_{1-\alpha}^{III}[Cr^{III}(CN)_6]_z \cdot nH_2O)$. ($C_y^I$ is a simplified formulation and covers more complex situations corresponding to different cationic or anionic species in the lattice to insure electroneutrality). To arrive at a simpler

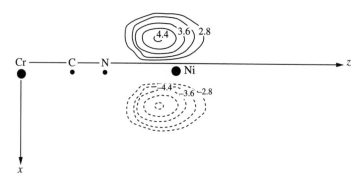

Figure 8. Projection on the xz-plane of the overlap density ρ between the t_{2g}–xz-type magnetic orbital centred on chromium and the e_g–z_2-type magnetic orbital centred on nickel.

representation of the phenomenon, we consider as a first step that the J value of the V^{II}–Cr^{III} pair is the same as the J value of the V^{III}–Cr^{III} pair (see Ferlay *et al.* 1999). In these conditions T_C follows equation (2.6) and depends in a straightforward manner upon the fraction of vanadium (II), α and upon the stoichiometry z:

$$T_C \propto z\sqrt{z}|J|\sqrt{C_{Cr^{III}}}\sqrt{C_V} \propto z\sqrt{z}\sqrt{C_{Cr^{III}}}\sqrt{[\tfrac{7}{4}\alpha - 2]}. \tag{2.6}$$

The surface giving $T_C = f(z, \alpha)$ is shown in figure 9 and demonstrates that the highest T_C is expected for $z = 1$ and $\alpha = 1$, i.e. for the compound $C^I V^{II}[Cr^{III}(CN)_6]$. This is a misleading conclusion of course, since $C^I V^{II}[Cr^{III}(CN)_6]$ is definitively an antiferromagnet due to the exact cancellation of the $\frac{3}{2}$ spins of the vanadium (II) and chromium (III). Nevertheless, a compound with a composition close to this one is expected to have a T_C close to the predicted maximum.

The above simple calculations and considerations not only allow one to put our simple qualitative orbital model on a firmer basis but also give new tools to improve the magnetic characteristics (T_C, magnetization at saturation, coercivity, etc.) of room-temperature magnets.

3. Attempts to improve the characteristics of room-temperature Prussian blue magnets

We will still stick to the [VCr] system. The first point to raise is that real chemistry is much more complex than any model: the expected $V_3^{II}[Cr^{III}(CN)_6]_2$ is an amorphous non-stoichiometric compound, a deep-blue mixture of V^{II} and V^{III}, $V_\alpha^{II}V_{(1-\alpha)}^{III}[Cr^{III}(CN)_6]_{0.86} \cdot 2.8H_2O$, with $\alpha = 0.42$, and very sensitive to air (see Ferlay *et al.* 1995). The magnetization at saturation is very weak ($0.15\mu_B$) in line with the proposed formulation. The coercive field is one of a very soft magnet (25 Oe at 10 K). To try to improve the properties and to obtain better characterized materials, our group and the groups of Girolami, Hashimoto and Miller undertook new syntheses. Many factors play an important role in the successful synthesis of [VCr] analogues with improved magnetic properties. Among them are the oxidation state of the vanadium ions, the stoichiometry and the structure (presence of counterions, solvent, etc.) and the crystallinity and the size of the particles. Today, a set of non-stoichiometric Prussian blue analogues $C_y^I V[Cr^{III}(CN)_6]_z \cdot nH_2O$ is available with T_C

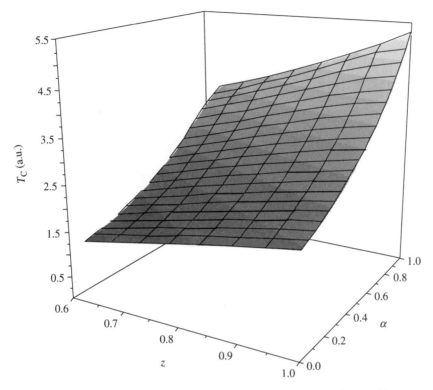

Figure 9. Computed values of T_C (in arbitrary units) when varying the stoichiometry z and the fraction α of vanadium (II) in the VCr system.

varying between 295 and 376 K (above 100 °C) and various magnetizations at saturation (Hashimoto *et al.*, personal communication).

We report here some of our endeavours in dealing with: (i) the influence of (an)aerobic conditions; (ii) the stoichiometry, controlled by the addition of counter-cations; (iii) the nature of the counteranion; and (iv) the nature of the solvent (see Dujardin *et al.* 1998) using the above theoretical analysis.

(i) Prepared in aerobic conditions, the [VCr] system gives a crystalline dark green product $(V^{IV}O)_3[Cr^{III}(CN)_6]_2$ with a much lower T_C (110 K) (see Ferlay *et al.* 1999). The use of anaerobic conditions appears a necessary condition to avoid the oxidation of V^{II}, even if sol–gel procedures seem to limit it (see Holmes & Girolami 1999).

(ii) Concerning stoichiometry and countercations, there is a warning: the compound with the highest computed T_C, V/Cr = 1, corresponds to an antiferromagnet. Indeed, the antiparallel alignment of the neighbouring spins in the magnetically ordered phase leads to a resulting total magnetization M_T which is the difference between the magnetization arising from the subset of chromium ions M_{Cr} and the magnetization from the subset of vanadium ions, M_V:

$$M_T = |M_{Cr} - M_V|. \tag{3.1}$$

It is then necessary to explore the role of the stoichiometry, tuned by countercations such as Cs^+, and to compute systematically the magnetization to be sure not to have

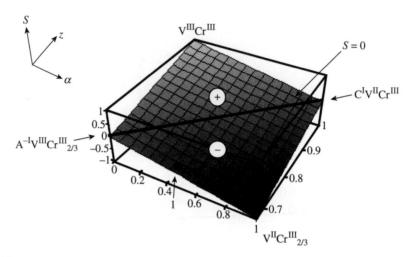

Figure 10. Variation of the spin values S versus z and α in mixed valency compounds,
$$C_y^I V_\alpha^{II} V_{1-\alpha}^{III} [Cr^{III}(CN)_6]_z, \; nH_2O.$$

a true antiferromagnet by exact cancellation of the spins ($M_{Cr} = M_V$). In the case of the $C_1^I V_\alpha^{II} V_{(1-\alpha)}^{III} [Cr^{III}(CN)_6]_z \cdot nH_2O$ system,

$$M_T = -(3z - \alpha - 2). \tag{3.2}$$

The corresponding plane is shown in figure 10, where α varies between 0 and 1 and z varies between $\frac{2}{3}$ and 1.

When $M_{Cr} > M_V$, the magnetic moments of the chromium ions are aligned parallel to an external applied field; when $M_V > M_{Cr}$, the magnetic moments of the vanadium ions now lie parallel to the field. We can verify that the first room-temperature magnet, with $M_S = 0.15\mu_B$, was close to the line $M_T = 0$, shown in bold in figure 10! We can also use this new predicting tool to prepare new compounds with enhanced magnetization. The two compounds formulated,

$$\mathbf{A} : V_{0.45}^{II} V_{0.53}^{III} (V^{IV}O)_{0.02} [Cr(CN)_6]_{0.69} (SO_4^{2-})_{0.23} (K_2SO_4)_{0.02} \cdot 3H_2O,$$
$$\mathbf{B} : Cs_{0.82} V_{0.66}^{II} (V^{IV}O)_{0.34} [Cr(CN)_6]_{0.92} (SO_4^{2-})_{0.20} \cdot 3 \cdot 6H_2O,$$

present similar T_C but a larger magnetization than in the first reported [VCr] ($0.36N_A\beta$ for $\mathbf{A}$ and $0.42N_A\beta$ for $\mathbf{B}$) (see Dujardin *et al.* 1998). The calculated M_T value is positive for $\mathbf{A}$ ($M_T = +0.36N_A\beta$) and negative for $\mathbf{B}$ ($M_T = -0.42N_A\beta$). The absolute values are in good agreement with the experimental values but the sign of the two M_T values is opposite. This intriguing situation can be revealed amazingly by X-ray magnetic circular dichroism (XMCD), a new X-ray spectroscopy developed with synchrotron radiation, which is an element and orbital selective magnetic local probe, able to give direct information about the local magnetic properties of the photon absorber (direction and magnitude of the local magnetic moment) whatever the state of the sample (crystals, powders, etc.). Besides the usefulness of XMCD to determine locally the spin orientation on each metal ion, the measurements explicitly showed

(1) the antiferromagnetic coupling between vanadium and chromium ions; and

(2) the change of the majority spin location caused by the modification of the chemical composition from vanadium in **A** to chromium in **B**.

The influence of the stoichiometry on the T_C has recently been beautifully shown (see Holmes & Girolami 1999), enabling T_C to reach 376 K in a compound formulated as $K^I V^{II}[Cr^{III}(CN)_6]$, displaying a weak magnetization, as expected.

(iii) and (iv) When looking at the structure (figure 2c), one realizes that vacancies and channels in Prussian blues can accommodate guest molecules: cations, as seen before, but also anions and solvent, which can either improve or compromise the short- and long-range structural organization of the material. We therefore varied the nature of the anions (I^-, SO_4^{2-}, Cl^-) and of the solvents (water, methanol, THF, etc.) under anaerobic conditions, according to the equation

$$3[V(H_2O)_6]A_2 + 2B_3[Cr(CN)_6] \xrightarrow{\text{solvent}} V_3[Cr(CN)_6]_2 \cdot xH_2O, \qquad (3.3)$$

where A (anions) $= I^-$ (weak Lewis base), SO_4^{2-} (weak Lewis base), Cl^- (stronger Lewis base), etc.; B $= K^+$, NH_4^+, NBu_4^+, etc.; and solvent $= H_2O$, MeOH, THF, DMSO, etc.

We obtained:

1 $V[Cr(CN)_6]_{0.73}(NBu_4^+)_{0.19}(I^-)_{0.03} \cdot 4H_2O$ (solvent, H_2O; starting materials, $(NBu_4)_3[Cr(CN)_3]$ and $[V(MeOH)_6]I_2$; $T_C = 326$ K);

2 $V[Cr(CN)_6]_{0.64}(NBu_4^+)_{0.04}(SO_4^{2-})_{0.15} \cdot 4 \cdot 4H_2O$ (solvent, H_2O; starting materials, $(NBu_4)_3[Cr(CN)_3]$ and $(NH_4)_2V(SO_4)_2 \cdot 6H_2O$; $T_C = 320$ K);

3 $V[Cr(CN)_6]_{0.69}(NBu_4^+)_{0.1}(Cl^-)_{0.14} \cdot 4H_2O$ (solvent, H_2O; starting materials: $(NBu_4)_3[Cr(CN)_3]$ and $VCl_2 \cdot 4H_2O$; $T_C = 308$ K).

We detected the Curie temperatures with a new device, which allowed the measurement of the permeability of the magnetic materials in their storage vessels. The T_C values obtained are indicated in figure 11. They are the same as those measured by SQUID.

In the IR spectra, we observe in the 2300–2000 cm^{-1} region two bands with different intensities for all compounds. The stronger band is centred around 2112 cm^{-1} and the weaker band is centred around 2162 cm^{-1}. We observe that the stronger band shifts to lower energy and the weaker band shifts to higher energy (ΔE is higher) when the compound has a smaller amount of V^{III} and its T_C is higher. Moreover, we observe that the weaker band not only shifts to higher energy (*ca.* 2170 cm^{-1}) but also its intensity increases when the compound is in a more oxidized state. We did not observe the band at 981 cm^{-1} corresponding to the $\nu_{as} V^{IV} = O$ stretching which confirms that there is no $V^{IV} = O$ in the compounds.

Counterions have large effects on T_C. Disorder in the Prussian blues' structures can be revealed at two levels: local order (coordination of anions which decreases the number of magnetic neighbours; distortion of the Cr—CN—V sequence, which decreases $|J|$); and long-range order (quality of the crystallization and size of the

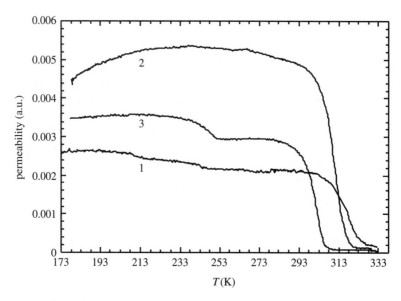

Figure 11. Permeability versus temperature measured for compounds **1**, **2** and **3**.

magnetic particles). The ions that do not distort the structures lead to better crystalline compounds with improved magnetic properties. The large size (NBu_4^+) and weakly coordinating ions, like I^-, induce less disorder and favour higher T_C.

We also carried out the reaction in organic solvents to avoid the oxidation of V^{II} by water and to study the influence of the solvent on the magnetic properties. In the same general conditions as for **1**, we got compounds **4** and **5** from reaction in methanol and THF, respectively:

    4 $V[Cr(CN)_6]_{0.69}(I^-)_{0.03} \cdot 1.5MeOH$ (solvent, methanol; starting materials, $(NBu_4)_3[Cr(CN)_3]$ and $[V(MeOH)_6]I_2$; $T_C = 200$ K);

    5 $V[Cr(CN)_6]_{0.68}(NBu_4^+)_{0.27}(I^-)_{0.33} \cdot 1.9H_2O$ (solvent, THF; starting materials, $(NBu_4)_3[Cr(CN)_3]$ and $[V(MeOH)_6]I_2$; $T_C = 290$ K).

Compounds **4** and **5** display the same IR bands as above. Figure 12 indicates the T_C of **1**, **4** and **5**.

We propose that the changes in the magnetic properties are mainly due to the changes in the structure induced by the kinetics of solvent exchange in $[V(solvent)_6]^{2+}$ (see Lincoln & Merbach 1995): when the exchange is fast, in water, the substitution of the solvent molecules by CN^- around V^{II} is more effective, the number of magnetic neighbours increases and so does T_C. In methanol, the kinetics of exchange is slow, hence the poor quality of the magnetism. In THF, with faster kinetics, the T_C of the bright blue compound reaches room temperature (17 °C).

4. Prospects

We give here three directions that we are exploring, among others: magnetic devices; magneto-optical properties; and conditions to get single-molecule magnets at room temperature.

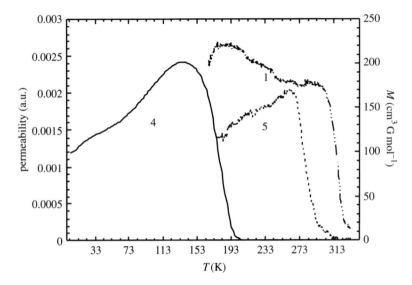

Figure 12. Permeability (compounds **1** and **5**) and magnetization
(compound **4**) versus temperature.

(*a*) *Devices from room-temperature magnets*

A molecule-based magnet, such as our [VCr] system, is a useful tool to illustrate easily, near room temperature, what a Curie temperature is. Figure 13 displays a demonstrator designed for this purpose.

[VCr] is sealed in a glass vessel under argon and suspended at the bottom of a pendulum (equilibrium, position (2) in absence of permanent magnet). It is then cycled between its two magnetic states: the three-dimensional ordered ferrimagnetic state, when $T < T_C$; and the paramagnetic state, when $T > T_C$. The three steps are as follows.

(i) The 'room-temperature magnet' is cold ($T < T_C$, ferrimagnetic state). It is attracted ($\rightarrow$) by the permanent magnet and deviates from the vertical direction towards position (1). The light beam is focused at position (1) above the permanent magnet. The light heats the sample.

(ii) When $T > T_C$, the hot 'room-temperature magnet' is in the paramagnetic state. It is no longer attracted and moves away from the magnet ($\leftarrow$) under the influence of its own weight. It is then air cooled and its temperature decreases.

(iii) When $T < T_C$, the cold 'room-temperature magnet' is attracted again by the permanent magnet ($\rightarrow$) and comes back to position (1). The system is ready for a new oscillation.

The demonstrator works well in our laboratory: millions of cycles have been accomplished without any fatigue. The demonstrator is an example of a thermo-dynamical machine working between two energy reservoirs with close temperatures (sun and shadow) allowing the conversion of light into mechanical energy. We are exploring other practical applications: thermal probes, magnetic switches, etc.

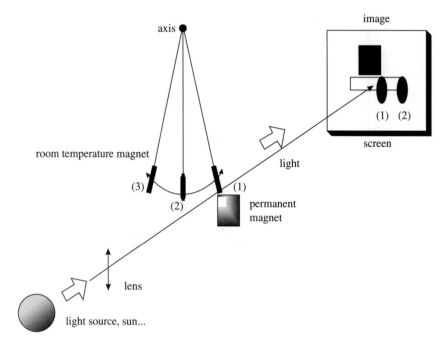

Figure 13. Demonstrator using the properties of a room-temperature magnet
to transform light into mechanical energy.

(b) Thin layers

Magnetic Prussian blue analogues display bright colours and transparency, among other interesting properties. To exploit these optical properties it is useful to prepare thin films, and a 1 μm thick film of vanadium–chromium magnet is indeed quite transparent. The best way to prepare thin films of these materials is electrochemical synthesis and various films have been prepared from hexacyanoferrates in recent years. To obtain [VCr] thin films, experimental conditions are extensively modified. One actually wants to produce and to stabilize the highly oxidizable ions. Thus, strongly negative potentials are applied at the working transparent semiconducting electrode. The deposition of the [VCr] film is obtained from aqueous solutions of $[Cr^{III}(CN)_6]^{3-}$ and V^{III} at fixed potential or by cycling the potential.

In passing, we point to a further interesting property of [VCr] thin films, in that they also exhibit electrochromism. The way is open for the preparation of electrochromic room-temperature magnets.

The magnetization of a transparent magnetic film, protected by a transparent glass cover, can then be probed by measuring the Faraday effect, corresponding to spin-dependent modifications of the polarization of the transmitted light. Spectroscopic measurements in the UV–vis range reveal information about the magnetization of the sample and, moreover, about its electronic structure. Moreover, observing the Faraday effect at room temperature in these compounds is the first step in demonstrating that these materials can be used in magneto-optical information storage. We succeeded in obtaining for the first time on a [VCr] room-temperature magnet a magneto-optical signal at room temperature through a transparent

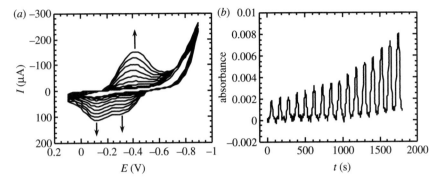

Figure 14. Cyclovoltammograms and electrochromism during the deposition
of thin layer [VCr] by cycling the applied potential.

semiconducting electrode. Further studies are needed to control the purity and the homogeneity of the layers, and to correlate the magneto-optical effects to the local magnetization of vanadium and chromium. They are underway, in a very promising area.

(c) *Towards room-temperature single-molecule magnets?*

The enhancement of the T_C values in Prussian blue magnets relies on the control of the interaction between the spin carriers. A brand new direction appeared some years ago when Sessoli *et al.* (1993) characterized the behaviour of a single-molecule magnet: a uniaxial anisotropic molecule presenting a ground state with a spin S and a zero-field splitting D displays an anisotropy barrier $E_a = DS_z^2$ and behaves as a magnet when $kT \ll E_a$. If the challenge in this field is the synthesis of a single-molecule magnet at room temperature, two conditions must be fulfilled: (i) the population of the high spin ground state at room temperature (and an interaction constant between the paramagnetic centres at *ca.* 400 K); and (ii) a uniaxial anisotropic barrier $E_a = DS_z^2$ larger than the thermal quantum at room temperature, i.e. *ca.* 400 K. The later condition can be reached, for example, with a total spin $S = 20$ and an anisotropy of the ground state $D = 1$ K. The challenge is difficult but not completely impossible to overcome. We are working in two directions: (i) the enhancement of the molecular spin, and we arrive at $S = \frac{27}{2}$ (see Mallah *et al.* 1995; Scuiller *et al.* 1996); and (ii) the improvement of the anisotropy of the system (see Scuiller 1999).

5. Conclusion

We have tried to show, through different examples essentially chosen from our own chemistry, that looking at old systems with new concepts can give rise to new systems, new properties and new applications. It is impossible to quote in a short text devoted to room-temperature magnets all the recent endeavours using cyanometallate precursors to build magnetic systems from zero to three dimensions. Bimetallic cyanide boxes and superboxes, complex frameworks using less symmetrical precursors than the hexacyano anions, thin layers, photomagnetic and multifunctional systems are appearing or near completion. We can expect in the near future important achievements in this field, where the flexibility of molecular precursors combines with the versatility of solid-state chemistry to build the new solid.

We thank the European Community (grants ERBCHRXCT92080, ERBFMBICT 972644 and ERBFMRXCT980181) and the European Science Foundation for financial support ('Molecular Magnets' Programme).

References

Babel, D. 1986 *Comments. Inorg. Chem.* **5**, 285.

Charlot, M. F. & Kahn, O. 1980 *Nouv. J. Chim.* **4**, 567.

Dujardin, E., Ferlay, S., Phan, X., Cartier dit Moulin, C., Sainctavit, P., Baudelet, F., Dartyge, E., Veillet, P. & Verdaguer, M. 1998 *J. Am. Chem. Soc.* **120**, 11 347.

Dunbar, K. R. & Heintz, R. A. 1997 In *Progress in inorganic chemistry* (ed. D. K. Kenneth), vol. 45, p. 282. Wiley.

Ferlay, S., Mallah, T., Ouahès, R., Veillet, P. & Verdaguer, M. 1995 *Nature* **378**, 701.

Ferlay, S., Mallah, T., Ouahès, R., Veillet, P. & Verdaguer, M. 1999 *Inorg. Chem.* **38**, 229.

Gadet, V., Mallah, T., Castro, I., Veillet, P. & Verdaguer, M. 1992 *J. Am. Chem. Soc.* **114**, 9213.

Gatteschi, D., Kahn, O., Miller, J. S. & Palacio, F. (eds) 1991 *Molecular magnetic materials.* NATO ASI Series E, vol. 198. Dordrecht: Kluwer.

Girerd, J. J., Journeaux, Y. & Kahn, O. 1981 *Chem. Phys. Lett.* **82**, 534.

Goodenough, J. B. 1963 *Magnetism and the chemical bond.* New York: Interscience.

Güdel, H. U. 1985 In *Magneto-structural correlation in exchange coupled systems* (ed. R. D. Willett, D. Gatteschi & O. Kahn). NATO ASI Series, p. 329. Dordrecht: Reidel.

Hay, P. J., Thibeault, J. C. & Hoffmann, R. 1975 *J. Am. Chem. Soc.* **97**, 4884.

Herpin, A. 1968 *Théorie du magnétisme.* Saclay: INSTN-PUF.

Holmes, S. M. & Girolami, S. G. 1999 *J. Am. Chem. Soc.* **121**, 5593.

Ito, A., Suenaga, M. & Ono, K. 1968 *J. Chem. Phys.* **48**, 3597.

Kahn, O. 1993 *Molecular magnetism.* New York: VCH.

Landrum, G. A. 1992 YAeHMOP: yet another extended Hückel molecular package. URL: http:// overlap.chem.cornell.edu:8080/yaehmop.html.

Larionova, J., Sanchiz, J., Gohlen, S., Ouahab, L. & Kahn, O. 1998 *Chem. Commun.*, p. 953.

Lincoln, S. F. & Merbach, A. E. 1995 *Adv. Inorg. Chem.* **42**, 1.

Ludi, A. & Güdel, H. U. 1973 *Struct. Bonding* **14**, 1.

Mallah, T., Thiébaut, S., Verdaguer, M. & Veillet, P. 1993 *Science* **262**, 1554.

Mallah, T., Auberger, C., Verdaguer, M. & Veillet, P. 1995 *J. Chem. Soc. Chem. Commun.*, p. 61.

Manriquez, J. M., Yee, G. T., McLean, R. S., Epstein, A. J. & Miller, J. S. 1991 *Science* **252**, 1415.

Miller, J. S. & Dougherty, D. A. (eds) 1989 *Proc. Conf. on Ferromagnetic and High Spin Molecular Based Materials. Mol Cryst. Liq. Cryst.* **176**.

Néel, L. 1948 *Ann. Phys. Paris* **3**, 137.

Sciuller, A. 1999 PhD thesis, Université Pierre et Marie Curie, Paris.

Sciuller, A, Mallah, T., Nivorozkhin, A., Verdaguer, M. & Veillet, P. 1996 *New J. Chem.* **20**, 1.

Sessoli, R., Tsai, H.-L., Schake, A. R., Wang, S., Vincent, J. B., Folting, K., Gatteschi, D., Christou, G. & Hendrickson, D. N. 1993 *J. Am. Chem. Soc.* **115**, 1804.

Design of novel magnets using Prussian blue analogues

By Kazuhito Hashimoto and Shin-ichi Ohkoshi

Research Center for Advanced Science and Technology, The University of Tokyo, 4-6-1 Komaba, Meguro-ku, Tokyo 153-8904, Japan

The main aim of this paper is to show how well we can design novel molecule-based magnets using Prussian blue analogues based on theory. We obtained various novel magnets, i.e. magnets containing both ferro- and antiferromagnetic interactions without spin frustration, a magnet having two compensation temperatures, coloured transparent magnets, a photomagnet and a photoinduced pole inversion magnet. These magnets were designed using simple theories, including molecular field theory.

Keywords: molecule-based magnet; Prussian blue; photomagnet; pole inversion; molecular field theory; supercharge interaction

1. Introduction

Control of magnetic properties is an attractive but very difficult problem in the field of magnetic materials. Although the theoretical analyses of various magnetic properties of materials are often successful in great detail, theoretical prediction for the production of novel magnets is difficult in general and especially for classical metal or metal oxide magnets. One of the main reasons is that various types of exchange and/or superexchange interactions exist among many spin sources. Moreover, metal or metal ion substitution often causes structural distortions. For molecule-based magnets, however, theoretical design could become more useful than for the classical magnets, because the molecular magnets can be obtained through a selection of appropriate spin sources (see Gatteschi *et al.* 1991; Kahn 1993; Miller & Epstein 1994). This is especially appropriate for Prussian blue analogues, some of which become ferromagnets, e.g. CsNi[Cr(CN)$_6$], $T_c = 90$ K (Entley & Girolami 1994), while others are ferrimagnets, e.g. V$_{1.16}$[Cr(CN)$_6$], $T_c = 315$ K (Ferlay *et al.* 1995), where the theoretical prediction of magnetic properties is very useful, as shown later. Let us consider the magnetic coupling of the metal centres in Prussian blue analogues. The coupling cannot be explained either by weak dipole–dipole interactions or by direct exchange interactions via overlapping metal orbitals. The coupling is described in terms of a superexchange mechanism through the cyanide ligands. The superexchange mechanism is summarized on the basis of the Goodenough–Kanamori rule that includes consideration of the bond angle and the symmetry of the metal and ligand orbitals concerned (see Goodenough 1958, 1959; Kanamori 1959). There are two mechanisms for superexchange interactions: the kinetic exchange mechanism (J_{KE}) and the potential exchange mechanism (J_{PE}) (figure 1*a*) (see Ginsberg 1971). On the one hand, kinetic exchange is mediated by a direct pathway of the overlapping orbitals, which connects the two interacting magnetic orbitals. It is antiferromagnetic in nature as a consequence of the Pauli principle, leading to an antiparallel spin ordering via a common covalent bond. On the other hand, potential exchange is effective between orthogonal magnetic orbitals with comparable orbital energy. In this case Hund's rule leads to a parallel spin alignment, i.e. a ferromagnetic interaction. In the case of Prussian blue analogues, the metal d-orbitals are split into t_{2g} and e_g sets by the CN ligands (figure 1b). Moreover, in most cases, they maintain their face-centred cubic (FCC) structure even after substituting the metal ions. Therefore, based on magnetic orbital symmetry, we can understand whether the orbital superexchange among each orbital on the metal ions is J_{KE} or J_{PE}. When the magnetic orbital symmetries of the metals are the same, the superexchange interaction might be J_{KE}. Conversely, when the magnetic orbital symmetries of the metals are different, the superexchange interaction might be J_{PE}. The total superexchange interaction is given by the sum of all the orbital exchange contributions between transition metal ions (see Ginsberg 1971; Nishino $et\ al.$ 1997).

For example, consider the case of the hexacyanochromate cyanide $A_y^{II}[Cr^{III}(CN)_6]$, with Cr^{III} being $(t_{2g})^3$ and $S_{Cr} = 3/2$. If each magnetic orbital of A^{II} has e_g symmetry, there is no overlap between the Cr^{III} and A^{II} magnetic orbitals. In this situation, the potential exchange mechanism becomes dominant, leading to a ferro-magnetic interaction between Cr^{III} and A^{II}. In fact, in $CsNi^{II}[Cr^{III}(CN)_6]$, with a high-spin state for Ni^{II} $((t_{2g})^6(e_g)^2$, $S_{Ni} = 1)$, a ferromagnetic interaction operates between Cr^{III} and Ni^{II}, as reported by Entley & Girolami (1994). However, when each A^{II} magnetic orbital has t_{2g} symmetry, the overlap between the $t_{2g}(A)$ and $t_{2g}(Cr)$ orbitals gives rise to kinetic exchange, leading to an antiferromagnetic interaction. If both t_{2g} and e_g electrons are present on A^{II}, the superexchange coupling constant (J_{AB}) is described as the sum of the ferromagnetic $(J_{PE} > 0)$ and antiferromagnetic $(J_{KE} < 0)$ orbital contributions. Kinetic exchange usually operates in preference to potential exchange, i.e. $|J_{KE}| > |J_{PE}|$ (see Anderson 1959). For example, Griebler & Babel (1982) showed that the interaction between Cr^{III} and Mn^{II} with a high-spin state for Mn^{II} $((t_{2g})^3(e_g)^2$, $S_{Mn} = 5/2)$ in $CsMn^{II}[Cr^{III}(CN)_6]$ is antiferromagnetic, and the compound is a ferrimagnet. In most Prussian blue analogues, therefore, we can readily understand the superexchange interactions.

2. Design of the superexchange interactions

Here, we will discuss the magnetic properties of Prussian blue analogues containing both ferromagnetic and antiferromagnetic interactions (see Ohkoshi $et\ al.$ 1997a, b). The magnets have both ferromagnetic and ferrimagnetic characteristics simulta-neously, and therefore could be called mixed ferro-ferrimagnets. It may be difficult to achieve such properties in classical metal oxide magnets, because various types of magnetic interactions involving the metal ions can operate in these systems, including superexchange interactions, direct exchange interactions, and dipole–dipole interac-tions. In Prussian blue analogues, however, their FCC structure and the relatively long distance between magnetic ions make it possible for ferromagnetic and antiferromagnetic interactions to exist independently. As prototypes exemplifying the mixed ferro-ferrimagnetism, we will discuss a series of $(Ni_x^{II}Mn_{1-x}^{II})_{1.5}[Cr^{III}(CN)_6]$ compounds, which can accommodate both ferromagnetic $(J_{NiCr} > 0)$ and antiferro-magnetic $(J_{MnCr} < 0)$ exchange interactions (figure 2).

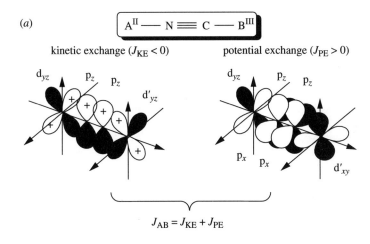

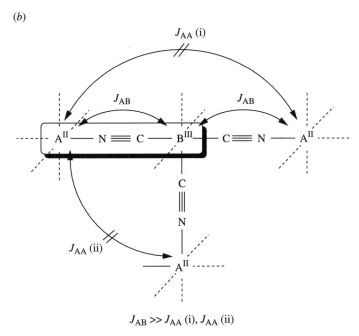

Figure 1. (*a*) The two basic mechanisms for the isotropic exchange in the magnetic coupling between the A^{II} and B^{III} ions in the CN-bridged complex. On the left is one of the significant kinetic exchange (J_{KE}) pathways ($d_{yz}\|\pi_z\|d_{yz'}$), and on the right is one of significant potential exchange (J_{PE}) pathways ($d_{yz}\|\pi_z \perp \pi_x\|d_{xy'}$). The superexchange coupling between A^{II} and B^{III} (J_{AB}) involves a superposition of J_{PE} and J_{KE}. (*b*) In the Prussian blue structure, superexchange interactions at a 180° angle between A^{II} and B^{III} are dominant over superexchange interactions of the second nearest-neighbour metals ($J_{AA(i)}$) and direct exchange interactions ($J_{AA(ii)}$).

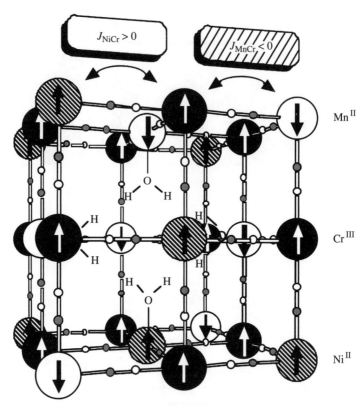

Figure 2. Schematic illustration of a ternary metal Prussian blue analogue containing both ferromagnetic and antiferromagnetic interactions. Cr^{III} and either Ni^{II} or Mn^{II}, which are randomly incorporated in the lattice, are linked in an alternating fashion.

Let us evaluate the theoretical saturation magnetization (I_s) values for materials. We first consider the model compounds, $Ni^{II}_{1.5}[Cr^{III}(CN)_6]$ ferromagnet and $Mn^{II}_{1.5}[Cr^{III}(CN)_6]$ ferrimagnet. The I_s values of these two compounds are expected to be $6\mu_B$ (due to parallel alignment of the spins, $S_{Ni} = 1$ and $S_{Cr} = 3/2$) and $4.5\mu_B$ (due to antiparallel alignment of the spins, $S_{Mn} = 5/2$ and $S_{Cr} = 3/2$), respectively. When powders of the two compounds are physically mixed, the total I_s will vary between 4.5 and $6\mu_B$, depending on the mixing ratio, as shown in figure 3 (dashed line). However, when the two compounds are mixed at an atomic level, parallel spins (Ni^{II} and Cr^{III}) and antiparallel spins (Mn^{II} and Cr^{III}) can partly or even completely cancel, depending on the mixing ratio, because Cr^{III} and either Ni^{II} or Mn^{II} are linked in an alternating fashion. In this manner, materials with I_s values anywhere in the range 0–$6\mu_B$ may be prepared. For the members of the series $(Ni^{II}_x Mn^{II}_{1-x})_{1.5}[Cr^{III}(CN)_6]$, I_s is given by

$$I_s = g\mu_B|S_{Cr} + 1.5[S_{Ni}x - S_{Mn}(1-x)]|. \tag{2.1}$$

The calculated dependence of I_s on x is shown in figure 3 (solid line). The I_s is predicted to vanish for $x = 3/7$ (0.428), and such a material should exhibit antiferromagnetic properties. Moreover, spin glass behaviour does not occur in this

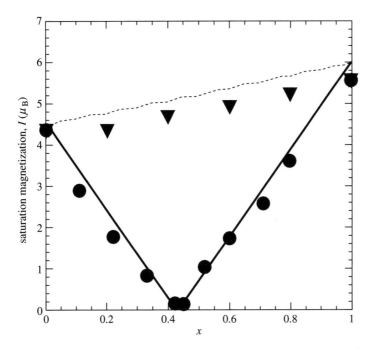

Figure 3. Calculated and experimentally observed saturation magnetizations as a function of x. Atomic-level mixture $(Ni_x^{II}Mn_{1-x}^{II})_{1.5}[Cr^{III}(CN)_6] \cdot zH_2O$: theory (——), observed (•). Macroscopic physical mixture $x \cdot Ni_{1.5}^{II}[Cr^{III}(CN)_6] + (1-x) \cdot Mn_{1.5}^{II}[Cr^{III}(CN)_6]$: calculated (– – – –); observed (▼).

series because the direction of each spin is consistent with the sign of J of the nearest-neighbour spin site.

These compounds can be prepared by reacting mixtures of $NiCl_2$ and $MnCl_2$ aqueous solutions (with a given molar ratio x_{mix}, ranging from zero to unity, of Ni^{2+} versus the total of Ni^{2+} and Mn^{2+}) with $K_3Cr(CN)_6$ aqueous solution to yield green precipitates. The x values in the resulting precipitates are in general in good agreement with the x_{mix} values used in the syntheses. The results for the X-ray powder diffraction measurements show that patterns for each member of the series are consistent with FCC structure, with the lattice constant decreasing from 10.787 to 10.467 Å with increasing x, and that the materials are not macroscopic mixtures of $Ni_{1.5}^{II}[Cr^{III}(CN)_6]$ and $Mn_{1.5}^{II}[Cr^{III}(CN)_6]$, but are ternary metal complexes in which Mn^{II} and Ni^{II} are randomly incorporated in the lattice, corresponding to the mixing ratio of $NiCl_2$ and $MnCl_2$. It is important to note that the carbon atoms of the cyano groups are always bonded to Cr^{III}, and the nitrogen atoms are always bonded to either Ni^{II} or Mn^{II}.

The observed I_s values of $Ni_{1.5}^{II}[Cr^{III}(CN)_6]$ and $Mn_{1.5}^{II}[Cr^{III}(CN)_6]$ are 5.57 and $4.38\mu_B$ for a given formula, respectively. When powders of the two compounds are physically mixed, the total I_s varied between 4.38 and $5.57\mu_B$, depending on the mixing ratio, as shown in figure 3 (triangles). Conversely, when the two compounds are mixed at an atomic level, those for the intermediate compositions varied in a systematic fashion as a function of x (figure 3, filled circles). Minimum values are obtained for x values close to $3/7$ (0.42 and 0.45); the molar I_s values are very close to zero, just at the point where parallel spins (Cr^{III} and Ni^{II}) and antiparallel spins (Mn^{II}) should completely cancel out. Thus it can be noticed that the molar I_s dependence on x follows equation (2.1) quite well. The Weiss temperature (θ_C) values increase monotonically from negative to positive values with increasing x, and this behaviour can be explained by considering that the weighted average of J_{NiCr} and J_{MnCr} changes as a function of x.

The other magnetic properties are also changed depending on x, but one of the most interesting aspects of the magnetic behaviour of this series is the thermodynamics of the magnetization. In general, ferromagnets exhibit monotonically increasing magnetization curves with decreasing temperature below T_C. However, in the case of ferrimagnets, Néel (1948) envisaged the possibility that saturation magnetization versus temperature curves could be classified into four types according to the shape, i.e. Q-, R-, P- and N-types. The Q- and R-type curves exhibit monotonic increases with decreasing temperature, while the P-type exhibits a single maximum and the N-type exhibits two maxima. For the present mixed ferro-ferrimagnets, all these four types of temperature dependence are observed with one series of compounds. In figure 4a are shown the magnetization versus temperature curves at 1000 G with various x values. These types of curves are consistent with the Néel classification, Type R ($x = 0$), Type N ($x = 0.38$), Type P ($x = 0.45$), and Type Q ($x = 1$). In addition, the magnetization of the compound for $0.33 < x < 0.43$ shows a negative value when the external magnetic field is lower than its coercive field (figure 5a). The temperature at which the magnetization becomes zero is called a compensation temperature (T_{comp}) (see Gorter 1954; Mathonière *et al.* 1994).

These temperature dependencies can be analysed using a molecular field (MF) theory (see Anderson 1964), considering only two types of superexchange couplings between the nearest-neighbour sites, one for Ni–Cr and the other for Mn–Cr, according to the model shown in figure 2. Those between the second nearest-neighbour sites (Mn–Ni and Cr–Cr) are neglected. The molecular fields H_{Ni}, H_{Mn} and H_{Cr} acting on the three sublattice sites in $(Ni_x^{II}Mn_{1-x}^{II})_{1.5}[Cr^{III}(CN)_6]$ can be expressed as follows:

$$H_{Mn} = H_0 + n_{MnCr}M_{Cr}, \qquad\qquad (2.2)$$

$$H_{Ni} = H_0 + n_{NiCr}M_{Cr}, \qquad\qquad (2.3)$$

$$H_{Cr} = H_0 + n_{CrMn}M_{Mn} + n_{CrNi}M_{Ni}, \qquad\qquad (2.4)$$

where H_0 is the external magnetic field, n_{ij} are the MF coefficients, and M_{Ni}, M_{Mn} and M_{Cr} are the sublattice magnetizations per unit volume for the Ni, Mn and Cr sites, respectively. The MF coefficients n_{ij} are related to the exchange coefficients (J_{ij}) by

$$n_{MnCr} = 2Z_{MnCr}J_{MnCr}/(\mu N(g\mu_B)^2), \qquad\qquad (2.5)$$

$$n_{NiCr} = 2Z_{NiCr}J_{NiCr}/(\mu N(g\mu_B)^2), \qquad\qquad (2.6)$$

$$n_{CrMn} = 2Z_{CrMn}J_{MnCr}/(\lambda(1-x)N(g\mu_B)^2), \qquad\qquad (2.7)$$

$$n_{CrNi} = 2Z_{CrNi}J_{NiCr}/(\lambda x N(g\mu_B)^2), \qquad\qquad (2.8)$$

where μ_B is the Bohr magneton, Z_{ij} are the numbers of the nearest-neighbour j-site ions surrounding an i-site ion, N is the total number of all types of metal ions per unit volume, and λ and μ represent the molar fractions for the A^{II} cations (total of the molar fractions for Mn^{II} and Ni^{II}) ions and for the Cr^{III} ions, respectively. When we designate the thermally averaged values of the spins of the Mn, Ni and Cr ions in their respective sites in the direction of each sublattice magnetization as $\langle S_{Mn} \rangle$, $\langle S_{Ni} \rangle$ and $\langle S_{Cr} \rangle$, the sublattice magnetization can be expressed as follows:

$$M_{Mn} = \lambda(1-x)Ng\mu_B \langle S_{Mn} \rangle, \tag{2.9}$$

$$M_{Ni} = \lambda x Ng\mu_B \langle S_{Ni} \rangle, \tag{2.10}$$

$$M_{Cr} = \mu Ng\mu_B \langle S_{Cr} \rangle. \tag{2.11}$$

Substituting (2.5)–(2.8) and (2.9)–(2.11) into (2.2)–(2.4), we have

$$H_{Mn} = H_0 + 2Z_{MnCr}J_{MnCr}\langle S_{Cr}\rangle/g\mu_B, \tag{2.12}$$

$$H_{Ni} = H_0 + 2Z_{NiCr}J_{NiCr}\langle S_{Cr}\rangle/g\mu_B, \tag{2.13}$$

$$H_{Cr} = H_0 + 2Z_{CrMn}(1-x)J_{MnCr}\langle S_{Mn}\rangle/g\mu_B + 2Z_{CrNi}xJ_{NiCr}\langle S_{Ni}\rangle/g\mu_B. \tag{2.14}$$

The magnitudes of $\langle S_{Mn} \rangle$, $\langle S_{Ni} \rangle$ and $\langle S_{Cr} \rangle$, setting $H_0 = 0$, are given by

$$\langle S_{Mn} \rangle = S_{Mn0}B_{SMn0}(2Z_{MnCr}J_{MnCr}S_{Mn0}\langle S_{Cr}\rangle/k_B T), \tag{2.15}$$

$$\langle S_{Ni} \rangle = S_{Ni0}B_{SNi0}(2Z_{NiCr}J_{NiCr}S_{Ni0}\langle S_{Cr}\rangle/k_B T), \tag{2.16}$$

$$\langle S_{Cr} \rangle = S_{Cr0}B_{SCr0}(2Z_{CrMn}(1-x)J_{MnCr}S_{Cr0}\langle S_{Mn}\rangle/k_B T \\ + 2Z_{CrNi}xJ_{MnCr}S_{Cr0}\langle S_{Ni}\rangle/k_B T), \tag{2.17}$$

where B_S is the Brillouin function, S_{Mn0}, S_{Ni0} and S_{Cr0} are the values of $\langle S_{Mn} \rangle$, $\langle S_{Ni} \rangle$ and $\langle S_{Cr} \rangle$ at $T = 0$ K and k_B is the Boltzmann constant. The $\langle S_{Mn} \rangle$, $\langle S_{Ni} \rangle$ and $\langle S_{Cr} \rangle$ can be calculated numerically. The total magnetization (M_{total}) is

$$M_{total} = -M_{Mn} + M_{Ni} + M_{Cr} \\ = Ng\mu_B[-\lambda(1-x)\langle S_{Mn}\rangle + \lambda x\langle S_{Ni}\rangle + \mu\langle S_{Cr}\rangle]. \tag{2.18}$$

The negative and positive signs are for antiparallel and parallel interactions, respectively. For $(Ni^{II}_x Mn^{II}_{1-x})_{1.5}[Cr^{III}(CN)_6] \cdot 7.5H_2O$, the numbers of the nearest neighbours Z_{ij} are $Z_{MnCr} = Z_{NiCr} = 4$; $Z_{CrMn} = 6(1-x)$, $Z_{CrNi} = 6x$; and other quantities are as follows: $\lambda = 1.5$; $\mu = 1$; $S_{Mn0} = 5/2$; $S_{Ni0} = 1$; $S_{Cr0} = 3/2$; and $g = 2$.

In order to calculate the temperature dependence, the J values of Mn^{II}–Cr^{III} and Ni^{II}–Cr^{III} must be estimated. These values can be obtained approximately from the observed T_c values for $Mn^{II}_{1.5}[Cr^{III}(CN)_6]$ and $Ni^{II}_{1.5}[Cr^{III}(CN)_6]$. The relationship between the T_c values for the Prussian blue analogues and the J values is expressed as follows:

$$T_c = 2(Z_{ij}Z_{ji})^{1/2}|J_{ij}|(S_i(S_i + 1)S_j(S_j + 1))^{1/2}/3k_B, \tag{2.19}$$

where $i =$ Ni or Mn and $j =$ Cr. Based on this equation, $J_{NiCr} = 5.6$ cm^{-1} and $J_{MnCr} = -2.5$ cm^{-1} are obtained, using T_c values of a 72 K for $Ni^{II}_{1.5}[Cr^{III}(CN)_6] \cdot 8H_2O$ and a 67 K for $Mn^{II}_{1.5}[Cr^{III}(CN)_6] \cdot 7.5H_2O$.

For the calculation of magnetization dependencies in the

$$(Ni^{II}_x Mn^{II}_{1-x})_{1.5}[Cr^{III}(CN)_6] \cdot zH_2O$$

series, the magnetizations of Ni and Cr are expected to be along the direction of the external magnetic field, because experimental curves shown here are obtained by a

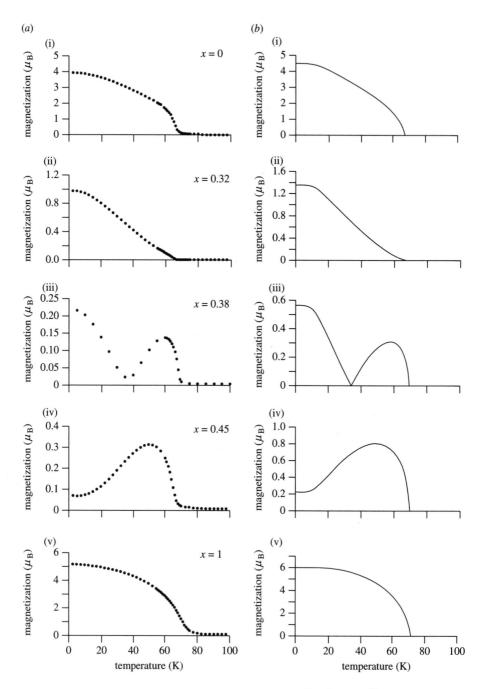

Figure 4. Magnetization versus temperature curves for $(Ni_x^{II}Mn_{1-x}^{II})_{1.5}[Cr^{III}(CN)_6] \cdot zH_2O$: (a) experimental points obtained at 1000 G and (b) calculated dependence of $|M_{total}|$ based on molecular field theory, assuming three sublattices, the two J coefficients ($J_{NiCr} = 5.6$ cm^{-1} and $J_{MnCr} = -2.5$ cm^{-1}), and the compositional parameter x.

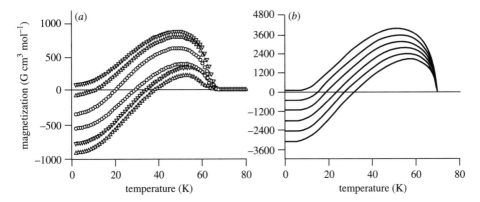

Figure 5. Magnetization versus temperature curves for $(Ni_x^{II}Mn_{1-x}^{II})_{1.5}[Cr^{III}(CN)_6] \cdot zH_2O$ ($x = 0.38$, 0.39, 0.40, 0.41, 0.42, 0.43, going from the lowest curve to the highest curve). (*a*) Experimental data obtained at 10 G and (*b*) calculated temperature dependences of magnetization M_{total} based on molecular field theory, with three sublattice sites (Ni, Mn, Cr), with J coefficients $J_{NiCr} = 5.6 \text{ cm}^{-1}$ and $J_{MnCr} = -2.5 \text{ cm}^{-1}$.

field cooling (10 G or 1000 G) method and the T_c value for $Ni_{1.5}^{II}[Cr^{III}(CN)_6]$ is higher than that for $Mn_{1.5}^{II}[Cr^{III}(CN)_6]$. Therefore, in this system, the signs of the magnetization of ferromagnetic sites (Ni and Cr) can be assumed to be positive. Note that the calculated spontaneous magnetizations are essentially the saturated values, because the MF theory does not consider the magnetization process. The negative magnetization below T_{comp} is compelled to invert along the external magnetic field direction when the external magnetic field is larger. Therefore, for simulations of observed temperature dependence curves at 10 G and 1000 G, the M_{total} and the $|M_{total}|$ curves are adopted, respectively.

Under these conditions, the temperature dependencies of the spontaneous magnetization are calculated for several different compositions. For $x = 0$, only J_{MnCr} appears in the equation, and hence the curve exhibits a monotonically increasing magnetization with decreasing temperature. For $0.33 < x < 3/7$ (0.429), the $|M_{total}|$ curves exhibit two maxima, with a minimum at T_{comp} (see curve for $x = 0.38$ in figure 4*b*). This T_{comp} shifts from T_c to 0 K with increasing x. On the other hand, M_{total} for $0.33 < x < 3/7$ exhibits negative magnetization with T_{comp} (figure 5*b*). For $x = 3/7$, T_{comp} was 0 K, and this value corresponds to that predicted by equation (2.1), in which the saturation magnetization disappears. For $x > 0.45$, the curve exhibits a single maximum. For $x = 1$, the curve exhibits a monotonic increase due to the fact that only J_{NiCr} appears. The calculated curves for $|M_{total}|$ and M_{total} qualitatively reproduce the experimental ones at 1000 G and 10 G, as shown in figures 4 and 5, respectively.

These various types of temperature dependence of the magnetization arise because the negative magnetization due to the Mn^{II} sublattice and the positive magnetizations due to the Ni^{II} and Cr^{III} sublattices have different temperature dependencies (figure 6*a*). For example, for $x = 0.38$, the relations of the magnitude among each sublattice magnetizations are as follows: $|M_{Ni}| + |M_{Cr}| > |M_{Mn}|$ at $T > T_{comp}$ and $|M_{Ni}| + |M_{Cr}| < |M_{Mn}|$ at $T < T_{comp}$, respectively (figure 6*b*). The close correspondence between the calculated and observed curves shows that the magnetic properties of the

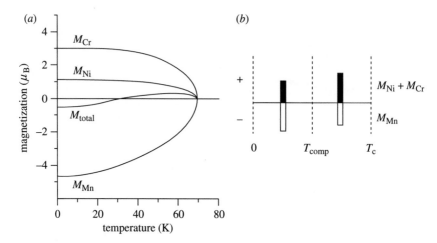

Figure 6. (*a*) Calculated temperature-dependence curves for each sublattice (M_{Mn}, M_{Ni}, M_{Cr}) and total magnetization (M_{total}) for $(Ni_{0.38}^{II}Mn_{0.62}^{II})_{1.5}[Cr^{III}(CN)_6]$ based on the three-sublattice molecular field theory. (*b*) Schematic diagram illustrating positive ($M_{Ni} + M_{Cr}$: ■) and negative magnetizations (M_{Mn}: □) versus the direction of the external magnetic field at $T > T_{comp}$ and $T < T_{comp}$, respectively.

$(Ni_x^{II}Mn_{1-x}^{II})_{1.5}[Cr^{III}(CN)_6] \cdot zH_2O$ series can be predicted using an MF model which considered only superexchange interactions between the nearest neighbours (Ni^{II}–Cr^{III} and Mn^{II}–Cr^{III}).

These results show that in Prussian blue analogues both the ferro- and antiferromagnetic interactions exist simultaneously without spin glass behaviour and their magnetic properties can be explained by MF theory.

3. A design of fickle magnet which switches pole direction twice with increasing temperature

Here we will show how well the MF theory works for the design of a novel magnet with Prussian blue analogues by designing a magnet that flips the orientation of its magnetic field twice as its temperature is increased. In other words, this magnet possesses two compensation temperatures (see Ohkoshi *et al.* 1999*a*).

The key objective here is to design Prussian blue analogues containing one ferromagnetic and two antiferromagnetic interactions simultaneously by incorporating four different metal ions into the lattice, three of them being randomly distributed, with the correct ratio being obtained by use of a calculation based on MF theory. We choose members of a new class of mixed ferro-ferrimagnets with the generic formula $(Ni_a^{II}Mn_b^{II}Fe_c^{II})_{1.5}[Cr^{III}(CN)_6] \cdot z\,H_2O$ ($a + b + c = 1$) (figure 7).

First, we evaluate theoretically the temperature dependencies of a series of compounds $(Ni_a^{II}Mn_b^{II}Fe_c^{II})_{1.5}[Cr^{III}(CN)_6] \cdot zH_2O$ using the MF theory, with the spin numbers for the four sublattice sites: $S_{Ni} = 1$; $S_{Mn} = 5/2$; $S_{Fe} = 2$ and $S_{Cr} = 3/2$. For this calculation, only the three types of superexchange couplings between the nearest-neighbour sites, Ni–Cr, Mn–Cr and Fe–Cr, are considered. The molecular fields H_{Ni},

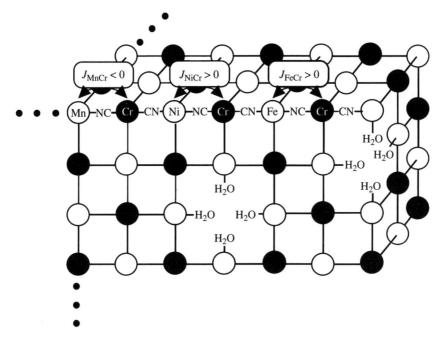

Figure 7. Schematic of mixed ferro-ferrimagnets composed of $(Ni_a^{II}Mn_b^{II}Fe_c^{II})_{1.5}[Cr^{III}(CN)_6] \cdot zH_2O$. Zeolitic water molecules in the unit cell are omitted for clarity. The ferromagnetic ($J_{NiCr} > 0$ and $J_{FeCr} > 0$) and antiferromagnetic ($J_{MnCr} < 0$) superexchange interactions can coexist without spin frustration, because the A^{II} (A = Ni, Mn or Fe) (○) and Cr^{III} (●) ions of the Prussian blue structure are linked in an alternating fashion. $(Fe_x^{II}Cr_{1-x}^{II})_{1.5}[Cr^{III}(CN)_6] \cdot zH_2O$.

H_{Mn}, H_{Fe} and H_{Cr} acting on the four sublattice sites can be expressed as follows:

$$H_{Mn} = H_0 + n_{MnCr}M_{Cr}, \tag{3.1}$$

$$H_{Ni} = H_0 + n_{NiCr}M_{Cr}, \tag{3.2}$$

$$H_{Fe} = H_0 + n_{FeCr}M_{Cr}, \tag{3.3}$$

$$H_{Cr} = H_0 + n_{CrMn}M_{Mn} + n_{CrNi}M_{Ni} + n_{CrFe}M_{Fe}, \tag{3.4}$$

where H_0, n_{ij}, J_{ij}, M_{Ni}, M_{Mn}, M_{Fe} and M_{Cr} are defined similar to those in §2. The sublattice magnetizations and total magnetization ($M_{total} = -M_{Mn} + M_{Ni} + M_{Fe} + M_{Cr}$) as a function of temperature can be evaluated using a Brillouin function. A J_{MnCr} value of -2.5 cm^{-1}, a J_{NiCr} value of $+5.6$ cm^{-1} and a J_{FeCr} value of $+0.9$ cm^{-1} are obtained from the experimental T_c values of

$$Mn_{1.5}^{II}[Cr^{III}(CN)_6] \cdot 7.5H_2O \qquad (T_c = 67 \text{ K}),$$

$$Ni_{1.5}^{II}[Cr^{III}(CN)_6] \cdot 8H_2O \qquad (T_c = 72 \text{ K}),$$

$$Fe_{1.5}^{II}[Cr^{III}(CN)_6] \cdot 7.5H_2O \qquad (T_c = 21 \text{ K}),$$

respectively. Using these J values and the compositional factors (a, b, c), the temperature dependencies are evaluated theoretically. Figure 8 shows the calculated temperature dependence curves for each sublattice and the total magnetization for the

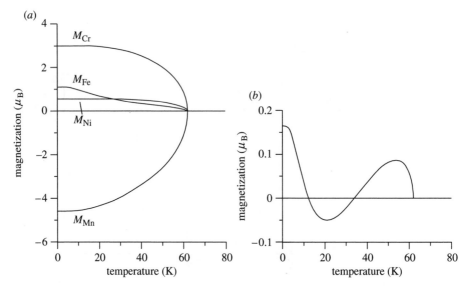

Figure 8. Calculated temperature-dependence curves for each sublattice and total magnetization for $(Ni_{0.20}^{II}Mn_{0.61}^{II}Fe_{0.19}^{II})_{1.5}[Cr^{III}(CN)_6] \cdot zH_2O$, based on molecular field theory, with four sublattice sites (Ni, Mn, Fe, Cr), with J coefficients $J_{NiCr} = +5.6$ cm^{-1}, $J_{FeCr} = +0.9$ cm^{-1} and $J_{MnCr} = -2.5$ cm^{-1}: (a) sublattice magnetization (M_{Mn}, M_{Ni}, M_{Fe}, M_{Cr}); and (b) total magnetization (M_{total}).

$(Ni_{0.20}^{II}Mn_{0.61}^{II}Fe_{0.19}^{II})_{1.5}[Cr^{III}(CN)_6] \cdot zH_2O$ system, which exhibits two compensation temperatures. On the basis of this theoretical prediction, we will synthesize members of this series.

The compounds can be prepared by reacting mixtures of $NiCl_2$, $MnCl_2$ and $FeCl_2$ aqueous solutions with $K_3Cr(CN)_6$ aqueous solution to yield light brown precipitates. Similar to the ternary metal Prussian blue analogues, Ni^{II}, Mn^{II} and Fe^{II} are randomly incorporated in sites where they are coordinated to the nitrogen ends of the cyano groups, and the Cr^{III} ions are always coordinated to the carbon ends of the cyano groups.

Figure 9 shows the field-cooled magnetization (FCM) versus temperature plots for the $(Ni_{0.22}^{II}Mn_{0.60}^{II}Fe_{0.18}^{II})_{1.5}[Cr^{III}(CN)_6] \cdot 7.6H_2O$ powder in an external magnetic field of 10 G. This compound shows two compensation temperatures ($T_{comp1} = 53$ K and $T_{comp2} = 35$ K). Its remnant magnetization versus temperature plots also show a similar behaviour. The theoretical calculation described above shows clearly that the two compensation temperatures for this compound are due to different temperature dependencies of the negative magnetization of the Mn^{II} sublattice and the positive magnetizations of the Ni^{II}, Fe^{II} and Cr^{III} sublattices. In the temperature range between 61 K (T_c) and 53 K, the positive magnetizations dominate. Between 53 K and 35 K, however, the negative magnetization outweighs the positive magnetizations. At temperatures below 35 K, the positive magnetizations again dominate due to the growth of the Fe^{II} sublattice contribution. Other magnetic properties also obey the theory of the mixed ferro-ferrimagnetism described in §2. For example, the I_s value for the composition $a = 0.24$, $b = 0.58$, $c = 0.18$ is $0.77\mu_B$, assuming the g factors for the metal ions to be 2.0. This observed value is close to the theoretical I_s value of

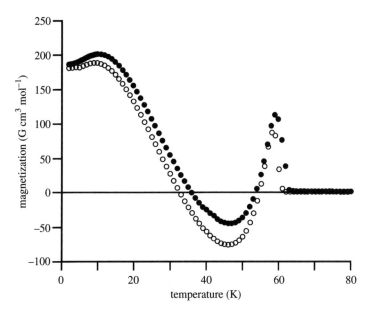

Figure 9. Experimental magnetization versus temperature curves for $(Ni^{II}_{0.22}Mn^{II}_{0.60}Fe^{II}_{0.18})_{1.5}$ $[Cr^{III}(CN)_6] \cdot 7.6H_2O$: ($\bullet$) field-cooled magnetization obtained with decreasing temperature $(80 \rightarrow 2\ K)$ in an external magnetic field of 10 G; ($\circ$) remnant magnetization obtained with increasing temperature $(2 \rightarrow 80\ K)$ after the temperature was first lowered in the applied magnetic field of 10 G.

$0.52\mu_B$. In addition, the coercive field (H_c) value of 570 G was larger than the H_c values for the respective binary compositions

$$Mn^{II}_{1.5}[Cr^{III}(CN)_6] \cdot 7.5H_2O \qquad (H_c = 6\ G),$$
$$Ni^{II}_{1.5}[Cr^{III}(CN)_6] \cdot 8H_2O \qquad (H_c = 120\ G),$$
$$Fe^{II}_{1.5}[Cr^{III}(CN)_6] \cdot 7.5H_2O \qquad (H_c = 200\ G).$$

This is because the H_c values for the mixed ferro-ferrimagnets are proportional to I_s^{-1}.

It is rather surprising that the compound, whose molecular composition is determined based on a simple magnetic theory (MF theory), gives such a complicated magnetic behaviour, possessing two compensation temperatures. This result shows that MF theory is useful for the design of novel magnets in the series of Prussian blue analogue.

4. Coloured magnetic thin films

Here we will show the design of various coloured transparent magnetic thin films (Ohkoshi *et al.* 1998). The key to this strategy is to control of the intervalence transfer bands of metal ions in a dye material exhibiting ferromagnetism. We prepared new classes of transparent magnetic thin films composed of $(Fe^{II}_x Cr^{II}_{1-x})_{1.5}[Cr^{III}(CN)_6] \cdot zH_2O$. Their colours could be controlled by controlling the compositional factor x $(= Fe^{II}/(Fe^{II} + Cr^{II}))$, e.g. colourless $(x = 0)$, violet $(x = 0.20)$, red $(x = 0.42)$ and

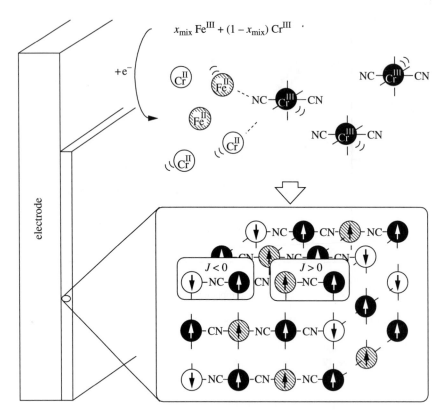

Figure 10. Schematic diagram illustrating electrochemical synthesis of magnetic thin films with both ferromagnetic ($J > 0$) and antiferromagnetic ($J < 0$) interactions. Cr^{III} (black spheres) and either Fe^{II} (hatched spheres) or Cr^{II} (white spheres), which are randomly incorporated in the lattice, are linked in an alternating fashion.

orange ($x = 1$). It should be emphasized that the strategy of this work is essentially different from the electrochromism. Moreover, their magnetic properties are also rich in variety depending on x, e.g. disappearance of saturation magnetization, compensation temperatures, and anomalous coercive fields.

The films of ternary metal Prussian blue can be prepared by reducing aqueous solutions containing three compounds $K_3[Cr(CN)_6]$, $CrCl_3$ and $FeCl_3$, where the mixing ratio x_{mix} ($= Fe^{III}/(Fe^{II} + Cr^{II})$) is controlled (figure 10). The x values of the $(Fe^{II}_x Cr^{II}_{1-x})_{1.5}[Cr^{III}(CN)_6] \cdot zH_2O$ series can be controlled either by the x_{mix} values of metal ions in the prepared solutions or by the electrode potential of reducing aqueous solutions.

The magnetic susceptibility and magnetization of obtained thin films depended strongly on the x values. The I_s values for $x = 0$ and $x = 1$ at fields up to 5 T, assuming $g = 2.0$ for metal ions, are determined to be $1.04\mu_B$ and $6.69\mu_B$, respectively. Those for the intermediate compositions vary in a systematic fashion as a function of x. Minimum values of the I_s are obtained with the film of x value close to 0.11. This is because the superexchange interaction between Cr^{III} and Fe^{II} are positive (which will be shown in §6) and that between Cr^{III} and Cr^{II} are negative. Therefore, the parallel

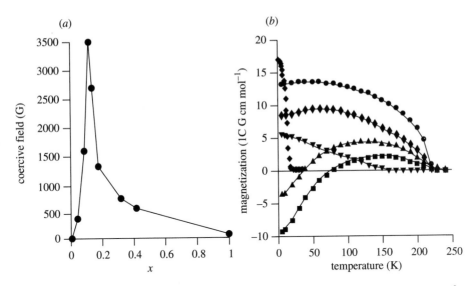

Figure 11. (a) Plots of H_c values versus x. (b) Magnetization versus temperature curves for $(Fe^{II}_x Cr^{II}_{1-x})_{1.5}[Cr^{III}(CN)_6] \cdot zH_2O$ (field of 10 G): (•) $x = 0$; (♦) $x = 0.08$; (▲) $x = 0.11$; (■) $x = 0.13$; (▼) $x = 0.42$; (♦) $x = 1$.

spins (Cr^{III} and Fe^{II}) and antiparallel spins (Cr^{III} and Cr^{II}) can be partly or even completely cancelled, depending on the x as shown in §2. Moreover, the H_c value for $x = 0.11$ was much larger than those at the other x values, e.g. 6 G ($x = 0$), 3500 G ($x = 0.11$), and 150 G ($x = 1$) (figure 11a). The particle sizes of deposited crystals for the whole x range are almost the same (200–300 nm diameter), so that the H_c values are theoretically expected to be proportional to I_s^{-1}. Therefore, the H_c value at minimum I_s value should become the largest. The magnetization versus temperature curves below T_c also exhibit various types of behaviour depending on x (field of 10 G), according to the theory described in §2 (figure 11b). Similar to the other ternary metal Prussian blue systems described previously, these temperature dependencies could be qualitatively reproduced using the molecular field theory, considering only two types of exchange couplings between nearest-neighbour sites, one for Fe^{II}–Cr^{III} ($J_{FeCr} = 0.9 \text{ cm}^{-1}$) and the other for Cr^{II}–Cr^{III} ($J_{CrCr} = -9.0 \text{ cm}^{-1}$).

The interesting aspect of the magnetic thin films that were obtained is their optical properties. The colours of $(Fe^{II}_x Cr^{II}_{1-x})_{1.5}[Cr^{III}(CN)_6] \cdot zH_2O$ transparent films are changed depending on x. For example, the film for $x = 0$ is colourless, that for $x = 0.20$ is violet, that for $x = 0.42$ is red, and that for $x = 1$ is orange. Their colours are due to the charge transfer (CT) band of Fe^{II} and Cr^{III} in the visible region (see Ludi & Güdel 1973), which is characteristic of mixed-valence compounds. As shown in figure 12, their CT bands shift from short to long wavelength in the visible region with decreasing x; e.g. $\lambda_{max} = 434$ nm ($x = 1$); 496 nm ($x = 0.42$); 506 nm ($x = 0.20$); 510 and 610 nm ($x = 0$) and hence their films exhibit various types of colour. In general, the wavelength of CT band depends on magnitudes of the vibronic coupling with asymmetrical distortion of each coordination sphere of metal ions and the intensity is proportional to the square of transfer integrals (see Robin & Day 1967; Hush 1967). Therefore, in a series of $(Fe^{II}_x Cr^{II}_{1-x})_{1.5}[Cr^{III}(CN)_6]$ films, magnitudes of vibronic

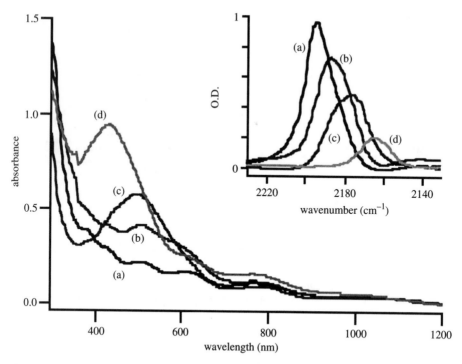

Figure 12. UV–vis and IR spectra of $(Fe^{II}_x Cr^{II}_{1-x})_{1.5}[Cr^{III}(CN)_6] \cdot zH_2O$: (a) $x = 0$; (b) $x = 0.20$; (c) $x = 0.42$; (d) $x = 1$. IR spectra are shown in the inset.

coupling parameters and transfer integrals are supposed to be continuously changed from those of Fe^{II}–Cr^{III} to those of Cr^{II}–Cr^{III} with decreasing x. In fact, frequencies of CN stretching and lattice constants also continuously shift, indicating that distances between metal ions of $(Fe^{II}_x Cr^{II}_{1-x})_{1.5}[Cr^{III}(CN)_6]$ are averaged values of the distance of Fe^{II}–Cr^{III} and Cr^{II}–Cr^{III} as a function of x.

We can thus design the various types of coloured magnetic thin films composed of Prussian blue analogues incorporating three or more types of metal ions. Here, note that our strategy to tune colour is essentially different from the electrochromism. Of course, when electrochemical reduction or oxidation of our thin films is performed on the electrode, each of thin films exhibits different types of colour and magnetic properties furthermore, e.g. red ($x = 0.42$) to dark blue at -1.0 V versus saturated calomel electrode (SCE). In general, the classical magnets show metallic lustre or are black. Here, note that even a yttrium iron garnet used as a photoisolater is seen to be black. Therefore, transparent and coloured magnets will enable one to develop new types of functional thin films.

5. Photomagnet

Here we demonstrate the photoinduced reversible change between paramagnet and ferrimagnet in a thin film of $K_{0.4}Co^{II}_{0.3}Co^{III}[Fe^{II}(CN)_6] \cdot 5H_2O$ (see Sato *et al.* 1996, 1997a, b; Einaga *et al.* 1997). The sodium complex of the cobalt iron cyanide thin film (of thickness *ca.* 0.05–0.1 μm), $Na_{1.4}Co^{II}_{1.3}[Fe^{II}(CN)_6] \cdot 5H_2O$, is first synthesized

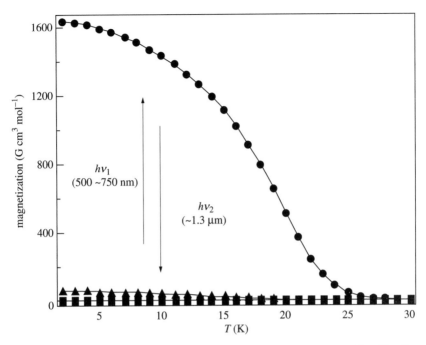

Figure 13. Field-cooled magnetization versus temperature curves for $K_{0.4}Co^{II}_{0.3}Co^{III}[Fe^{II}(CN)_6]\cdot$ $5H_2O$ at $H = 5$ G; ■, before illumination; ●, after visible light illumination; ▲, after near-IR light illumination.

electrochemically on a Pt electrode by reducing $K_3Fe^{III}(CN)_6$ in the presence of $Co^{II}NO_3$ in an aqueous solution of $NaNO_3$ electrolyte. Then this film is electro-chemically oxidized in KCl solution, resulting in a formation of $K_{0.4}Co^{II}_{0.3}Co^{III}[Fe^{II}(CN)_6]\cdot 5H_2O$. The absorption spectrum shows a strong, broad absorption band around 550 nm, assigned to the CT band from Fe^{II} to Co^{III}.

This compound exhibits an electron transfer from Fe^{II} to Co^{III} involving a spin transition around 340 K. The product of the molar magnetic susceptibility and temperature $(\chi_M T)$ versus T plot decreases slightly from 1.3 cm^3 mol^{-1} K at 300 K to 0.8 cm^3 mol^{-1} K at 20 K. Plots of FCM versus T at a magnetic field of 5 G do not exhibit abrupt breaks in the temperature range between 340 K and 2 K (figure 13). Long-range magnetic ordering is prevented by the presence of a large amount of diamagnetic components of Fe^{II} and Co^{III}, whereas the paramagnetic components of Co^{II} are responsible for the paramagnetic character of the compound.

When a red light is used to excite the CT band at 5 K, an increase of the magnetization value is observed. The FCM versus T plots after illumination for 10 min at 5 K show an abrupt break around 26 K (figure 13), indicating three-dimensional, long-range magnetic ordering. The increase in the value of T_c is explained by an increase in the number of magnetic neighbours. The plots of magnetization versus external field at 2 K before illumination do not show hysteresis loops, but those after illumination show the clear presence of hysteresis loops with a remnant magnetization of about 3800 cm^3 mol^{-1} G and H_c of about 6000 G at 2 K (figure 14). These results demonstrate that the paramagnetic material is converted to

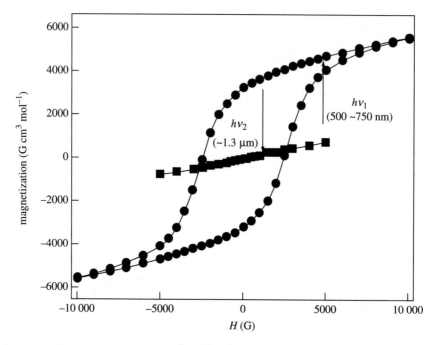

Figure 14. Hysteresis loops for $K_{0.4}Co^{II}_{0.3}Co^{III}[Fe^{II}(CN)_6] \cdot 5H_2O$ at 2 K; ■, before illumination; ●, after visible light illumination; ▲, after near-IR light illumination.

a magnetic one by light illumination. The change persists for periods of several days at 5 K. When the temperature of the sample is raised to 150 K, the magnetic properties quickly relax to almost the initial state.

After illumination, a new absorption peak appears at 400 nm and the peak intensity at 550 nm decreases, while the IR spectrum at 14 K shows that a peak at 2162 cm^{-1} appears and the peak at 2135 cm^{-1} nearly disappears. The Mössbauer spectrum at 25 K before the illumination shows a singlet absorption peak (figure 15a) indicating the presence of only low spin Fe^{II} (figure 15b). After illumination, however, a doublet absorption peaks appear, showing that Fe^{II} (low spin) is oxidized to Fe^{III} (low spin) during illumination. Taking into consideration of the above results, the change in the electronic states induced by light illumination is expressed as follows:

$$Co^{III}(t^6_{2g}, S = 0)-NC-Fe^{II}(t^6_{2g}, S = 0)$$
$$\rightarrow Co^{II}(t^5_{2g}e^2_g, S = 3/2)-NC-Fe^{III}(t^5_{2g}, S = 1/2). \quad (5.1)$$

Note that the ratio of nitrogen (CN) coordination and oxygen (H$_2$O) coordination to Co depends on the stoichiometry, by which the ligand field of Co is modified. Furthermore, the uptake of K$^+$ makes possible an interaction between K$^+$ and the Fe–CN–Co framework.

The compound, which is characterized by the Fe^{III}–CN–Co^{II} structure, has absorption peaks corresponding to the d–d transition for Fe^{III} and Co^{II} around 400 nm and 1300 nm, respectively. By the excitation of the near-IR band, the magnetization value decreases. In figure 13 are also shown the FCM plots after light illumination at 5 K. The enhanced magnetization resulting from visible light illumination can be

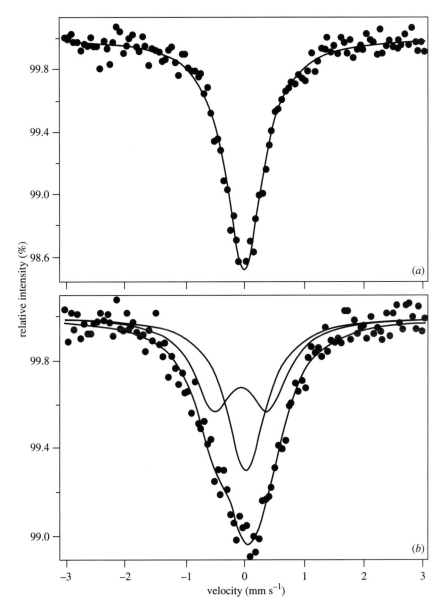

Figure 15. ^{57}Fe Mössbauer spectra of $K_{0.5}Co_{1.25}[Fe(CN)_6] \cdot 3.6H_2O$ at 25 K (*a*) before and (*b*) after visible light illumination.

almost completely reversed back to the original condition. Similarly, the hysteresis loop almost completely disappears as shown in figure 14. Changes in IR spectra are observed in which the CN stretching peak at 2162 cm^{-1} decreases and a peak around 2135 cm^{-1} increases in intensity, indicates that the reverse process of (3.1) is induced by the near-IR band illumination. The magnetization decreased by the near-IR illumination can then be increased again with visible light. Such a magnetization

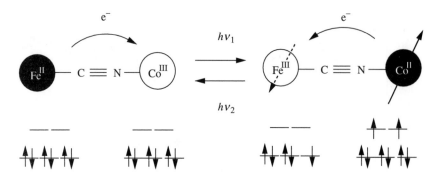

Figure 16. Schematic illustration of photomagnetic behaviour of a Fe–Co Prussian blue analogue
based on an electron-transfer mechanism.

increase and decrease by visible light, or near IR-light illumination can be reversibly
repeated, showing that the magnetic properties of cobalt iron cyanide thin films can
be switched between paramagnetic and ferrimagnetic by a photoinduced electron-
transfer process (figure 16).

It is interesting to note that such a photomagnet is obtained only with the
compounds whose Co and Fe ratio is $ca.$ 1.4–1.2. Neither $Co_{1.5}[Fe(CN)_6] \cdot 6H_2O$
(Co:Fe = 1.5:1) nor $KCoFe(CN)_6$ (Co:Fe = 1:1), which is a well-known cobalt–iron
cyanide, shows the photoeffects. The Mössbauer and other measurements suggest that
the former 1.5:1 compound takes the electronic state of $Co^{II}(t_{2g}^5 e_g^2, S = 3/2)$–NC–$Fe^{III}$
($t_{2g}^5, S = 1/2$) even without illumination. Conversely, the electronic state of the latter
1:1 compound is Co^{III} ($t_{2g}^6, S = 0$)–NC–$Fe^{II}(t_{2g}^6, S = 0)$, which is the same as that of
the photomagnet before illumination, $K_{0.4}Co_{1.3}[Fe(CN)_6] \cdot 5H_2O$, but it does not
change on illumination. These results suggest that the energy level of Co^{III}–NC–Fe^{II} is
situated close to that of Co^{II}–NC–Fe^{III} in cobalt iron cyanide, and that they are
changed by varying the ratio of Co and Fe. In other words, energy levels can be tuned
by introducing defect sites (sites coordinated by water molecules) in the three-
dimensional network. For $K_{0.4}Co_{1.3}[Fe(CN)_6] \cdot 5H_2O$, the energy of Co^{II}–NC–Fe^{III} is
slightly higher than that of Co^{III}–NC–Fe^{II}, forming a metastable state. This makes it
possible to change the electronic state from one to the other reversibly by external
stimulation.

6. Photo-induced magnetic pole inversion

We here design a novel magnet exhibiting photoinduced magnetic pole inversion by
the following strategy. We showed in §2 that we can control T_{comp} by tuning the
mixing ratio of mixed ferro-ferrimagnets composed of Prussian blue analogues. We
also showed in §5 that the I_s value could be changed by photo irradiation for some of
the Prussian blue analogues. By introducing such a photosensitive ferro- (or ferri-)
magnetic site into a mixed ferro-ferrimagnet showing a negative magnetization, the
T_{comp} will be controlled by photostimuli (see Ohkoshi et $al.$ 1997c; Ohkoshi &
Hashimoto 1999b). That is, if the photoinduced magnetization changes proceeds at
the either ferromagnetic or ferrimagnetic site in a mixed ferro-ferrimagnet, the balance
of magnetization between ferromagnetic site and ferrimagnetic site will be upset by

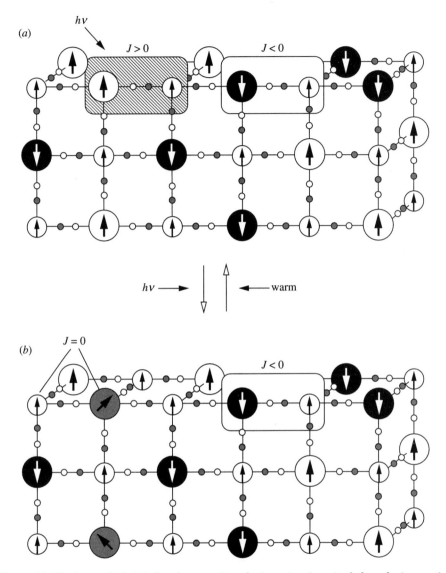

Figure 17. Strategy of photoinduced magnetic pole inversion in mixed ferro-ferrimagnetic Prussian blue analogue, $(Fe_x^{II}Mn_{1-x}^{II})_{1.5}Cr^{III}(CN)_6$: (*a*) before irradiation (*b*) after irradiation. The carbon (small grey spheres) ends of the cyano groups are bonded to Cr^{III} (middle small spheres), and the nitrogen (small white spheres) ends are bonded to either Fe^{II} (white spheres) and photoreactive Fe^{II} (grey spheres) or Mn^{II} (black spheres).

photoirradiation, with the result that the photoinduced magnetic pole inversion is realized at particular temperatures (figure 17).

On the basis of this strategy, we designed an $(Fe_xMn_{1-x})_{1.5}[Cr(CN)_6] \cdot 7.5H_2O$ magnet containing both a ferromagnetic (Fe–Cr) and antiferromagnetic (Mn–Cr) interactions. The $Mn_{1.5}[Cr(CN)_6] \cdot 7.5H_2O$ for an antiferrimagnetic site has no absorption in the visible region and hence does not show any change of magnetization by the visible light irradiation. Conversely, the magnetization of $Fe_{1.5}[Cr(CN)_6] \cdot 7.5H_2O$ for a ferromagnetic sites is reduced by the visible light irradiation.

Let us first show the photoinduced magnetization change of this ferromagnetic site. The ferromagnetic interaction between Fe^{II} and Cr^{III} cannot be explained by the orbital symmetry rule described in §1. However, the FCM plots of $Fe_{1.5}[Cr(CN)_6] \cdot$ $7.5H_2O$ at $H = 10$ G show an abrupt break at $T_c = 21$ K and the saturated magnetization observed at 5 K is $6.6\mu_B$. Moreover, the θ_c value estimated from the susceptibility values in the paramagnetic region is 27 K. All these data indicate that this complex is a ferromagnet (Ohkoshi et al. 1999b). M. Verdaguer (personal communication) suggested that this could be explained by the double-exchange mechanism between these metal ions. However, the estimation of J_{AF} and J_F values between Cr^{III} and Fe^{II} through the CN bond for Prussian blue analogues suggests that $|J_{AF}|$ is almost equal to $|J_F|$. Therefore, contrary to metal oxides, the value of superexchange interaction between Cr^{III} and Fe^{II} could become positive for a Prussian blue analogue even though it contains J_{AF} terms (Ohkoshi & Hashimoto 1999a).

As for optical properties, this compound shows the metal (Fe^{II}) to metal (Cr^{III}) CT band in the visible region ($\lambda_{max} = 454$ nm). By irradiation with a filtered blue light (360–450 nm) at 5 K, the magnetization of $Fe_{1.5}[Cr(CN)_6] \cdot 7.5H_2O$ decreases gradually under the external magnetic field (10 G, 50 G, 500 G, 7 T). About 10% of magnetization is decreased after eight hours irradiation at 10 G. This reduced magnetization persists for a period of several days at 5 K after turning off the light. Note that, in the dark, the magnetization at low temperatures at 10 G does not vary at all by a slight change of the temperature, indicating that this optically induced magnetization change is a photon mode not a photothermal mode. The magnetic property of this illuminated sample returns to the initial one when the temperature of the sample is raised above 30 K, showing that the magnetization can be reduced by a photon mode and recovered by a thermal mode repeatedly. The ^{57}Fe Mössbauer spectra of the $Fe_{1.5}[Cr(CN)_6] \cdot 7.5H_2O$ before and after irradiation indicate that the spin state of the iron is not changed by irradiation and that neither electron transfer nor spin transition occurs photochemically, but the spins of Fe^{II} and Cr^{III} are no longer aligned after the irradiation. This mechanism is different from that of the Fe–Co photomagnet described in §5. These results suggest that a ferromagnetic state is converted into a paramagnetic state without changing the valences of the Cr^{III} as a result of the irradiation. This decrease in photoinduced magnetization is explained as follows. The photoexcited state is a mixed valence state of Cr^{III}–CN–Fe^{II} and Cr^{II}–CN–Fe^{III}. This state would relax to a metastable state in which the magnetic interaction is too weak to maintain the spins' ordering. This metastable state is probably stabilized by structural distortion, and returns to the original ferromagnetic state above 40 K.

For the preparation of the Prussian blue analogues incorporating three different metal ions, $(Fe_xMn_{1-x})_{1.5}[Cr(CN)_6] \cdot zH_2O$, an aqueous solution containing both $FeCl_2$ and $MnCl_2$ is added to a concentrated aqueous solution of $K_3[Cr(CN)_6]$, yielding a light-brown-coloured microcrystalline powder. The fraction of Fe^{II} versus $(Fe^{II} + Mn^{II})$ in the above diluted aqueous solution is varied from zero to one, keeping the total metal ion concentration constant.

The magnetic properties of this series agree well with the mixed ferro-ferrimagnetism theory described in §2. For example, the I_s values at 5 K show a systematic change

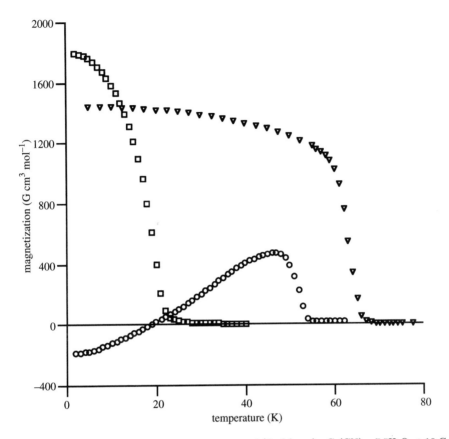

Figure 18. Magnetization versus temperature curves of $(Fe_xMn_{1-x})_{1.5}Cr(CN)_6 \cdot 7.5H_2O$ at 10 G. ($\triangledown$) $x = 0$; ($\circ$) $x = 0.40$; ($\square$) $x = 1$. The temperature dependence of $x = 0.40$ is due to a sum of the positive magnetization of the Mn^{II} sublattice and two negative magnetizations of the Fe^{II} and Cr^{III} sublattices which have different temperature dependencies.

as a function of x, and the values of materials metal ions. In other words, the ferromagnetic coupling between Fe^{II} and around $x \approx 0.4$ is nearly zero. This is because parallel spins (Cr^{III}, $S = 3/2$ and Fe^{II}, $S = 2$) and antiparallel spins (Mn^{II}, $S = 5/2$) cancel depending on the mixing ratio. The T_c values decrease from 67 to 21 K with increasing x. In addition, the temperature dependencies of magnetization are rich in variety. Particularly, the material prepared with $x = 0.40$ shows a negative magnetization at low temperature (figure 18).

When the compound for $x = 0.40$ is irradiated at 16 K below T_{comp} with the filtered blue light (360–450 nm), the negative magnetization at 16 K changes gradually to a positive one under the existing external field of 10 G. Simultaneously, the T_{comp} value also shifts to a smaller value and then that disappears, as shown in figure 19. This indicates that the magnetic pole below 19 K is inverted by the optical stimuli. This inverted magnetic pole persists for a period of several days at 16 K after turning off the light. This phenomenon can be explained by the fact that the ratio of the ferromagnetic part (Fe–Cr site) to the antiferromagnetic part (Mn–Cr site) of magnetization of this mixed ferro-ferrimagnet changes due to the decrease of magne-

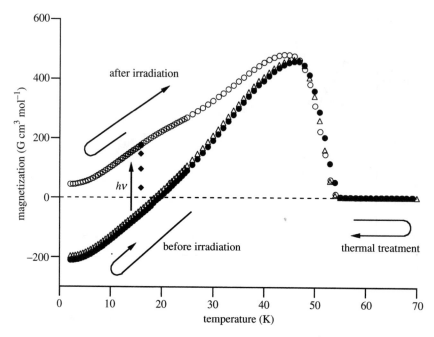

Figure 19. Magnetization versus temperature curves of $Fe_{0.40}Mn_{0.60}Cr(CN)_6 \cdot 7.5H_2O$ in the field of $H = 10$ G before (•) and after (○) light irradiation at 16 K for 72 h. The magnetic pole inversion is observed below the compensation temperature of 19 K by the light irradiation and the pole is reversed back again by the thermal treatment above 80 K (△). Magnetic measurement sequence: 70 K → • → 2 K → • → 16 K (light irradiation for 6, 24, 48, 72 h; (◆) → ○ → 2 K → ○ → 70 K → 80 K (thermal treatment) → 70 K → △ → 2 K.

tization in the ferromagnetic sites. In addition, the magnetization versus temperature curve is recovered by warming to 80 K, indicating that the magnetic pole inversion can be induced repeatedly by alternate optical and thermal stimulation as shown in figure 20.

The phenomenon of photoinduced magnetic pole inversion is used in a photomagnetic recording device made of TbFe, in which the magnetic material is heated above its critical temperature by photoirradiation in the presence of a reverse external magnetic field. During the cooling process below T_c, the magnetization arises parallel to the external field, resulting in a pole inversion. However, this phenomenon is induced by a photothermal process, not by a pure photo process. In the present molecule-based magnet, however, the pole inversion is induced by a pure photo process in the absence of an external magnetic field, and thus this process is completely different from that reported previously.

7. Conclusion

We have demonstrated various novel magnets, which were designed based on simple theories including a molecular field theory. The reasons why those simple theories can be applicable for the class of Prussian blue analogues are summarized as follows. Firstly, only the superexchange interactions between the nearest-neighbour metal ions

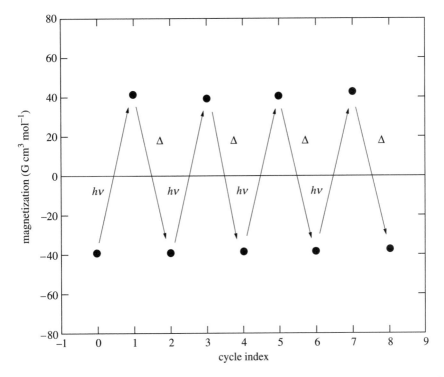

Figure 20. Reversible magnetic pole inversions are demonstrated by alternate stimuli: the magnetization at 15 K after the light illumination ($h\nu$) for 3 h and the magnetization at 15 K after the thermal treatment ($\triangle$) above 80 K.

have to be considered. In other words, contributions from the second nearest-neighbour sites can be neglected due to the relatively long distances between metal ions, and hence spin frustration does not occur. Secondly, the face-centred cubic (FCC) crystal structure is always maintained, and the substitution of the metal ion does not change the crystal structure significantly. Thirdly, the character of superexchange interactions can be easily predicted from a simple orbital symmetry rule, because the directions of the symmetry axes of the spin source d-orbitals for all of the metal ion sites and all of the ligand p-orbitals are identical due to their FCC structure. Moreover, a room temperature magnet has been obtained with a Prussian blue analogue (Ferlay *et al.* 1995). We thus believe that Prussian blue analogues will open new avenues in the field of novel magnets.

References

Anderson, E. E. 1964 Improved molecular field solutions for yttrium and gadolinium iron garnets. In *Proc. Int. Conf. on Magnetism, Nottingham*, p. 660.

Anderson, P. W. 1959 New approach to the theory of superexchange interactions. *Phys. Rev.* **115**, 2.

Einaga, Y., Sato, O., Iyoda, T., Kobayashi, Y., Ambe, F., Hashimoto, K. & Fujishima, A. 1997 Electronic states of cobalt iron cyanides studied by [57]Fe Mössbauer spectroscopy. *Chem. Lett.*, p. 289.

Entley, W. R. & Girolami, G. S. 1994 New three-dimensional ferrimagnetic materials: $K_2Mn[Mn(CN)_6]$,$Mn_3[Mn(CN)_6]_2 \cdot 12H_2O$, and $CsMn[Mn(CN)_6] \cdot \frac{1}{2}H_2O$. *Inorg. Chem.* **33**, 5165.

Ferlay, S., Mallah, T., Ouahes, R., Veillet, P. & Verdaguer, M. 1995 A room-temperature organometallic magnet based on Prussian blue. *Nature* **378**, 701.

Gatteschi, D., Kahn, O., Miller, J. S. & Palacio, F. 1991 *Magnetic molecular materials.* NATO ASI Series E, vol. 198. Dordrecht: Kluwer.

Ginsberg, A. P. 1971 Magnetic exchange in transition metal complexes: aspects of exchange coupling in magnetic cluster complexes. *Inorg. Chim. Acta Rev.* **5**, 45.

Goodenough, J. B. 1958 An interpretation of the magnetic properties of the perovskite-type mixed crystals. *J. Phys. Chem. Solids* **6**, 287.

Goodenough, J. B. 1959 Theory of the role of covalence in the perovskite-type maganites $[La,M(II)]MnO_3$. *Phys. Rev.* **100**, 564.

Gorter, E. W. 1954 Saturation magnetization and crystal chemistry of ferrimagnetic oxides. *Philips Res. Rep.* **9**, 295, 321, 403.

Griebler, W. D. & Babel, D. Z. 1982 Rontgenographische und magnetische Untersuchungen an perowskitverwandten Cyanoverbindungen $CsM^{II}M^{III}(CN)_6$. *Naturforsch.* Teil B **87**, 832.

Hush, N. S. 1967 Intervalence-transfer absorption. Part 2. Theoretical considerations and spectroscopic data. *Prog. Inorg. Chem.* **8**, 391.

Kahn, O. 1993 *Molecular magnetism.* New York: VCH.

Kanamori, J. 1959 Superexchange interaction and symmetry properties of electron orbitals. *J. Phys. Chem. Solids* **10**, 87.

Ludi, A. & Güdel, H. U. 1973 In *Structure and bonding* (ed. J. D. Dunitz), vol. 14, p. 1. Springer.

Mathonière, C., Nuttal, C. J., Carling, S. G. & Day, P. 1994 Molecular-based mixed valency ferrimagnets $(XR_4)Fe^{II}Fe^{III}(C_2O_4)_3$ (X = N, P; R = *n*-propyl, *n*-butyl, phenyl): anomalous negative magnetisation in the tetra-n-butylammonium derivative. *J. Chem. Soc. Chem. Commun.*, p. 1551.

Miller, J. S. & Epstein, A. J. 1994 Organic and organometallic molecular magnetic materials— designer magnets. *Angew. Chem. Int. Ed. Engl.* **33**, 385.

Néel, L. 1948 Proprietes magnetiques des ferrites; ferrimagnetisme et antiferromagnetisme. *Ann. Phys.* **3**, 137.

Nishino, M., Kubo, S., Yoshioka, Y., Nakamura, A. & Yamaguchi, K. 1997 Theoretical studies on magnetic interactions in Prussian blue analogs and active controls of spin states by external fields. *Mol. Cryst. Liq. Cryst.* **305**, 109.

Ohkoshi, S. & Hashimoto, K. 1999*a* Ferromagnetism of cobalt–chromium polycyanides. *Chem. Phys. Lett.* (In the press.)

Ohkoshi, S. & Hashimoto, K. 1999*b* Design of novel magnet exhibiting photo-induced magnetic pole inversion based on molecular field theory. *J. Am. Chem. Soc.* **121**. (In the press.)

Ohkoshi, S., Sato, O., Iyoda, T., Fujishima, A. & Hashimoto, K. 1997*a* Tuning of superexchange couplings in a molecule-based ferroferrimagnet: $(Ni^{II}_x Mn^{II}_{1-x})_{1.5}[Cr^{III}(CN)_6]$. *Inorg. Chem.* **36**, 268.

Ohkoshi, S., Iyoda, T., Fujishima, A. & Hashimoto, K. 1997*b* Magnetic properties of mixed ferro-ferrimagnets composed of Prussian blue analogs. *Phys. Rev.* B **56**, 11 642.

Ohkoshi, S., Yorozu, S., Sato, O., Iyoda, T., Fujishima, A. & Hashimoto, K. 1997*c* Photoinduced magnetic pole inversion in a ferro-ferrimagnet: $(Fe^{II}_{0.40} Mn^{II}_{0.60})_{1.5} Cr^{III}(CN)_6$. *Appl. Phys. Lett.* **70**, 1040.

Ohkoshi, S., Fujishima, A. & Hashimoto, K. 1998 Transparent and colored magnetic thin films: $(Fe^{II}_x Cr^{II}_{1-x})_{1.5}[Cr^{III}(CN)_6]$. *J. Am. Chem. Soc.* **120**, 5349.

Ohkoshi, S., Abe, Y., Fujishima, A. & Hashimoto, K. 1999*a* Design and preparation of a novel magnet exhibiting two compensation temperatures bed on molecular field theory. *Phys. Rev. Lett.* **82**, 1285.

Ohkoshi, S., Fujishima, A. & Hashimoto, K. 1999*b* Magnetic properties and optical control of iron-chromium polycyanides. *J. Electroanal. Chem.* **473**, 245.

Robin, M. B. & Day, P. 1967 Mixed valence chemistry—a survey and classification. *Adv. Inorg. Chem. Radiochem.* **10**, 247.

Sato, O., Iyoda, T., Fujishima, A. & Hashimoto, K. 1996 Photoinduced magnetization of a cobalt iron cyanide. *Science* **272**, 704.

Sato, O., Einaga, Y., Iyoda, T., Fujishima, A. & Hashimoto, K. 1997*a* Cation driven electron transfer involving a spin transition at room temperature in a cobalt iron cyanide thin film. *J. Phys. Chem.* **10**, 3903.

Sato, O., Einaga, Y., Iyoda, T., Fujishima, A. & Hashimoto, K. 1997*b* Reversible photoinduced magnetization. *J. Electrochem. Soc.* **144**, 11.

Magnetic anisotropy in molecule-based magnets

By Olivier Kahn

Laboratoire des Sciences Moléculaires, Institut de Chimie de la Matière Condensée de Bordeaux, UPR CNRS No. 9048, 33608 Pessac, France

The goal of this paper is to introduce the dimension 'magnetic anisotropy' in the field of molecule-based magnets. For that, we have focused on two aspects, namely the design of a hard magnet through the incorporation of magnetically anisotropic spin carriers, and the determination of the magnetic phase diagrams of two strongly anisotropic ferromagnets. The hard magnet contains three kinds of spin carriers, Co^{II} and Cu^{II} ions as well as radical cations. It structure is very peculiar; it consists of two perpendicular honeycomb-like networks which interpenetrate in such a way that each hexagon belonging to one of the networks is interlocked with a hexagon belonging to the perpendicular network. The coercive field depends on the grain size. It can be as large as 25 kOe at 5 K. The two strongly anisotropic ferromagnets are cyano-bridged $Mn^{II}Mo^{III}$ compounds synthesized from the $[Mo^{III}(CN)_7]^{4-}$ precursor. Mo^{III} has a low-spin configuration, with a local spin $S_{Mo} = 1/2$, and a strongly anisotropic **g** tensor. One of the compounds, $Mn_2(H_2O)_5Mo(CN)_7 \cdot 4H_2O$, has a three-dimensional structure. The other one, $K_2Mn_3(H_2O)_6[Mo(CN)_7]_2 \cdot 6H_2O$, has a two-dimensional structure. For both compounds, we have succeeded to grow well-shaped single crystals suitable for magnetic anisotropy measurements, and we have investigated the magnetic properties as follows: first, we have determined the magnetic axes by looking for the extremes of the magnetization in the three crystallographic planes, ab, bc, and ac. Then, we have measured the temperature and field dependencies of the magnetization in the DC mode along the three magnetic axes. These measurements have revealed the existence of several magnetically ordered phases for the three-dimensional compound, and of field-induced spin reorientations for both compounds. For the very first time in the field of molecular magnetism, we believe we have been able to determine the magnetic phase diagrams.

Keywords: coercivity; magnetic anisotropy; magnetic phase diagrams; molecule-based magnet; molecular material; supramolecular chemistry

1. Introduction

The first two molecular compounds exhibiting a spontaneous magnetization below a certain critical temperature, T_c, were described in 1986 (Miller *et al.* 1986; Pei *et al.* 1986). These reports have opened a new field of research, that of molecule-based magnets, and in the last decade quite a few new compounds of that kind have been synthesized (Miller *et al.* 1987; Kahn *et al.* 1988; Caneschi *et al.* 1989; Broderick *et al.* 1990; Yee *et al.* 1991; Nakazawa *et al.* 1992; Tamaki *et al.* 1992; Chiarelli *et al.* 1993; Miller & Epstein 1994; Gatteschi 1994; Inoue & Iwamura 1994; Decurtins *et al.* 1994*a, b*; Ohba *et al.* 1994; Inoue *et al.* 1996; Mathonière *et al.* 1996; Iwamura *et al.*

1998). What characterizes this field of research is its deeply multidisciplinary nature; it brings together synthetic organic, organometallic, and inorganic chemists along with theoreticians and physicists as well as material and life science people.

To a large extent, the field so far has been governed by the race toward high critical temperature, T_c. In that respect, T_c above room temperature is a must. This goal has been reached with a compound of formula $V^{II}_{0.42}V^{III}_{0.58}[Cr^{III}(CN)_6]_{0.86} \cdot 2.8H_2O$ (Ferlay *et al.* 1995). This compound belongs to the vast family of the Prussian blue-like phases with the general formula $A_k[B(CN)_6]_1 \cdot nH_2O$, where A is a high spin metal ion and B a low spin one. In contrast, not much effort has been devoted so far to the study of the phenomena arising from the magnetic anisotropy. The goal of this paper is to introduce this 'magnetic anisotropy' dimension in the field of molecule-based magnets. To do so we will focus on two aspects, namely, the design of a hard magnet through the incorporation of magnetically anisotropic spin carriers, and the determination of the magnetic phase diagrams of two strongly anisotropic ferromagnets synthesized from the $[Mo(CN)_7]^{4-}$ precursor.

2. Design of a very hard molecule-based magnet

In 1993, we reported on a compound containing three spin carriers, two metal ions and an organic radical, whose structure consisted of two interpenetrating quasi-perpendicular honeycomb-like networks (Stumpf *et al.* 1993, 1994*a*). This compound is a soft magnet, with a very small coercive field in the magnetically ordered state, below $T_c = 22.5$ K. The problem we were then faced with was the following. What chemical modifications should be made to introduce a significant memory effect? How to transform this compound to end up with a hard magnet? Here, we report on the solution we found. For that we compare two compounds, of formula

$$(Etrad)_2Mn_2[Cu(opba)]_3(DMSO)_{0.5} \cdot 0.25H_2O \qquad (1)$$

and

$$(Etrad)_2Co_2[Cu(opba)]_3(DMSO)_{1.5} \cdot 0.25H_2O \qquad (2),$$

respectively. Etrad$^+$ is the radical cation shown below,

in which an unpaired electron is equally shared between the two nitroxide groups, opba is the ligand *ortho*-phenylenebis(oxamato), and S stands for solvent molecules. These compounds are isostructural. Their structure consists of two nearly perpendicular graphite-like networks with edge-sharing hexagons. The corners of each hexagon

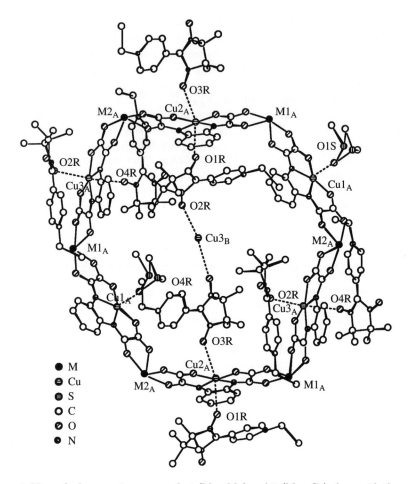

Figure 1. View of a hexagon in compounds **1** (M = Mn) and **2** (M = Co) along with the copper atom belonging to the nearly perpendicular network, located nearby the centre of the hexagon. This view also shows the presence of Cu^{II}–Etrad$^+$ chains connecting the two networks.

are occupied by M^{II} (Mn^{II} or Co^{II}) ions, and the middles of the edges by Cu^{II} ions. The two networks are interlocked, the centre of each hexagon being occupied by a Cu^{II} ion belonging to a nearly perpendicular hexagon. The compounds contain three kinds of spin carriers, M^{II} and Cu^{II} ions antiferromagnetically coupled through oxamato bridges, and Etrad$^+$ cations, bridging the Cu^{II} ions through the nitronyl nitroxide groups, and forming Cu^{II}–Etrad$^+$ chains which further connect the two networks. Figure 1 shows the structure of a hexagon, and figure 2 the interlocking of the two graphite-like networks.

The magnetic properties of **1** and **2** were investigated in detail. Both compounds may be considered as ferrimagnets. The S_M (M = Mn or Co) spins align along the field direction, and the S_{Cu} and S_{Etrad} spins along the opposite direction. The temperature dependencies of the field-cooled magnetization (FCM) and the remnant magnetization (REM) are represented in figure 3. The critical temperatures were determined as the

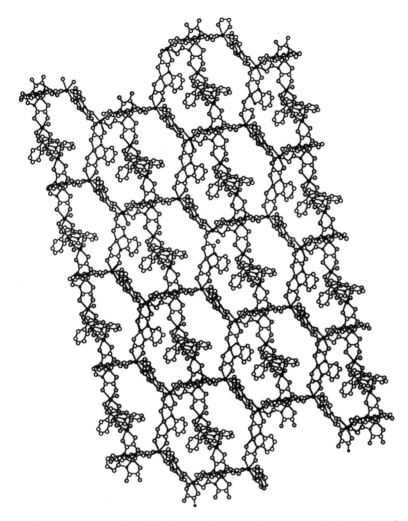

Figure 2. Interlocking of the two nearly perpendicular networks in compounds **1** and **2**.

extrema of the derivatives $dFCM/dT$, and found as $T_c = 22.8$ K for **1** and $T_c = 37$ K for **2**.

The main difference between the two compounds concerns the field dependencies of the magnetization in the magnetically ordered state, below T_c. **1** is a soft magnet. The field dependence of the magnetization at 5 K reveals a coercive field smaller than 10 Oe. In contrast, **2** is a very hard magnet. Figure 4 shows the hysteresis loop at 6 K for two samples consisting of crystals of different sizes. The black points were obtained with rather large crystals, and the coercive field is found as 8.5 kOe. The white points were obtained with crystals whose volume is roughly 50 times smaller, and the coercive field is of the order of 25 kOe.

The coercivity of a magnet depends on both chemical and structural factors (Stumpf *et al.* 1994*b*; Vaz *et al.* 1999). As far as the chemical factors are concerned,

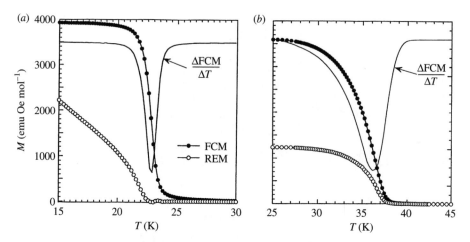

Figure 3. FCM and REM versus T curves for compound **1** (a) and **2** (b). The applied magnetic field is 20 Oe. The figure also shows the dFCM/dT derivatives.

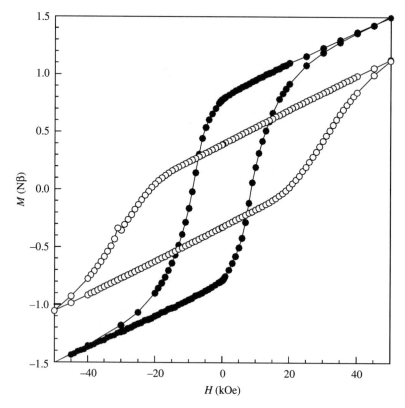

Figure 4. Field dependencies of the magnetization for two samples of compound **2**: (•) largest crystals; (○) smallest crystals.

the key role is played by the magnetic anisotropy of the spin carriers. The main structural factors are related to the size and shape of the particles. Here these two types of factors are illustrated. The high-spin Co^{II} ion in a distorted octahedral environment has a strong magnetic anisotropy, while the Mn^{II} ion in the same environment is almost isotropic. This chemical difference explains why compound **2** is much more coercive than compound **1**. The coercivity, however, is not an intrinsic property. It also depends on structural factors such as grain size and shape.

3. Magnetic phase diagrams for two cyano-bridged bimetallic ferromagnets synthesized from the $[Mo(CN)_7]^{4-}$ precursor

One of the appealing aspects of the magnetic studies dealing with Prussian blue phases (Babel 1986; Gadet *et al.* 1992; Mallah *et al.* 1993; Entley & Girolami 1994, 1995) resides in the fact that it is possible to predict the nature, and to estimate *a priori* the value of the critical temperature, using simple theoretical models based on the symmetry of the singly occupied orbitals (Kahn 1993). This is due to the fact that the symmetry of both the A and B metal sites is strictly octahedral, so that the e_g and t_{2g} orbitals do not mix. The design of the room temperature magnet mentioned in the introduction does not arise from serendipity, but was achieved in a rational way (Ferlay *et al.* 1995; Kahn 1995). The cubic symmetry of the Prussian blue phases, however, has a cost. These compounds are structurally, and hence magnetically, isotropic, and many interesting features associated with the structural and magnetic anisotropies cannot be observed. It may be noticed that nobody so far has succeeded in growing single crystals of Prussian blue phases suitable for detailed magnetic measurements. This situation is not too embarrassing as no anisotropy is expected, except perhaps a weak shape anisotropy for thin film samples. On the other hand, the thorough investigation of the properties of molecule-based magnets of low symmetry requires obviously to work with well-shaped single crystals.

Recently, we initiated a project concerning bimetallic compounds synthesized from the $[Mo(CN)_7]^{4-}$ precursor. The choice of this precursor was motivated by three reasons. (i) As for the Prussian blue phases, the presence of cyano ligands can lead to extended lattices. (ii) These networks should be of low symmetry; as a matter of fact, the heptacoordination of the precursor is not compatible with a cubic symmetry. (iii) In $KNa_2[Mo(CN)_7] \cdot 2H_2O$, the Mo^{III} ion is in a low-spin pentagonal bipyramid environment. The orbitally degenerate ground state, $^2E_1''$, is split into two Kramers doublets by the spin orbit coupling, and the ground Kramers doublet is strongly anisotropic (Rossman *et al.* 1973; Hursthouse *et al.* 1980; Young 1932).

(a) The three-dimensional compound $Mn_2(H_2O)_5Mo(CN)_7 \cdot 4H_2O$, α phase (3)

The slow diffusion of two aqueous solutions containing $K_4[Mo(CN)_7] \cdot 2H_2O$ and $[Mn(H_2O)_6](NO_3)_2$, respectively, affords two kinds of single crystals, with elongated plate (α phase) and prism (β phase) shapes (Larionova *et al.* 1998*a, b*). Here, we restrict ourselves to $Mn_2(H_2O)_5Mo(CN)_7 \cdot 4H_2O$, α phase (**3**). There is one molybdenum site along with two manganese sites, denoted as Mn1 and Mn2. The molybdenum atom is surrounded by seven –C–N–Mn linkages, four of them involving a Mn1 site and three of them a Mn2 site. The geometry may be described as a slightly distorted pentagonal bipyramid. Both Mn1 and Mn2 sites are in distorted octahedral

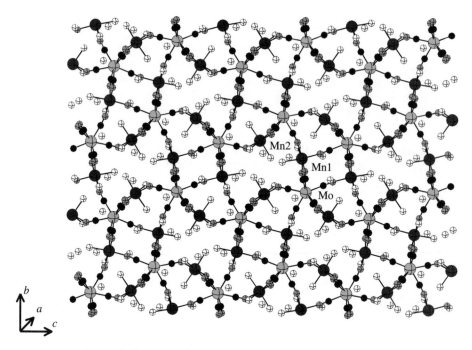

Figure 5. Structure of compound **3** viewed along the *a*-direction.

surroundings. Mn1 is surrounded by four –N–C–Mo linkages and two water molecules in *cis* conformation. Mn2 is surrounded by three –N–C–Mo linkages and three water molecules in a *mer* conformation. The three-dimensional organization for **3** may be described as follows. Edge-sharing lozenge motifs (MoCNMn1NC)$_2$ form bent ladders running along the *a*-direction. Each ladder is linked to four other ladders of the same kind along the [011] and [01$\bar{1}$] directions through cyano bridges. These ladders are further connected by the Mn$_2$(CN)$_3$(H$_2$O)$_3$ groups. Mn2 is linked to a Mo site of one of the ladders and to two Mo sites of the adjacent ladder. The structure as a whole viewed along the *a*-direction is represented in figure 5.

We first checked that in the three crystallographic planes *ab*, *bc* and *ac*, the extremes of the magnetization are obtained when the field is aligned along the *a*-, *b*- and *c**-axes. Therefore, these axes are the magnetic axes of the compound. It may be noticed that the twofold axis of the monoclinic lattice, *b*, was necessarily one of the magnetic axes (Wooster 1973). All the magnetic measurements were carried out on single crystals with the external field successively applied along these three axes. We first measured the temperature dependencies of the magnetization, *M*, under a field of 5 Oe. The three curves are shown in figure 6. They reveal that the material is anisotropic, the magnetization along the easy magnetization axis, *b*, being about twice as large as along the *a*-axis. Moreover, these curves exhibit a break with an inflexion point at $T_{1c} = 50.5$ K, along with another anomaly, more visible along the *a*-axis, at $T_{2c} = 43$ K.

The most accurate technique to determine transition temperatures is the measure of the heat capacity as a function of temperature. In the present case, the heat capacity curve shows a lambda peak at 50.5 K, but nothing is detected at 43 K.

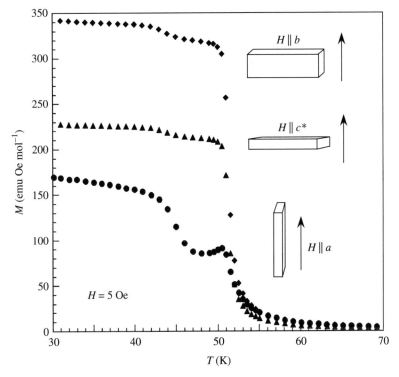

Figure 6. Temperature dependencies of the magnetization along the *a*-, *b*- and *c**-directions, using an external field of 5 Oe, for compound **3**.

In order to obtain more insights on the magnetic anomaly detected at 43 K, visible essentially along the *a*-direction, we measured the magnetization along this direction with an external field varying from 1 up to 100 Oe. The results are displayed in figure 7. T_{2c} is shifted toward higher temperatures as the magnetic field increases, and eventually, for a field of *ca.* 100 Oe, merges with the transition at 50.5 K. For each field, T_{2c} was determined as the inflexion point of the $M = f(T)$ curve.

We then measured the field dependencies of the magnetization at 5 K along the *a*-, *b*- and *c**-directions (see figure 8). The curves are strictly identical when increasing and decreasing the field; the compound exhibits no coercivity. Along the easy magnetization direction, *b*, the saturation is reached with *ca.* 1 kOe. The saturation magnetization is found to be equal to 11 Nβ. This value corresponds to what is expected for one $S_{Mo} = 1/2$ and two $S_{Mn} = 5/2$ local spins aligned along this direction. The interaction between adjacent Mo^{III} and Mn^{II} ions is ferromagnetic. Along the *c**-direction, the magnetization increases progressively when applying the field, and even at 50 kOe the saturation is not totally reached. Along the *a*-direction, the *M* versus *H* curve is peculiar; it shows an inflexion point for a critical field, H_c, of about 2.2 kOe. We are faced with a field-induced spin reorientation phenomenon (Salgueiro da Siva *et al.* 1995; Garcia-Landa *et al.* 1996; Mendoza & Shaheen 1996; Cao *et al.* 1997; Kou *et al.* 1997).

Let us examine in more detail what happens when applying the field along the *a*-axis. In zero field, the resulting moment of 11 N is essentially aligned along *b*. When

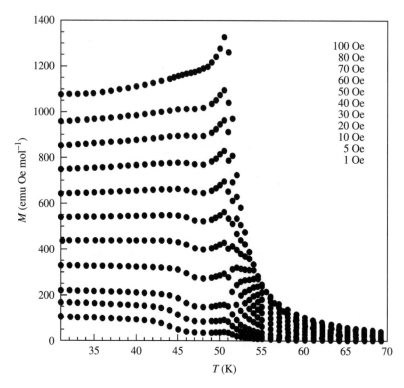

Figure 7. Temperature dependencies of the magnetization along the a-direction, using different values of the external field, for compound **3**.

applying the field, first this moment hardly rotates from b to a, then for a critical value of the field the rotation becomes much easier. Finally, for a saturation value of the field, H_{sat}, the moment is aligned along a. The spin reorientation is a nonlinear phenomenon. It is difficult to unhook the moment from the b-axis. When the field is large enough this unhooking is realized, and then a weak increase of the field induces a strong rotation until the moment is collinear with a. To determine the H_c and H_{sat} versus T curves, we measured the field dependence of the magnetization along a every 5 K in the 5–51 K range. The critical field at each temperature was determined as the field for which the dM/dH derivative is maximum. The saturation field at each temperature was determined as the weakest field for which the saturation is reached.

The temperature dependencies of H_c and H_{sat} for a field applied along the a-axis are used to determine the magnetic phase diagram of the compound. This diagram, shown in figure 9, presents four domains. Domain I is the paramagnetic domain in which the spins are either randomly oriented, or aligned along the field direction. Domains II and III are ferromagnetic domains in which the spins are essentially aligned along the b-direction. Domain III is limited to the 43–51 K temperature range and the 0–100 Oe field range. Domain IV, finally, is limited by the $H_c = f(T)$ and $H_{sat} = f(T)$ curves, and corresponds to a mixed domain in which the spins may rotate easily from the b- to the a-direction. It is worth mentioning that the

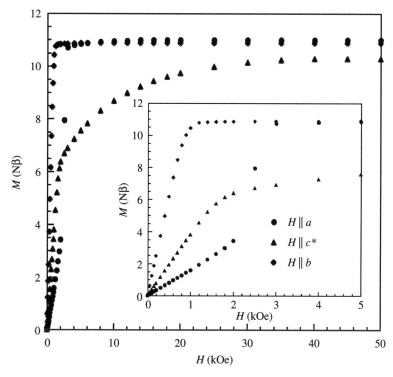

Figure 8. Field dependencies of the magnetization at 5 K along the *a*-, *b*- and *c**-directions for compound **3**.

boundary between domains II and IV corresponds to a first-order transition, while that between domains I and IV corresponds to a second-order transition.

The magnetic data suggest that the main difference between domains II and III concerns the component of the resulting moment along *a*. Assuming that the magnetic symmetry of the low-temperature domain II is lower than that of the high-temperature domain III, we may assume that domain III is a perfectly ferromagnetic domain with the magnetic moments aligned along *b* in zero field, and that a small canting occurs as *T* is lowered below 43 K, with a weak component of the resulting moment along *a*. Alternatively, the two domains might be weakly canted, with a component of the moment along *a*, the degree of canting being slightly more pronounced in domain II.

(b) *The two-dimensional compound* $K_2Mn_3(H_2O)_6[Mo(CN)_7]_2 \cdot 6H_2O$ (4)

When the slow diffusion between aqueous solutions containing $K_4[Mo(CN)_7] \cdot 2H_2O$ and $[Mn(H_2O)_6](NO_3)_2$, respectively, takes place in the presence of an excess of K^+ ions, a two-dimensional compound of formula $K_2Mn_3(H_2O)_6[Mo(CN)_7]_2 \cdot 6H_2O$ is obtained (Larionova *et al.* 1999). The structure again contains a unique molybdenum site along with two manganese sites, denoted as Mn1 and Mn2. The molybdenum atom is surrounded by six –C–N–Mn linkages and a terminal –C–N ligand. The geometry may be described as a strongly distorted pentagonal bipyramid, and both

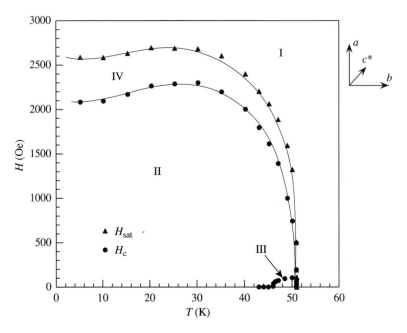

Figure 9. Magnetic phase diagram for compound **3**; the full lines are just eye-guides.

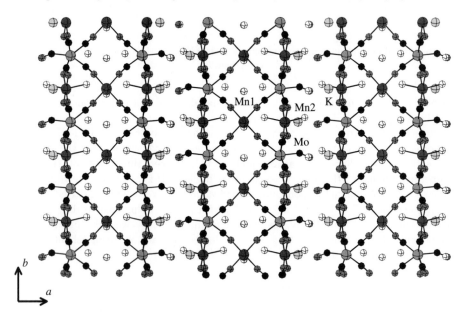

Figure 10. Structure of the compound **4** in the *ab*-plane.

the Mo–C–N and C-N–Mn bridging angles deviate significantly from 180°. Both the Mn1 and Mn2 sites are surrounded by four –N–C–Mo linkages and two water molecules in *trans* conformation. The two-dimensional structure is made of anionic double-sheet layers parallel to the *bc*-plane, and K$^+$ and non-coordinated water

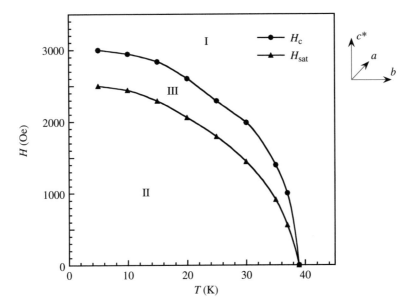

Figure 11. Magnetic phase diagram for the compound **4**; the full lines are just eye-guides.

molecules situated between the layers, as shown in figure 10. Each sheet is a kind of grid in the bc-plane made of edge-sharing lozenges [MoCNMn$_2$NC]$_2$. Two parallel sheets of a layer are further connected by Mn1(CN)$_4$(H$_2$O)$_2$ units situated between the sheets. The thickness of a double-sheet layer is 8.042 Å, and that of the gap between two layers is 7.263 Å.

We first checked that a, b and c^* were the magnetic axes of the compound, b being the easy magnetization axis, then we studied the temperature and field dependencies of the magnetization along the three axes. No hysteresis was observed along a and b, while a narrow hysteresis, of about 125 Oe, was observed along c^* at 5 K. These measurements revealed that the compound exhibits a long-range ferromagnetic ordering at $T_c = 39$ K, and that below T_c a field-induced spin reorientation occurs along the c^*-axis. We determined the critical and saturation fields every 5 K below T_c when applying the field along c^*. The H_c and H_{sat} versus T curves shown in figure 11 define the magnetic phase diagram for the compound when the magnetic field is applied along c^*. This diagram is simpler than that of figure 9. It presents only three domains. Domain I corresponds to the paramagnetic, or saturated paramagnetic domain. Domain II corresponds to the ferromagnetically ordered domain in which the spins are essentially aligned along the b-axis. Domain III, finally, is a spin-reorientation domain in which the spins rotate from the b- to the c^*-direction as the field increases from H_c to H_{sat}.

So far, we have only spoken of magnetic measurements recorded in the DC mode, i.e. with a static external field. Additional information can be obtained by working in the AC mode (Palacio *et al.* 1989). We report here on two experiments of this kind. First, we measured the temperature dependencies of the in-phase, χ'_{AC}, and out-of-phase, χ''_{AC}, magnetic susceptibilities under a zero static field. Along the three directions, the in-phase responses exhibit a break at $T_c = 39$ K. The χ'_{AC} values along the b-axis below T_c are much higher than along the other two directions. The

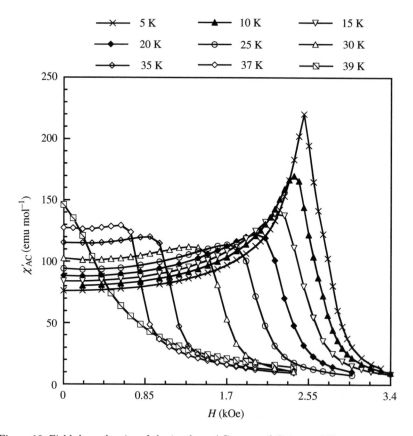

Figure 12. Field dependencies of the in-phase AC susceptibilities at different temperatures along the c^*-axis for the compound **4**.

out-of-phase response along the b-axis is zero down to 39 K, then presents an abrupt break as T is lowered below this temperature, and reaches a maximum around 34 K. Along the other two directions, χ''_{AC} is negligibly weak down to 2 K.

When the field is applied along the easy magnetization axis, b, both the displacement of the domain walls and the rotation of the magnetic moments contribute to the AC magnetic response. On the other hand, when the field is applied along a hard magnetization axis, only the rotation of the magnetic moments contributes to the AC response (Kou *et al.* 1996). In the present case, the very high response along b as compared with the responses along a and c^* indicates that the domain walls move very easily, which is in line with the quasi-absence of hysteresis in the $M = f(H)$ curves.

The second experiment consisted of measuring the field dependence of χ'_{AC} along the c^*-axis every 5 K in the magnetically ordered phase. The results are displayed in figure 12. At each temperature χ'_{AC} first increases as the field increases, reaches a maximum, then tends to zero at high field. The maximum of χ'_{AC} determines the critical field, H_c, and the extreme of the derivative $d\chi'_{AC}/dH$ determines the saturation field, H_{sat}. The $H_c = f(T)$ and $H_{sat} = f(T)$ curves deduced from this experiment are strictly similar to those shown in figure 11.

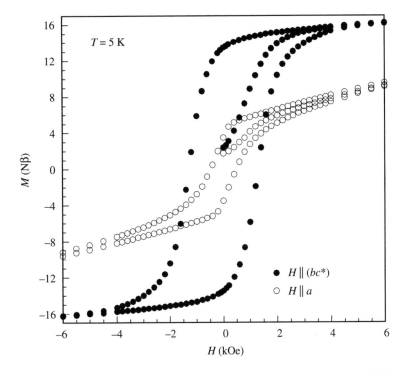

Figure 13. Hysteresis loops in the bc^*-plane and along the a-axis at 10 K for a partly dehydrated crystal of compound **4**.

(c) *Modification of the magnetic properties through partial dehydration*

The magnetic properties of $K_2Mn_3(H_2O)_6[Mo(CN)_7]_2 \cdot 6H_2O$ can be dramatically modified through partial dehydration. When the non-coordinated water molecules of a single crystal are released under vacuum, the external shape of the crystal is not modified. Magnetic measurements suggest that the crystallographic directions are retained. We then investigated the temperature dependencies of the magnetization along these directions. The $M = f(T)$ curves along b and c^* are not distinguishable. After dehydration, the bc-plane may be considered as an easy magnetization plane, even in low field. The spin reorientation is suppressed. Both in the bc-plane and perpendicular to this plane, the magnetization shows a break at $T_c = 72$ K, while the critical temperature for the non-dehydrated compound is 39 K. We measured the field dependencies of the magnetization at 10 K both in the bc^*-plane and along the a-direction. The results are displayed in figure 13. In the easy magnetization plane, the saturation value of 17 Nβ corresponding to the parallel alignment of all the spins is obtained under *ca.* 5.0 kOe. On the other hand, along the a-axis, the saturation is not reached yet under 50 kOe. Magnetic hystereses are observed along both directions, with coercive fields of 1.3 kOe in the bc-plane, and 0.55 kOe along a.

We can notice here that the Prussian blue phases are also hydrated. It would be interesting to see whether their partial dehydration also modifies their magnetic properties.

$$Mo^{III}(CN)_7$$

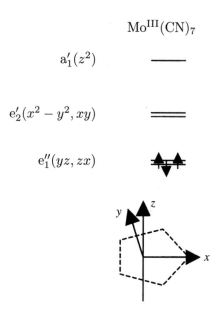

$$a_1'(z^2)$$

$$e_2'(x^2 - y^2, xy)$$

$$e_1''(yz, zx)$$

Scheme 1.

(d) Why is the Mo^{III}–C–N–Mn^{II} interaction ferromagnetic?

One of the striking features concerning the compounds described in this article is the ferromagnetic nature of the interaction between low-spin Mo^{III} and high-spin Mn^{II} ions through the cyano bridge. For several decades, quite a few studies have been devoted to the microscopic mechanisms of the interaction between spin carriers, more particularly to those of the mechanisms favouring a parallel alignment of the electron spins (Kahn 1993; Kollmar & Kahn 1993). Three situations have been found to stabilize the parallel spin state, namely the following. (i) *The strict orthogonality of the magnetic orbitals.* Such a situation is achieved when all the singly occupied orbitals centred on a spin carrier (also called magnetic orbitals) are orthogonal (i.e. give a zero overlap integral) with all the singly occupied orbitals centred on the adjacent spin carrier. This situation of strict orthogonality of the magnetic orbitals is certainly the most efficient way to achieve a ferromagnetic interaction. (ii) *The electron transfer from a singly occupied orbital on a site toward an empty orbital on an adjacent site, or from a doubly occupied orbital on a site toward a singly occupied orbital on an adjacent site.* This mechanism is probably less efficient than the previous one, and more difficult to control. (iii) *The interaction between a zone of negative spin density on a fragment and a zone of positive spin density on an adjacent fragment.* This mechanism, first suggested in the context of assemblies of organic radicals (McConnell 1963), may apply for transition metal species as well.

We carefully examined the relevance of each of these mechanisms in the case of Mo^{III}–C–N–Mn^{II}. The strict orthogonality of the magnetic orbitals is not realized, and the absorption spectra of the compounds synthesized from the $[Mo(CN)_7]^{4-}$ precursor do no reveal any metal–metal charge transfer band of relatively low energy. Therefore, we are wondering whether the ferromagnetic interaction could arise from a close contact between negative and positive spin density zones.

The orbital energy diagram for low-spin Mo^{III} in pentagonal bipyramid symmetry (D_{5h}) is represented in scheme 1.

At the self-consistent field (SCF) approximation, the $e'_2(x^2 - y^2, xy)$ and $a'_1(z^2)$ orbitals with a predominant 4d character are empty. It follows that at this approximation level, there is no spin density along both equatorial and axial Mo–C–N directions. It is now well established that such a view is oversimplified. The $^2E''_1$ SCF ground state may couple with SCF excited states of the same symmetry in which electrons have been promoted from the doubly occupied and bonding e'_2 and a'_1 orbitals with a predominant cyano character to the empty and antibonding e'_2 and a'_1 orbitals with a predominant metal character. This configuration interaction gives rise to a negative spin density in the σ orbitals of the cyano ligands, along the Mo–C–N directions. This spin polarization effect has been experimentally observed in the hexacyanometallates involving 3d ions, such as $[Cr(CN)_6]^{3-}$ and $[Fe(CN)_6]^{3-}$, from polarized neutron diffraction experiments (Faggis *et al.* 1987, 1993). The crucial point is that the stronger (i.e. the more covalent) the M—(CN) bond is, the more pronounced the spin polarization. Owing to the diffuseness of the 4d orbitals compared with the 3d orbitals, the Mo^{III}—(CN) bond is significantly stronger than the Cr^{III}—(CN) or Fe^{III}—(CN) bond. It follows that the negative spin density along the Mo–C–N directions of the $Mo(CN)_7$ fragment might be particularly important. If it was so, the interaction between this negative spin density with a σ character and the σ singly occupied orbitals of Mn^{II} might favour the parallel alignment of the S_{Mo} and S_{Mn} spins.

(e) What is the origin of the anisotropy?

Let us list the various factors affording magnetic anisotropy. We begin with the two local factors: the anisotropy of the **g** tensors and the zero-field splitting of the local spin states for the magnetic ions with a local spin higher than $1/2$. In addition, there are several many-body (or collective) factors. These are (i) the anisotropic interactions, resulting from the synergistic effect of the local spin-orbit coupling for an ion and the interaction between the excited states of this ion and the ground state of the adjacent ion; (ii) the antisymmetric interaction whose origin is similar to that of anisotropic interaction, but in addition requires a low lattice symmetry; (iii) the dipolar interactions which may become important for lattices of low symmetry, and/or for high local spins (for instance, $S_{Mn} = 5/2$); (iv) the shape anisotropy, finally, depending on the shape and size of the single crystals or particles used for the magnetic measurements (Bencini & Gatteschi 1989). Applying an external field H along the easy magnetization axis results in an internal field H_i related to H through

$$H_i = H - NM,$$

where N is the demagnetizing factor depending on the shape anisotropy, and NM the demagnetizing field. The field dependencies of M shown in figure 8 are not corrected of the demagnetizing field.

Except when all the local spins are $1/2$, the local zero-field splittings are usually more important than the anisotropic interactions. As for the antisymmetric interaction, it leads to the spin canting, which may superimpose to both an antiferromagnetic or a ferromagnetic state. In this latter case, it gives rise to the phenomenon of weak ferromagnetism.

What are the relevant factors in the case of the compounds described in this article? Mn^{II} in octahedral surroundings has a $^6A_1$ ground state, with a very weak zero-field spltting. On the other hand, the **g** tensor for the $[Mo^{III}(CN)_7]$ chromophore is expected to be strongly anisotropic (Hursthouse *et al.* 1980). In other respects, the lattice symmetries are very low, even for the three-dimensional compounds. It follows that the two main anisotropy factors are the anisotropy of the **g** tensor for the $[Mo^{III}(CN)_7]$ chromophore along with the dipolar interactions. The spin reorientation might be due to a competition between these two factors. In zero (or low) external field, the ferromagnetically coupled local spins tend to align along the direction of the Mo–C– N–Mn–C–N infinite linkages (for instance, the *b*-axis for $Mn_2(H_2O)_5Mo(CN)_7 \cdot 4H_2O$, α phase), which minimizes the dipolar energy. When the magnetic field reaches a certain value, the **g** anisotropy for Mo^{III} favours the spin alignment along another direction.

4. Conclusion

In this paper, we have explored two facets of the magnetic anisotropy in the field of molecule-based magnets. First, we have been concerned by the coercivity. The possibility to design molecule-based magnets displaying wide magnetic hysteresis loops was not obvious, at least for us. Actually, some time ago, we thought that the softness of the molecular state would prevent us from synthesizing very coercive magnets. We were not right; there is no contradiction between soft lattices and hard magnets. The key factor of the coercivity for compound **2** is the presence of the very anisotropic Co^{2+} spin carrier. This factor, however, is not the only one. The Co^{2+}-containing two-dimensional magnets of formula $cat_2Co_2[Cu(opba)]_3 \cdot S$ display coercive fields at 6 K weaker than 5 kOe (Stumpf *et al.* 1994*b*). The three-dimensional character of **2** resulting from the interlocking of two quasi-perpendicular graphite-like networks also contributes to the coercivity.

The second part of this paper deals with the peculiar magnetic phase diagrams arising from the low symmetry. In the field of molecule-based magnets, this low symmetry may lead to very interesting physical phenomena. Of course, the thorough investigation of these phenomena and their correct interpretation require work on single crystals, which is time-consuming and sometimes not trivial. However, accepting to do so may be very rewarding. To the best of our knowledge, the spin reorientation phenomenon had only been found so far for ferromagnetic intermetallic compounds, and not for insulating ferromagnets. In other respects, the magnetic phase diagrams of figures 9 and 11 are the very first for magnetic materials synthesized from molecular precursors.

This work was partly funded by the TMR Research Network ERBFMRXCT980181 of the European Union, entitled 'Molecular Magnetism: from Materials toward Devices'.

References

Babel, D. 1986 *Comments Inorg. Chem.* **5**, 285.
Bencini, A. & Gatteschi, D. 1989 *EPR of exchange coupled systems.* Springer.
Broderick, W. E., Thompson, J. A., Day, E. P. & Hoffman, B. M. 1990 *Science* **249**, 410.
Caneschi, A., Gatteschi, D., Sessoli, R. & Rey, P. 1989 *Acc. Chem. Res.* **22**, 392.
Cao, G., McCall, S. & Crow, J. E. 1997 *Phys. Rev.* B **55**, R672.

Chiarelli, R., Nowak, M. A., Rassat, A. & Tholence, J.-L. 1993 *Nature* **363**, 147.

Decurtins, S., Schmalle, H. W., Oswald, H. R., Linden, A., Ensling, J., Gütlich, P. & Hauser, A. 1994*a Inorg. Chim. Acta* **216**, 65.

Descurtins, S., Schmalle, H. W., Schneuwly, P., Ensling, J. & Gütlich, P. 1994*b J. Am. Chem. Soc.* **116**, 9521.

Entley, W. R. & Girolami, G. S. 1994 *Inorg. Chem.* **33**, 5165.

Entley, W. R. & Girolami, G. S. 1995 *Science* **268**, 397.

Ferlay, S., Mallah, T., Ouahès, R., Veillet, P. & Verdaguer, M. 1995 *Nature* **378**, 701.

Figgis, B. N., Forsyth, J. B. & Reynolds, P. A. 1987 *Inorg. Chem.* **26**, 101.

Figgis, B. N., Kucharski, E. S. & Vrtis, M. 1993 *J. Am. Chem. Soc.* **115**, 176.

Gadet, V., Mallah, T., Castro, I. & Verdaguer, M. 1992 *J. Am. Chem. Soc.* **114**, 9213.

Garcia-Landa, B., Tomey, E., Fruchart, D., Gignoux, D. & Skolozdra, R. 1996 *J. Magn. Magn. Mater.* **157–158**, 21.

Gatteschi, D. 1994 *Adv. Mater.* **6**, 635.

Hursthouse, M. B., Maijk, K. M. A., Soares, A. M., Gibson, J. F. & Griffith, W. P. 1980 *Inorg. Chim. Acta* **45**, L81.

Inoue, K. & Iwamura, H. 1994 *J. Am. Chem. Soc.* **116**, 3173.

Inoue, K., Hayamizu, T., Iwamura, H., Hashizume, D. & Ohashi, Y. 1996 *J. Am. Chem. Soc.* **118**, 1803.

Iwamura, H., Inoue, K. & Koga, N. 1998 *New J. Chem.* **10**, 201.

Kahn, O. 1993 *Molecular magnetism.* New York: VCH.

Kahn, O. 1995 *Nature* **378**, 667.

Kahn, O., Pei, Y., Verdaguer, M., Renard, J.-P. & Sletten, J. 1988 *J. Am. Chem. Soc.* **110**, 782.

Kollmar, C. & Kahn, O. 1993 *Acc. Chem. Res.* **26**, 259.

Kou, X. C., Grössinger, R., Hischer, G. & Kirchmayr, H. R. 1996 *Phys. Rev.* B **54**, 6421.

Kou, X. C., Dahlgren, M., Grössinger, R. & Wiesinger, G. 1997 *J. Appl. Phys.* **81**, 4428.

Larionova, J., Sanchiz, J., Golhen, S., Ouahab, L. & Kahn, O. 1998*a Chem. Commun.*, p. 953.

Larionova, J., Clérac, R., Sanchiz, J., Kahn, O., Golhen, S. & Ouahab, L. 1998*b J. Am. Chem. Soc.* **120**, 13 088.

Larionova, J., Kahn, O., Gohlen, S., Ouahab, L. & Clérac, R. 1999 *J. Am. Chem. Soc.* **121**, 3349.

McConnell, H. M. 1963 *J. Chem. Phys.* **39**, 1910.

Mallah, T., Thiebaut, S., Verdaguer, M. & Veillet, P. 1993 *Science* **262**, 1554.

Mathonière, C., Nuttall, C. J., Carling, S. & Day, P. 1996 *Inorg. Chem.* **35**, 1201.

Mendoza, W. A. & Shaheen, S. A. 1996 *J. Appl. Phys.* **79**, 6327.

Miller, J. S. & Epstein, A. J. 1994 *Angew. Chem. Int. Ed.* **33**, 385.

Miller, J. S., Calabrese, J. C., Epstein, A. J., Bigelow, R. W., Zang, J. H. & Reiff, W. M. 1986 *J. Chem. Soc. Chem. Commun.*, p. 1026.

Miller, J. S., Calabrese, J. C., Rommelman, H., Chittipedi, S. R., Zang, J. H., Reiff, W. M. & Epstein, A. J. 1987 *J. Am. Chem. Soc.* **109**, 769.

Nakazawa, Y., Tamura, M., Shirakawa, N., Shiomi, D., Takahashi, M., Kinoshita, M. & Ishikawa, M. 1992 *Phys. Rev.* B **46**, 8906.

Ohba, M., Maruono, N. & Okawa, H. 1994 *J. Am. Chem. Soc.* **116**, 11 566.

Palacio, F., Lazaro, F. J. & Duyneveldt, A. J. 1989 *Mol. Crys. Liq. Cryst.* **176**, 289.

Pei, Y., Verdaguer, M., Kahn, O., Sletten, J. & Renard, J.-P. 1986 *J. Am. Chem. Soc.* **108**, 428.

Rossman, G. R., Tsay, F. D. & Gray, H. B. 1973 *Inorg. Chem.* **12**, 824.

Salgueiro da Siva, M. A., Moreira, J. M., Mendes, J. A., Amaral, V. S., Sousa, J. B. & Palmer, S. B. 1995 *J. Phys.: Condens. Mater.* **7**, 9853.

Stumpf, H. O., Ouahab, L., Pei, Y., Grandjean, D. & Kahn, O. 1993 *Science* **261**, 447.

Stumpf, H. O., Ouahab, L., Pei, Y., Bergerat, P. & Kahn, O. 1994*a J. Am. Chem. Soc.* **116**, 3866.

Stumpf, H. O., Pei, Y., Michaut, C., Kahn, O., Renard, J. P. & Ouahab, L. 1994*b Chem. Mater.* **6**, 257.

Tamaki, H., Zhong, Z. J., Matsumoto, N., Kida, S., Koikawa, S., Achiwa, S., Hashimoto, Y. & Okawa, H. 1992 *J. Am. Chem. Soc.* **114**, 6974.

Vaz, M. G. F., Pinheiro, L. M. M., Stumpf, H. O., Alcântara, A. F. C., Gohlen, S., Ouahab, L., Cador, O., Mathonière, C. & Kahn, O. 1999 *Chem. Eur. J.* **5**, 1486.

Wooster, W. A. 1973 *Tensor and group theory for the physical properties of crystals.* Oxford: Clarendon Press.

Yee, G. T., Manriquez, J. M., Dixon, D. A., McLean, R. S., Groski, D. M., Flippen, R. B., Narayan, K. S., Epstein, A. J. & Miller, J. S. 1991 *Adv. Mater.* **3**, 309.

Young, R. C. 1932 *J. Am. Chem. Soc.* **54**, 1402.

Multifunctional coordination compounds: design and properties

By Silvio Decurtins

Departement für Chemie und Biochemie, Universität Bern,
Freiestrasse 3, CH-3012 Bern, Switzerland

Cleverly designed molecular building blocks provide chemists with the tools of a powerful molecular-scale construction set. They enable them to engineer materials that have a predictable order and useful solid-state properties. Hence, it is in the realm of supramolecular chemistry to follow a strategy for synthesizing materials that combine a selected set of properties, for instance from the areas of magnetism, photophysics and electronics. As a possible approach, host–guest solids that are based on extended anionic, homo- and bi-metallic oxalato-bridged transition-metal compounds with two- and three-dimensional connectivities are investigated. In particular, we report in detail about their structural properties and their multifunctional characteristics in the area of molecular magnetism and photophysics.

Keywords: coordination compounds; network compounds; oxalate complexes;
molecular magnetism; energy transfer

1. Introduction

Interest in functional materials based on molecular crystals persists, owing to the potential for manipulating solid-state properties by systematic variation of the molecular structures as well as of the stoichiometries and properties of the molecular components. However, until now, the crystal engineering of molecular materials has been frustrated by the absence of reliable and general structural paradigms that are needed for systematic design of crystal lattices with predictable structure and desirable functions. Therefore, learning how to create large supramolecular units and the elucidation of rules mediating their macroscopic organization into multifunctional materials will offer a fascinating prospect for technology (Zaworotko 1997; Zimmerman 1997; Desiraju 1996; Lehn 1995).

The purpose of the present work is to set an example of a supramolecular system exhibiting molecular self-organization. Thereby, we aim to exploit host–guest solids, where each component will contribute its own physical characteristics. In general, the two entities could behave independently, resulting in composite properties, or they might interact synergetically, potentially leading to new physical properties, e.g. in the fields of molecular magnetism and photophysics. Particularly, the present work focuses on the transition-metal oxalate system, its fascinating structural versatility, and its potential as a multifunctional material mainly in the areas of molecule-based magnetism and photophysics (figure 1). Starting with a detailed discussion about the structural topology of this supramolecular coordination compound, we will subsequently present some experimental results in the areas of magnetism and

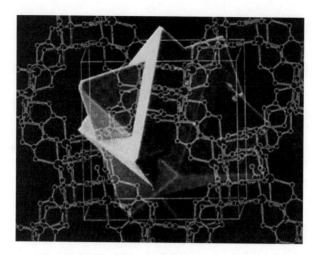

Figure 1. Single crystal of a chiral three-dimensional molecular network compound with transition metals as spin carriers.

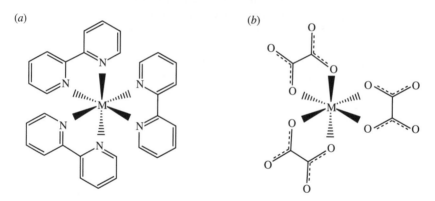

Figure 2. Schematic representation of the two chiral (the Λ-isomers are shown) preorganized cationic (*a*) and anionic (*b*) coordination entities. M = transition-metal ion as spin carrier centre.

photophysics; altogether this exemplifies a strategy we explored in order to attempt a multifunctional material.

2. Crystal engineering

Our laboratory has reported on the synthesis and structure determination of coordination solids based on transition-metal oxalates, which, typically, behave as host–guest compounds with different lattice dimensionalities (Schmalle *et al.* 1996; Pellaux *et al.* 1997; Decurtins *et al.* 1993, 1994*a*, *b*, 1996*b*). Thereby, the idea of a reasonable strategy looks simple: mix metal ions with a preference for a particular coordination geometry with bridging ligand systems and under the right conditions— and it is important to control both the kinetics and thermodynamics of the assembly process—a crystalline network will nucleate and grow. Clearly, the strategy relies on

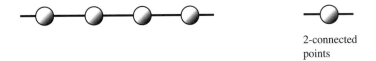

2-connected
points

Figure 3. An atom or a molecular complex forming two bonds comprises
a structural unit for a one-dimensional compound.

the robustness of some coordination subunits and of the supramolecular motif as a whole. Figure 2 depicts two preorganized transition metal complexes that act as mutually complementary molecular partners. Due to their specific coordination properties and their similar size and shape, they are predisposed on one hand to act as template molecules and on the other to form extended two- and three-dimensional network structures.

In the following, we will present a detailed discussion about the topics of lattice dimensionality and chirality whereby, first, we will rely on some distinct topological rules and, in a second step, we will refer to the appropriate chemical building blocks. Quite surprisingly, there exist simple topological rules (Wells 1984) that will determine the connectivity and structural dimensionality for many of these molecular assemblies, ranging from infinitely extended one-dimensional to three-dimensional motifs (compare figures 3–5). In accordance with this structural concept, one easily recognizes that any pattern that repeats regularly in one, two, or three dimensions consists of units that join together when repeated in the same orientation, that is, all units are identical and related only by translation. The repeat unit may be represented by a single point or a group of connected points, and it must have at least two, four or six free links available for attachment to its neighbours. Consequently, molecular subunits with two free links, represented by 2-connected points, will combine to form one-dimensional molecular chains (figure 3). Analogously, as illustrated in figure 4, the formation of two-dimensionally linked assemblies affords subunits possessing four free links, which corresponds to two non-parallel lines. Evidently, either 4-connected points or a combination of two 3-connected points will form the appropriate building blocks. In the latter case, the honeycomb lattice type results.

Along this line, extended three-dimensionally linked assemblies rely on building blocks comprising three non-coplanar lines as free links as represented either by 6-connected points, by a combination of two 4-connected points, or by four 3-connected points. Figure 5 shows that, in each case and very strictly, distinct three-dimensional lattice types are created. Most interestingly, the subunits that will form a 3-connected three-dimensional net must contain four 3-connected points in order to obtain the necessary number of six free links. Similarly oriented subunits must be joined together through the free links, so that decagon circuits are formed. Hence, the structure represents a uniform net in the sense that the shortest path, starting from any point along any link and returning to that point along any other link, is a circuit of 10 points.

In the actual case of this report, the molecular subunits are anionic tris-chelated transition-metal oxalato complexes $[M^{z+}(ox)_3]^{(6-z)-}$, where $ox = C_2O_4^{2-}$ (see figure 2b). Therefore, as a consequence of this $[M(L^\wedge L)_3]$ type of connectivity, each coordinated metal ion represents a chiral centre with D_3 point-group symmetry, showing either Δ- or Λ-helical chirality. Evidently, this property adds a further aspect to our discussion about molecular topology. Now, if such building blocks of different chirality, while

(a)

(b)

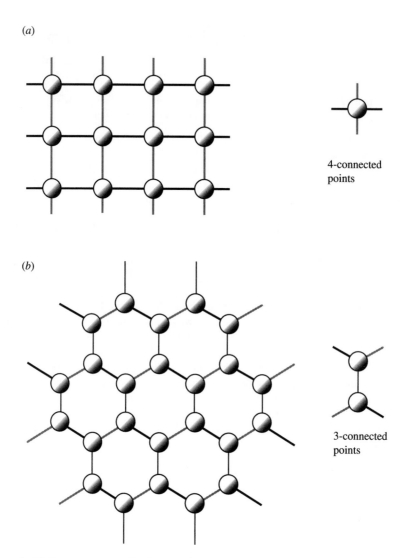

4-connected
points

3-connected
points

Figure 4. (a) An atom or a molecular complex forming four bonds comprises a structural unit for a two-dimensional compound. (b) The repeat unit consists of a pair of 3-connected points.

still corresponding to 3-connected points, are alternately linked, the bridged metal ions are confined to lie within a plane, as illustrated in figure 6. Consequently, a layered structure motif will result. In contrast, as also depicted in figure 6, an assembling of building blocks of the same chiral configuration will lead to a three-dimensional framework structure. It remains to apply the topological rules described above in order to define the number of 3-connected subunits that are needed to build closed circuits, hence, extended framework motifs. Figure 7 illustrates the way that two dimeric subunits may be combined to form the planar honeycomb network. In an analogous manner, it can be seen from figure 8 that two tetrameric subunits are

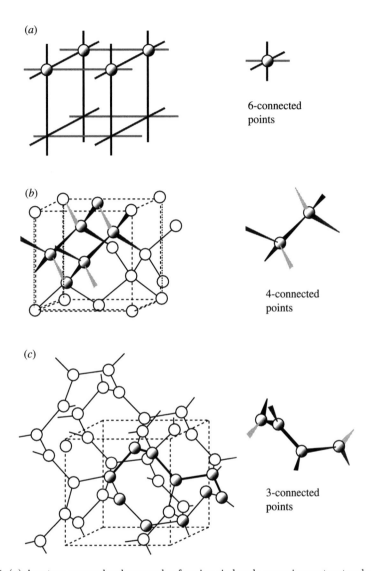

Figure 5. (*a*) An atom or a molecular complex forming six bonds comprises a structural unit for a three-dimensional compound. (*b*) The repeat unit consists of a pair of 4-connected points, leading to a three-dimensional (6,4) network. (*c*) The repeat unit consists of a tetramer of 3-connected points, leading to a three-dimensional (10,3) network.

needed to build closed circuits composed of 10 metal centres, which, in sum, define the chiral three-dimensional decagon framework structures.

Although the topological rules give an understanding of the different structural possibilities, the synthetic chemists still need to find the optimal reaction and crystallization conditions for each specific material. In the actual case of these tris-chelated transition metal oxalato complexes, the discrimination between the formation and crystallization of either a two-dimensional or a three-dimensional framework structure

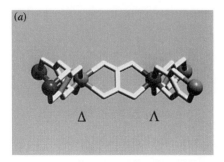
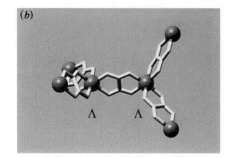

Figure 6. Chiral $[M^{z+}(ox)_3]^{(6-z)-}$ building blocks assembled with
(a) alternating chiral configuration, (b) equal chiral configuration.

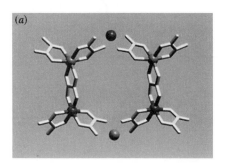
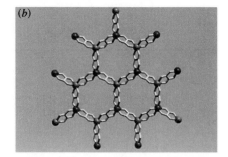

Figure 7. (a) Two dimeric units of the alternating chirality type are necessary to form a
closed hexagon ring; (b) the resulting planar network motif.

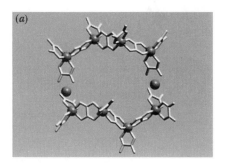
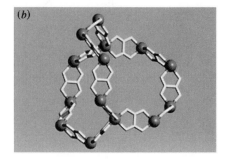

Figure 8. Two tetrameric units of the same chirality type are necessary to form a closed decagon
ring. (b) A fragment of the three-dimensional chiral framework.

relies on the choice of the templating counterion. Evidently, the template cation
determines the crystal chemistry. In particular, $[XR_4]^+$ (X = N, P; R = phenyl,
n-propyl, n-butyl, n-pentyl) cations initiate the growth of two-dimensional layer
structures containing $[M^{II}M^{III}(ox)_3]_n^{n-}$ (M^{II} = V, Cr, Mn, Fe, Co, Ni, Cu, Zn;
M^{III} = V, Cr, Fe) network stoichiometries. The structures consist of anionic two-
dimensional honeycomb networks that are interleaved by the templating cations.
Although these two-dimensional compounds are not chiral, they express a structural
polarity due to the specific arrangement of the templating cations (see figure 9). These
organic cations, which are located between the anionic layers, determine the interlayer

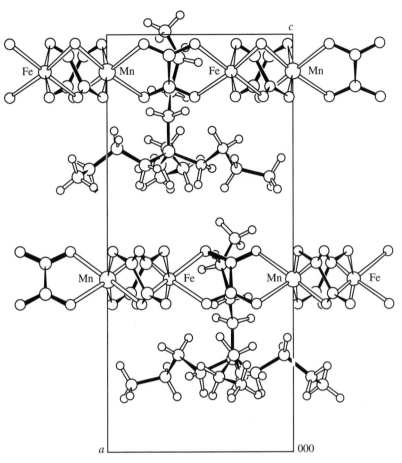

Figure 9. [010] projection of $\{[\mathrm{N}(n\text{-}\mathrm{C}_4\mathrm{H}_9)_4][\mathrm{Mn}^{\mathrm{II}}\mathrm{Fe}^{\mathrm{III}}(\mathrm{C}_2\mathrm{O}_4)_3]\}_n$.

separations. From single-crystal X-ray studies, these distances have been determined to have the values 9.94 Å, 9.55 Å, 8.91 Å and 8.20 Å for the *n*-pentyl, phenyl, *n*-butyl and *n*-propyl derivatives, respectively (Decurtins *et al.* 1994a; Pellaux *et al.* 1997; Mathonière *et al.* 1996).

Remarkably, the cationic tris-chelated transition-metal diimine complexes,

$$[\mathrm{M}(\mathrm{bpy})_3]^{2+/3+}$$

(see figure 2a), act as templates for the formation and crystallization of the three-dimensional decagon framework structures. As outlined above, the topological principle implies that, for the three-dimensional case, only subunits of the same chiral configuration are assembled. Consequently, the uniform anionic three-dimensional network type with stoichiometries like

$$[\mathrm{M}_2^{\mathrm{II}}(\mathrm{ox})_3]_n^{2n-}, \quad [\mathrm{M}^{\mathrm{I}}\mathrm{M}^{\mathrm{III}}(\mathrm{ox})_3]_n^{2n-} \quad \text{or} \quad [\mathrm{M}^{\mathrm{II}}\mathrm{M}^{\mathrm{III}}(\mathrm{ox})_3]_n^{n-}$$

is chiral, as it is composed of $2n$ centres exhibiting the same kind of chirality. Naturally, this chiral topology is in line with the symmetry elements that are present

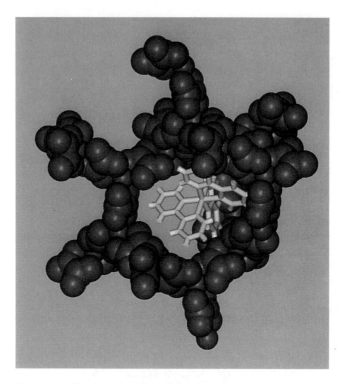

Figure 10. The three-dimensional host–guest compound. Only one
guest molecule is shown within the chiral framework.

in the crystalline state of the three-dimensional frameworks, which, in sum, constitute
either one of the enantiomorphic cubic space groups $P4_3 32$ or $P4_1 32$ for the former
and the cubic space group $P2_1 3$ for the latter bi-metallic stoichiometries. Thereby, the
$2n$ metal ions occupy special sites with a threefold symmetry axis. Figure 10 depicts a
view of the three-dimensional host–guest assembly.

Overall, with a straightforward synthetic method that is mainly based on the
function of appropriate molecular templates and specific molecular building blocks,
extended two-dimensional and three-dimensional supramolecular materials are
accessible. Thereby, some general topological rules elegantly describe the resulting
distinctive architectures.

3. Multifunctionality: molecular magnetism

The magnetic properties of molecule-based materials have become an important focus
of scientific interest in the last few years (Miller & Epstein 1994, 1995; Kahn 1993;
Ferlay et al. 1995), whereby the search for molecule-based ferromagnets that order at
or above room temperature is a major driving force moving this field (Ferlay et al.
1995). Moreover, the synthesis of materials combining two or more functional
properties, e.g. from the area of magnetism and photophysics or magnetism and
superconductivity, represents a current challenge for the preparative chemists.

When did the research activities in the area of molecule-based magnetism start?
The first genuine molecular compound displaying a ferromagnetic transition was

described by Wickman *et al.* as early as 1967 (Wickman *et al.* 1967). This compound, a chlorobis(diethyldithio-carbamato)iron(III) complex, exhibiting an intermediate spin $S = 3/2$, orders ferromagnetically at 2.46 K. Since then, over the years, a scientific community has been established throughout the world, focusing on the aspects of molecular magnetism, and an increasing number of international conferences have turned up a lot of new chemistry and brought synthetic chemists into close contact with physics and material science.

Notice that there are several features of potential practical impact that distinguish magnetic materials based on molecules from their analogues consisting of continuous ionic or metallic lattices. Examples would include the search for materials combining two or more functional properties, e.g. magnetism and transparency for magneto-optical applications or the design of mesoscopic molecules possessing large magnetic moments. Synthetic methods will also be quite different, and, consequently, magnetic thin films might be deposited with methods such as solvent evaporation.

A brief comment on the characteristics of the elements that constitute a magnetic molecular solid will give us a feeling of the extension, complexity and wide diversity of magnetic phenomena that can be found in these materials. Generally, a magnetic molecular solid can be formed by free radicals, transition-metal ions, rare earth ions and diamagnetic ligands. Any combination of these components is possible, although only the free radicals can form a magnetic molecular solid by themselves. In the following, we will concentrate only on coordination compounds, where the importance of transition-metal ions as spin carrier centres stems from at least three main reasons:

(i)   transition metal to ligand interactions are extremely variable, thus, the building up of novel higher-dimensional architectures can profit very much from the coordination algorithm of the metal ions as well as from the availability of various bridging ligand systems;

(ii) transition metals are prone to quick and reversible redox changes; hence, supramolecular functions like energy- and charge-transfer processes can benefit from these properties; and

(iii) the collective features of components bearing free spins may result in supra-molecular assemblies exhibiting molecule-based magnetic behaviour, whereby the critical role of the dimensionality of the compounds is simultaneously taken into account.

Accordingly, molecular precursors implying transition-metal ions entail the synthesis of ferromagnetic and antiferromagnetic systems with a tuneable critical temperature.

In addition, any synthetic strategy aimed at designing molecular magnets has to answer the questions of

(i)   how to control the interaction between the nearest neighbouring magnetic spins; and

(ii) how to control parallel alignment of the magnetic spin vectors over the three-dimensional lattice.

Naturally, if the compounds assume a two-dimensional layer structure, the magnetic properties depend on the nature of both the intra- and inter-layer magnetic interactions.

With respect to the first question, it is well known that the oxalate bridge is a good mediator in both antiferromagnetic and ferromagnetic interactions between similar and dissimilar metal ions; therefore, it has been widely used to construct polynuclear compounds in the search for new molecular-based magnets (Kahn 1993). Naturally, in search for an answer to the second question, effort has to be made to investigate the magnetic ordering behaviour of the above-described two-dimensional and three-dimensional systems.

Along this line, a successful molecular design of two-dimensionally extended metal-complex magnets that are based on trioxalatochromium(III) building blocks (see figure 2b) has been reported in Tamaki et al. (1992). Within a series of layered oxalate-bridged bi-metallic compounds, ferromagnetic ordering behaviour has been shown to occur at temperatures below 14 K. Since then, a variety of analogous two-dimensional bi-metallic assemblies, also with mixed-valency stoichiometries, have been prepared and characterized (Mathonière et al. 1994, 1996; Nuttall et al. 1995; Carling et al. 1996). Overall, many of these layered compounds exhibit ferromagnetic, ferrimagnetic or antiferromagnetic long-range ordering behaviour, and, in some cases, they show evidence for at least short-range interactions.

In contrast with the large body of experimental results that has been published from magnetic susceptibility and magnetization measurements with molecule-based magnetic materials, very limited experience has been gained so far from elastic neutron-scattering experiments aimed at elucidating the spin structure in the magnetically ordered phase. Therefore, in the following, a brief account of the current state of the ongoing investigations by means of the neutron-scattering technique will be given. Thereby, complementary to the above-mentioned two-dimensional systems, a three-dimensional molecular network compound will be chosen for a brief discussion.

Thus, we briefly report on neutron diffraction experiments, performed on a three-dimensional polycrystalline sample with stoichiometry

$$[Fe^{II}(d_8\text{-bpy})_3]_n^{2+} \, [Mn_2^{II}(ox)_3]_n^{2n-},$$

with the goal of determining the magnetic structure of this helical supramolecule in the antiferromagnetically ordered phase, thus, below $T_N = 13$ K (Decurtins et al. 1996a). The existence of this magnetically ordered phase has formerly been suggested from magnetic DC-susceptibility measurements, which revealed a rounded maximum at about 20 K in the χ_M versus T curve (thus, $T_N < 20$ K) as well as a Weiss constant Θ of -33 K in the $1/\chi_M$ versus T plot (Decurtins et al. 1994b). Accordingly, a magnetic AC-susceptibility experiment revealed an ordering temperature around 15 K. This long-range magnetic ordering basically originates from the exchange interaction between neighbouring Mn^{2+} ions, mediated by the bridging oxalate ligands.

As anticipated from these bulk measurements, an increase of the intensities due to long-range antiferromagnetic ordering of the Mn^{2+} ions could be detected with the neutron diffraction measurements performed in the temperature range from 30 K to 1.8 K. Figure 11 illustrates the observed (difference $I(1.8\ \text{K}) - I(30\ \text{K})$), calculated and difference magnetic neutron-diffraction patterns. Thereby, it has to be noted that

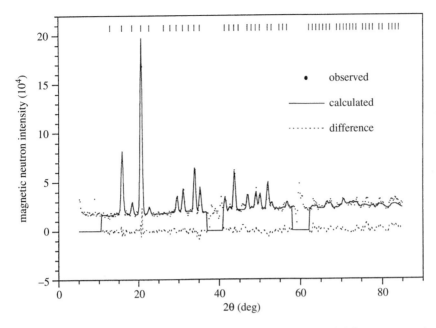

Figure 11. Observed (difference $I(1.8 \text{ K}) - I(30 \text{ K})$), calculated and difference magnetic neutron-diffraction patterns of a polycrystalline sample of $[Fe^{II}(d_8\text{-bpy})_3]_n^{2+} [Mn_2^{II}(ox)_3]_n^{2n-}$.

the increase of the intensities corresponds to a propagation vector $\boldsymbol{K} = \boldsymbol{0}$, i.e. a magnetic unit cell being equal to the chemical cell. The temperature dependence of the dominant magnetic intensity (210) at $2\theta = 21.1°$ indicates an ordering temperature $T_N = 13(0.5)$ K, in good agreement with the magnetic susceptibility experiments.

Furthermore, it remains to discuss the determined magnetic moment configuration of this three-dimensionally linked Mn^{2+} network. With respect to space group $P4_132$, the Mn^{2+} ions occupy sites $8c; x, x, x$ with $x = 0.649\,07$. The best agreement between observed and calculated magnetic neutron intensities was achieved with a collinear antiferromagnetic arrangement of Mn^{2+} moments according to the three-dimensional irreducible representation τ_4, which is derived from the enantiomorphic pair of the chiral cubic crystallographic space groups $P4_332/P4_132$ (Sikora 1994). Thereby, the ordered magnetic moment at 1.8 K amounts to $\mu_{Mn} = 4.6(1)\mu_B$, where μ_B is the electron *Bohr* magneton. The saturation magnetization M_S is related to the equation $M_S = g\mu_B NS$, where S is the spin quantum number, N the Avogadro number and g the electron g-factor. Thus, a gS value corresponding to the number of unpaired electrons of 4.6 is obtained, which is compatible with the expected five unpaired electrons ($g = 2$) from the Mn^{2+} ions. Naturally, in the present experiment, no information about a preferred direction of the magnetic moments with respect to the crystallographic axes can be gained from the polycrystalline sample with cubic symmetry.

Figure 12 shows the pattern of the magnetic structure within the three-dimensional manganese(II) network. Despite the three-dimensional helical character of the framework structure incorporating the magnetic ions, a two-sublattice antiferromagnetic spin arrangement has been proven to occur, hence, no helimagnetic structure has

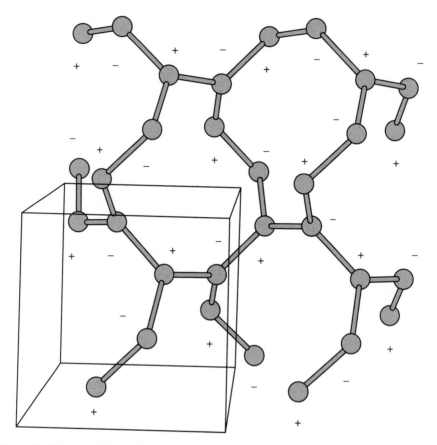

Figure 12. Scheme of the antiferromagnetic collinear configuration of the magnetic moments originating from the Mn^{2+} ions, which constitute the chiral three-dimensional network compound.

shown up. After all, the behaviour is in accordance with the typical isotropic character of the *Heisenberg* ion Mn^{2+}.

This example of the extended three-dimensional Mn^{2+} coordination solid illustrates the potential of the elastic neutron-scattering technique to elucidate the spin structures within the magnetically ordered phases occurring in such polymeric molecular materials.

4. Multifunctionality: photophysics

In this section, we will comment on some observations of excitation energy-transfer processes within the three-dimensional supramolecular host–guest compounds. Depending upon the relative energies of the excited states of the chromophores, energy transfer is observed either from the guest system with the tris-bipyridine cations as donors to the host system where the oxalate backbone acts as acceptor sites or *vice versa*. In addition, energy transfer between identical chromophores occurs within the

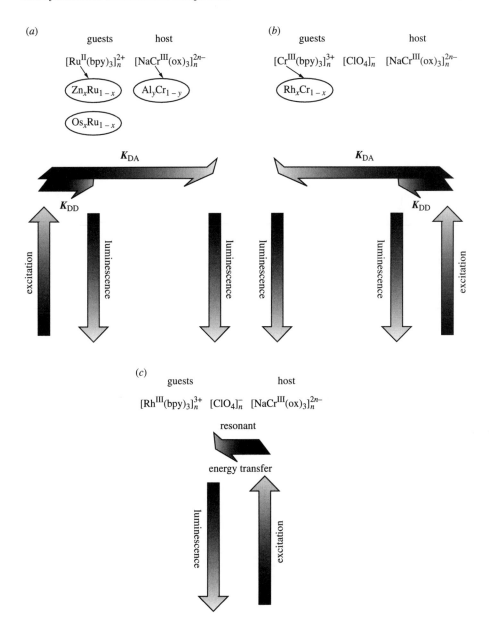

Figure 13. Schematic representation of energy-transfer processes for different stoichiometries: (*a*) excitation into the guest system; (*b*) and (*c*) excitation into the host system.

host as well as within the guest system (Decurtins *et al.* 1996*b*; Hauser *et al.* 1996; von Arx *et al.* 1996).

Figure 13 illustrates these energy-transfer processes schematically with different

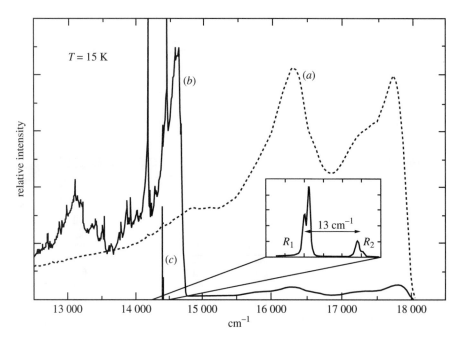

Figure 14. Luminescence spectra at $T = 15$ K of (a) [Ru(bpy)$_3$][NaAl(ox)$_3$];
(b) [Ru$_{1-x}$Os$_x$(bpy)$_3$][NaAl(ox)$_3$], $x = 1\%$; (c) [Ru(bpy)$_3$][NaCr(ox)$_3$]; $\lambda = 476$ nm.

host–guest stoichiometries, and, particularly, one can point out that chemical
variation and combination of metal ions of different valencies in the oxalate backbone
as well as in the tris-bipyridine cation offer unique opportunities for studying a large
variety of photophysical phenomena. Naturally, the sensitizer can be incorporated
into the oxalate backbone or the tris-bipyridine cation, either in low concentration as
dopant, at higher concentration in mixed crystals, or fully concentrated in neat
compounds.

As an example of the energy-transfer processes, figure 14 shows the luminescence
spectra of three representative compounds. For instance, if Al^{3+} is replaced by Cr^{3+},
the [Ru(bpy)$_3$]$^{2+}$ luminescence from the spin-forbidden MLCT transition is completely quenched, and the sharp luminescence bands characteristic for the zero-field
components of the ^{2}E $\rightarrow$ ^{4}A$_2$ transition of octahedrally coordinated and trigonally
distorted Cr^{3+} are observed at 14 400 cm^{-1}. This is a clear indication of very efficient
energy transfer from the initially excited [Ru(bpy)$_3$]$^{2+}$ to [Cr(ox)$_3$]$^{3-}$. However, not
only acceptors on the oxalate backbone may quench the [Ru(bpy)$_3$]$^{2+}$ luminescence.
Replacing a fraction of the [Ru(bpy)$_3$]$^{2+}$ with [Os(bpy)$_3$]$^{2+}$ results in luminescence
from [Os(bpy)$_3$]$^{2+}$ and a quenching of the [Ru(bpy)$_3$]$^{2+}$ luminescence too. Indeed, the
energy transfer to [Os(bpy)$_3$]$^{2+}$ is even more efficient than to [Cr(ox)$_3$]$^{3-}$.

Alternatively, as shown schematically in figure 13b, irradiating into the spin-
allowed ^{4}A$_2$ $\rightarrow$ ^{4}T$_2$ absorption band of [Cr(ox)$_3$]$^{3-}$ from the host system results in
intense luminescence from the ^{2}E state of [Cr(bpy)$_3$]$^{3+}$ within the guest system, again
demonstrating a rapid energy-transfer process (Decurtins et $al.$ 1996b). Finally,
according to figure 13c, the stoichiometry [Rh(bpy)$_3$]$_n^{3+}$[ClO$_4$]$_n^-$[NaCr(ox)$_3$]$_n^{2n-}$ allows

us to study the energy transfer within the R_1 line of the ^2E state of Cr^{3+}. In that case, from a fluorescence-line-narrowing experiment, clear evidence for a resonant energy-transfer process could be gained (Hauser *et al.* 1996; von Arx *et al.* 1996).

5. Summary

Nowadays, the synthetic chemists have turned their attention to an ambitious architectural goal: the assembly of relatively simple molecules into complex polymeric structures. The big challenge of building new ever-more-intricate molecules is to learn how to control the ordering of the component molecules so that the supramolecular assembly has the desired structure, stability and properties. Accordingly, the results presented in this paper show a straightforward concept for the synthesis of two-dimensional and three-dimensional network structures. Thereby, the oxalate ion, although a fairly ubiquitous ligand, plays a key role in the formation of a whole class of transition-metal-based supramolecular host–guest systems. Ongoing studies focus on the elucidation of the magnetic structures for different magnetically ordered phases and on the investigation of the fascinating photophysical behaviour. Overall, we are looking for synergistic properties within this class of multifunctional materials.

Gratitude is expressed to the Swiss National Science Foundation and to the European-TMR Network '3MD' project for financial support.

References

Carling, S. G., Mathonière, C., Day, P., Malik, K. M. A., Coles, S. J. & Hursthouse, M. B. 1996 *J. Chem. Soc. Dalton Trans.*, p. 1839.

Decurtins, S., Schmalle, H. W., Schneuwly, P. & Oswald, H. R. 1993 *Inorg. Chem.* **32**, 1888.

Decurtins, S., Schmalle, H. W., Oswald, H. R., Linden, A., Ensling, J., Gütlich, P. & Hauser, A. 1994*a Inorg. Chim. Acta* **216**, 65.

Decurtins, S., Schmalle, H. W., Schneuwly, P., Ensling, J. & Gütlich, P. 1994*b J. Am. Chem. Soc.* **116**, 9521.

Decurtins, S., Schmalle, H. W., Pellaux, R., Huber, R., Fischer, P. & Ouladdiaf, B. 1996*a Adv. Mater.* **8**, 647.

Decurtins, S., Schmalle, H. W., Pellaux, R., Schneuwly, P. & Hauser, A. 1996*b Inorg. Chem.* **35**, 1451.

Desiraju, G. R. (ed.) 1996 The crystal as a supramolecular entity. In *Perspectives in supramolecular chemistry*, vol. 2. Wiley.

Ferlay, S., Mallah, T., Ouahès, R., Veillet, P. & Verdaguer, M. 1995 *Nature* **378**, 701.

Hauser, A., Riesen, H., Pellaux, R. & Decurtins, S. 1996 *Chem. Phys. Lett.* **261**, 313.

Kahn, O. 1993 *Molecular magnetism*. Weinheim: VCH.

Lehn, J.-M. 1995 *Supramolecular chemistry*. Weinheim: VCH.

Mathonière, C., Carling, S. G., Yusheng, D. & Day, P. 1994 *J. Chem. Soc. Chem. Commun.*, p. 1551.

Mathonière, C., Nuttall, C. J., Carling, S. G. & Day, P. 1996 *Inorg. Chem.* **35**, 1201.

Miller, J. S. & Epstein, A. J. 1994 *Angew. Chem. Int.* **33**, 385.

Miller, J. S. & Epstein, A. J. (eds) 1995 *Proc. IV Int. Conf. on Molecule-Based Magnets: 'Molecular Crystals and Liquid Crystals', London, 1995*, vols 271–274. London: Gordon & Breach.

Nuttall, C. J., Bellitto, C. & Day, P. 1995 *J. Chem. Soc. Chem. Commun.*, p. 1513.

Pellaux, R., Schmalle, H. W., Huber, R., Fischer, P., Hauss, T., Ouladdiaf, B. & Decurtins, S. 1997 *Inorg. Chem.* **36**, 2301.

Schmalle, H. W., Pellaux, R. & Decurtins, S. 1996 *Z. Kristallogr.* **211**, 533.

Sikora, W. 1994 PC program MODY. In *Collected abstracts of a 'Workshop on magnetic structures and phase transitions', Krakow, Poland.*

Tamaki, H., Zhong, Z. J., Matsumoto, N., Kida, S., Koikawa, M., Achiwa, N., Hashimoto, Y. & Okawa, H. 1992 *J. Am. Chem. Soc.* **114**, 6974.

von Arx, M. E., Hauser, A., Riesen, H., Pellaux, R. & Decurtins, S. 1996 *Phys. Rev.* B **54**, 15 800.

Wells, A. F. 1984 *Structural inorganic chemistry.* Oxford: Clarendon.

Wickman, H. H., Trozzolo, A. M., Williams, H. J., Hull, G. W. & Merritt, F. R. 1967 *Phys. Rev.* **155**, 563.

Zaworotko, M. J. 1997 *Nature* **386**, 220.

Zimmerman, S. C. 1997 *Science* **276**, 543.

Discussion

M. VERDAGUER (*Laboratoire de Chimie Inorganique et Matériaux Moléculaires, Université Pierre et Marie Curie, Paris, France*). Professor Decurtins has shown an impressive number of chiral three-dimensional systems.

Was the synthesis and the crystallization performed with racemic or resolved reactants?

Starting with racemic reactants, were 50/50 mixtures of A and B enantiomers obtained? Was enantiomeric excess in the crystallization observed in some cases? I am aware of such a possibility with some systems.

Were Λ or Δ, three-dimensional systems, obtained at will starting with Λ or Δ resolved reactants?

S. DECURTINS. The synthesis and crystallization has been performed with racemic reactants.

We only investigated the solid-state products in that respect by means of single-crystal X-ray diffraction experiments. This means that either one of the enantiomeric forms are observed, since the crystallization leads to a distinct enantiomeric form for each crystal.

I think the outcome is obvious, and, in any case, one gets at will three-dimensional single crystals, each one with a distinct chirality.

Ferrimagnetic and metamagnetic layered cobalt(II)-hydroxides: first observation of a coercive field greater than 5 T

Institut de Physique, Chimie et Matériaux de Strasbourg,
23 rue du Loess, 67037 Strasbourg Cedex, France

The synthesis, characterization by XRD, UV–vis, IR and TGA and the magnetic
properties of four layered compounds, namely

1   $Co_5(OH)_8(H_2O)_2(NO_3)_2$;

2   $Co_5(OH)_8(O_2CC_6H_4CO_2) \cdot 2H_2O$;

3   $Co_2(OH)_3(NO_3)$;

4   $Co_4(OH)_2(O_2CC_6H_4CO_2)_3 \cdot (NH_3)_{1.5}(H_2O)_{2.5}$,

are reported. **1** and **2** are characterized by triple-deck layers consisting of both
octahedral and tetrahedral cobaltous ions and **3** and **4** contain single-deck layers of
only octahedral cobaltous ions. **1** and **2** behave as ferrimagnets which are
characterized by a minimum in the χT versus T plot, spontaneous magnetization,
imaginary AC-susceptibility and hysteresis loop. The Curie temperatures are 30 and
40 K and the coercive fields at 4.2 K are 850 and 1750 Oe, respectively. **3** and **4** behave
as metamagnets which are characterized by maxima in the susceptibility at the Néel
temperature of 10 and 38 K, respectively, and by a critical field (antiferromagnetic ↔
paramagnetic) of *ca.* 1 kOe for **3** at 4.5 K and greater than 50 kOe for **4** at 2 K. The
tricritical temperature, separating the region of reversible and non-reversible M versus
H, for **4** is established at 22.5 K. The long-range magnetic ordering for the two
structural types is discussed on the basis of dipolar interactions between layers. The
intralayer interactions are ferromagnetic in all cases whereas the interlayer interactions are ferromagnetic for **1** and **2** and antiferromagnetic for **3** and **4**. The results
indicate that the Curie or Néel temperature is weakly dependent on the interlayer
distance and its observation does not depend on the existence of covalent bonds
between the layers. The large coercive fields observed are due to the alignment of the
moments perpendicular to the layers and the synergy between crystalline shape and
single-ion anisotropies.

Keywords: cobalt; magnet; layer; clay; hydroxide

---

1. Introduction

Among the many infinite assemblies of magnetic moments relevant to magnetism,
those consisting of two-dimensional arrays are of particular importance (De Jongh
1990; Kahn 1993; Heinrich & Bland 1994). An important commercial demand is

alignment of moments perpendicular to the layers, a feature of great value for magnetic recording devices with high-density information-storage capacity (Andrä *et al.* 1991). This class of materials also invites interesting theoretical and experimental questions (De Jongh 1990). One of the first is the existence of long-range ordering for two-dimensional Heisenberg systems. Spin-frustration in triangular lattices is another issue. Due to the existence of weak interlayer interactions through van der Waal or hydrogen bonds and via dipolar through space, an important question often asked is 'when can a crystalline material be classified as two dimensional'. These fundamental questions have consequently stimulated an enormous amount of study in the last twenty years (De Jongh & Miedma 1974; De Jongh 1990; Kahn 1996). From a chemical point of view they are additionally attractive as one is able to conserve the structure within the two-dimensional sheets and systematically tune that in the third dimension (Schollhorn 1984; Mitchell 1990; Clearfield 1991; Jacobsen 1992; O'Hare 1993; Meyn *et al.* 1993; Carlino 1997; Newman & Jones 1998).

From a magnetic point of view the problem is reduced to one dimension as principally the interlayer magnetic exchange interactions are modified. Such studies have been documented for several families of molecular compounds which include: (*a*) the layered perovskite, A_2MX_4, where A is an alkali metal or an alkyl- or aryl-ammonium and M is Cu^{II} (Miedema *et al.* 1963; De Jongh *et al.* 1969) or Cr^{II} (Fair *et al.* 1977; Bellitto & Day 1992; Day 1997); (*b*) the heterometallic layered oxalates, $AM^{II}M^{III}(C_2O_4)_3$ (Tamaki *et al.* 1992*a*, *b*; Atovmyan *et al.* 1993; Mathonière *et al.* 1996; Carling *et al.* 1996; Clemente-Leon *et al.* 1997; Day 1997; Decurtins *et al.* 1998; Nuttall & Day 1998); and (*c*) the metal dihalides (Schieber 1967; Day 1988; Aruga Katori *et al.* 1996) having the $CdCl_2$ or CdI_2 structure (Wells 1984) and some phosphates (Carling *et al.* 1995; Bellitto *et al.* 1998; Fanucci *et al.* 1998) of divalent metals. The first class is ferromagnetic due to the orthogonal alignment of the magnetic orbitals brought about by the Jahn–Teller distortion for metals with d^4 and d^9 electronic configurations. The second class are either ferromagnetic or ferrimagnetic due to uncompensated moments between the two sublattices (M^{II} and M^{III}) of the honeycomb structure. Compounds of the third class are antiferromagnets except for a few cases of canting. Similar behaviour has been reported for the phosphates.

Recent observations of long-range magnetic ordering in the series of transition metal hydroxides (Rabu *et al.* 1993; Fujita & Awaga 1996; Rouba 1996; Laget *et al.* 1998; Kurmoo *et al.* 1999) and the high level of magnetic hardness (Kurmoo *et al.* 1999) in the cobalt compounds prompted us to explore these systems further, on the one hand, by tuning the interlayer spacing and, on the other, by inserting electronically or optically active molecules or ions in the galleries to generate dual-property (electronic–magnetic or magneto-optic) compounds (Kurmoo *et al.* 1995; Nicoud 1994). The possibility of inserting neutral and charged species is also appealing since one is able to tune the magnetic interactions by electrostatic means. There are some controversial reports regarding the types of long-range ordering for these systems; for instance, for the copper(II) compounds ferromagnetic (Laget *et al.* 1998), antiferro-magnetic and weak ferromagnetic ground states (Fujita & Awaga 1996) have been proposed. For the cobalt compounds the multiphase nature and the lack of structural data have hampered development. In an attempt to clarify some of these questions, we have developed layered metal hydroxides with a range of anions having different coordinating groups, such as alkyl- or aryl-carboxylates, -dicarboxylates, -sulphonates (Kurmoo *et al.* 1999), -sulphates, and also polycyanide anions such as dicyanamide and tricyanomethanide. In this paper, we present the characterization and the magnetic results of two structural types of cobalt compounds, one having only octahedral sites and the other containing both octahedral and tetrahedral sites. In each structural type we have employed the nitrate and the terephthalate ions; the second can provide covalently bonded bridges (Burrows *et al.* 1997; Cano *et al.* 1997; Cotton *et al.* 1998; Fogg *et al.* 1998*a, b*; Li *et al.* 1998) between the layers.

For cobalt hydroxide materials with nitrate as an anion three compounds have been reported: $Co(OH)(NO_3)$ is a pink double chain compound (Rouba 1996), whereas $Co_2(OH)_3(NO_3)$ and $Co_7(OH)_{12}(NO_3)_2$ are pink and green layered compounds (Laget *et al.* 1996), respectively. The structures of the latter two are thought to be built up of brucite, $Mg(OH)_2$, layers. The first two are reported to behave as meta-magnets and the third as a ferromagnet (Laget *et al.* 1996). In a systematic study, we have recently (Kurmoo *et al.* 1999) shown that it is possible to create pillared clay-like compounds by using long-chain alkyl sulphonates using exchange for the copper salt and a one-pot synthesis for the cobalt and nickel. The copper compound, $Cu_2(OH)_3C_{12}H_{25}SO_3 \cdot H_2O$, exhibits short-range antiferromagnetic interactions; the nickel compound, $Ni_2(OH)_3C_{12}H_{25}SO_3 \cdot H_2O$, is a ferromagnet ($T_C = 18$ K, coercive field is 400 Oe at 4.2 K); and the cobalt compound, $Co_5(OH)_8(C_{12}H_{25}SO_3)_2 \cdot 5H_2O$, is a ferrimagnet ($T_C = 50$ K, coercive field is 19 000 Oe at 2 K). The most unusual result in this series is the observation of long-range magnetic ordering for metal layers sep-arated by as much as 25 Å. The long-range magnetic ordering at such a high temper-ature was explained by a dipolar mechanism (Drillon & Panissod 1998) whereby the correlation length in the layer diverges as the Curie temperature is approached from above, resulting in clusters with large effective spin which interact via dipolar interac-tions. Furthermore, the wide hysteresis loop classifies $Co_5(OH)_8(C_{12}H_{25}SO_3)_2 \cdot 5H_2O$ as one of the hardest metal–organic magnets known (Kurmoo & Kepert 1998). The magnetic hardness results from the alignment of the moments perpendicular to the layers and the synergy of crystalline shape and single-ion anisotropies (Chikazumi 1978).

One particular objective for this study was to establish the magnetic ground states of the different structural types of cobalt layers. The second point was to understand the role of the bridging ligand on the exchange interactions. It is anticipated that such studies may shed some light on the confusion regarding the various magnetic ground states reported. Therefore, we have reinvestigated the nitrate-containing com-pounds and report an extensive magnetic study of $Co_2(OH)_3(NO_3)$ and of a novel phase, $Co_5(OH)_8(H_2O)_2(NO_3)_2$. We also present the results of two new compounds, $Co_4(OH)_2(O_2CC_6H_4CO_2)_3 \cdot (NH_3)_{1.5}(H_2O)_{2.5}$ and $Co_5(OH)_8(O_2CC_6H_4CO_2) \cdot 2H_2O$ containing bridging terephthalate.

2. Experimental section

(*a*) *Syntheses*

$Co_5(OH)_8(H_2O)_2(NO_3)_2$, **1**, was obtained as a bright green powder by adding 3 ml of aqueous ammonia (30%) drop by drop to a warm (40 °C) solution of $Co(H_2O)_6(NO_3)_2$ (3 g) and $NaNO_3$ (3 g) in 250 ml of a 1:1 mixture of water and absolute ethanol. The precipitate was filtered, washed and dried in air. Yield: 1.0 g.

For $Co_5(OH)_8(O_2CC_6H_4CO_2)\cdot 2H_2O$, **2**, a procedure similar to that described above for **1** was employed, using terephthalic acid (1 g) in the place of the sodium nitrate. Yield: 1.1 g.

$Co_2(OH)_3(NO_3)$, **3**, was prepared as a pink solid by a slight modification to a procedure reported previously (Rouba 1996): a solution of NaOH (1.2 g in 20 ml of water) was added, drop by drop, to a refluxing solution of $Co(H_2O)_6(NO_3)_2$ (5.8 g in 20 ml of water) under a flow of argon for 24 h. The precipitate was washed with cold water and acetone and air dried. Yield: 2.1 g.

$Co_4(OH)_2(O_2CC_6H_4CO_2)_3\cdot(NH_3)_{1.5}(H_2O)_{2.5}$, **4**, was prepared by mixing Co $(H_2O)_6(NO_3)_2$ (3 g), terephthalic acid (1 g) and 3 ml of aqueous ammonia (30%) in a warm (40°C) solution of 250 ml of water and absolute ethanol (1:1 mixture). A blue green precipitate was produced which was transformed into the titled compound by reducing the solution volume to 60 ml at 80°C. The pale pink powder was washed with water, ethanol and dried in air. Yield: 1.1 g.

The compositions of the compounds were confirmed by chemical and thermogravimetric analyses.

(b) Thermogravimetric analyses

Data were collected in the temperature range 20–1100 °C by warming the sample at a rate of 6 °C min^{-1} in air on a SETARAM TGA 92 calorimeter.

(c) X-ray diffraction

X-ray powder diffraction were recorded at room temperature on a SIEMENS D500 diffractometer employing Co-$K\alpha1$ (1.789 Å) and a flat plate geometry. Data were recorded in the 2θ range from 2–72° with a step of 0.02°.

(d) Infrared and UV–vis spectroscopies

Infrared spectra were obtained from fine powder of the compounds dispersed onto a KBr plate with the use of a MATTSON FTIR spectrometer. UV–vis spectra were recorded in the range 300–900 nm by transmission through thin films of the compounds dispersed in paraffin oil. The scattering due to the powdered samples was compensated by a piece of tissue paper soaked with the oil and placed in the reference beam.

(e) Magnetic properties

(i) *Faraday balance*

The temperature-dependence (4–300 K) of the magnetization of each sample was measured on a home-built Faraday-type pendulum in a field up to 13 000 Oe. The temperature of the sample in the continuous flow cryostat was controlled by either an Oxford Instruments ITC4 or a BT 400. The sample was held in a gelatine capsule and the data were corrected for its diamagnetism and those of the atomic diamagnetism using Pascal's constants.

(ii) *SQUID*

The temperature (2–300 K) and field (up to 5 T) dependence of the magnetization of the samples were recorded on a MPMS-XL magnetometer. AC susceptibilities were measured on the same instrument in a field of 1 Oe oscillating at 20 Hz. The samples were held in gelatine capsules. For hysteresis measurements in high field the samples were fixed in the capsules to prevent rotation. Any special protocol of each experiment is detailed in the relevant section.

(iii) *Vibrating sample magnetometer*

Isothermal magnetization of the compounds was recorded at several temperatures ($T > 4.2$ K) in fields up to ± 1.8 T by use of a Princeton Applied Research Model 155. The temperature of the sample in a bath cryostat was controlled by an Oxford Instruments ITC4. Samples were mounted in 5 mm diameter perspex containers.

3. Results and discussion

Before describing the results it is worthwhile for the understanding of this paper to give a basic description of metamagnetism (Herpin 1968), for it is rarely observed. Furthermore, the large coercive field exhibited by the metamagnet, **4**, described below, opens up a new class of magnetic materials. A metamagnet is an antiferromagnet with strong anisotropy. Below T_N, application of a magnetic field along the easy axis reverses the moments to the parallel orientation without passing through a spin-flop state and resulting in a paramagnetic state. Strong correlation between the moments is thought to create domains which give rise to hysteresis, in some cases, below a tricritical or a bicritical end-point temperature (Stryjewski & Giordano 1977). Metamagnetism is usually observed in layered compounds, where there exists a strong structural anisotropy and, consequently, anisotropy in and competitions between the exchange interactions. Some of the most studied compounds are the halides (Cl, Br and I) (Stryjewski & Giordano 1977; Aruga Katori & Katsumata 1996; Day 1988; Schieber 1967) and hydroxides (Takada *et al.* 1966*a*, *b*) of the divalent iron group ions. They adopt the CdI_2 or $CdCl_2$ structure with strong covalent bonds within the layer and weak hydrogen bonds between layers (Wells 1984). The nearest-neighbour intralayer exchange interactions between metals are ferromagnetic and interlayer interactions are antiferromagnetic and generally, dipolar in nature (Stryjewski & Giordano 1977). It has been demonstrated, experimentally and theoretically, that when the easy axis is perpendicular to the layers (e.g. $FeCl_2$), hysteresis is more pronounced than when it is parallel (e.g. $CoCl_2$ and $NiCl_2$). Therefore, $FeCl_2$ behaves, in some ways, like a ferromagnet with domains and $CoCl_2$ and $NiCl_2$ behave as paramagnets.

(a) *Syntheses*

The synthesis of a particular phase depends on the choice of transition metals: copper and nickel behave differently to cobalt due to the fact that copper and nickel adopt only octahedral geometry, whereas cobalt atoms can have either only octahedral geometry or both octahedral and tetrahedral geometries. The pertinent

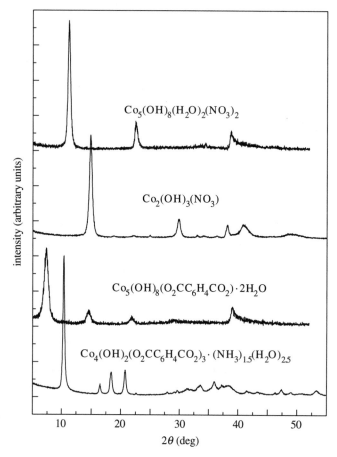

Figure 1. Powder X-ray diffraction of **1**, **2**, **3** and **4** (Co-$K\alpha$, $\lambda = 1.789$ Å).

reaction conditions depend on the concentration of the ions and the pK_a of the anions. For cobalt, the reactions with nitrate and terephthalate proceed as follows:

$$\text{Co}_5(\text{OH})_8(\text{H}_2\text{O})_2(\text{NO}_3)_2 \rightarrow \text{Co}_2(\text{OH})_3(\text{NO}_3)$$
$$\rightarrow \text{Co(OH)(NO}_3) \cdot \text{H}_2\text{O}$$
$$\rightarrow \beta\text{-Co(OH)}_2$$
$$\text{Co}_5(\text{OH})_8(\text{O}_2\text{CC}_6\text{H}_4\text{CO}_2) \cdot 2\text{H}_2\text{O} \rightarrow \text{Co}_4(\text{OH})_2(\text{O}_2\text{CC}_6\text{H}_4\text{CO}_2)_3 \cdot (\text{NH}_3)_{1.5}(\text{H}_2\text{O})_{2.5}$$

(b) X-ray diffraction

The X-ray diffraction patterns of the four compounds are shown in figure 1. **1** and **2** display only a progression of the 00ℓ in their diffraction patterns and two broad features at d-spacing of 2.70 and 1.55 Å. As mentioned in the introduction, the structures of these compounds are related to those observed for monoclinic $\text{Zn}_5(\text{OH})_8(\text{H}_2\text{O})_2(\text{NO}_3)_2$ (Stählin & Oswald 1970) and hexagonal $\text{Zn}_5(\text{OH})_8\text{Cl}_2 \cdot \text{H}_2\text{O}$ (Allmann 1968). However, the available information is not enough to differentiate

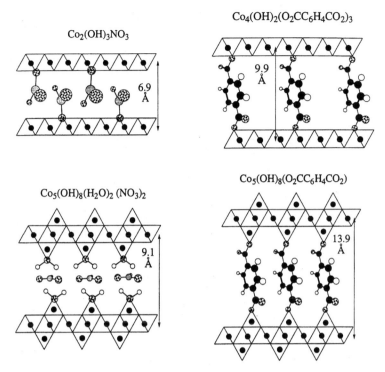

Figure 2. Idealized single-decker layer structures for **2** and **4** and triple-layer structures for **1** and **3** showing a single layer of octahedral cobalt sandwiched by two layers of tetrahedral cobalt and the terephthalate in a bridging position.

between the two possibilities. The interlayer separation of 9.14 Å for **1** and the infrared data (*vide supra*) suggest that the nitrate is not bonded to the cobalt. The 13.87 Å spacing for **2** is as expected for a terephthalate di-anion acting as a bridge between two layers (Burrows *et al.* 1997; Newman & Jones 1998; Cano *et al.* 1997; Cotton *et al.* 1998; Fogg *et al.* 1998b). The calculated distance between the two oxygen atoms of the two carboxylate groups is 7.4 Å, giving a thickness of 6.5 Å for the inorganic triple-deck layer. This value can be compared to an estimation of 3 Å for the pink single-deck layered compounds. A unique set of unit-cell parameters for **3** was found using TREOR (Werner *et al.* 1985) and they were refined within the pattern matching module of FULLPROF (Rodriguez-Carvajal 1997) to give $a = 5.53(1)$, $b = 6.31(1)$, $c = 6.95(1)$ Å, $\beta = 93.1°$, $V = 242$ Å^3, which are in good agreement with those in the literature (Rouba 1996). Similar procedures give $a = 9.97(1)$, $b = 11.24(1)$, $c = 6.29(1)$ Å, $\beta = 95.95°$ for **4**. The a parameter corresponds to the interlayer separation, the b parameter to $4r \sin \frac{1}{3}\pi$ and the c parameter to $2r$, where r is the Co–Co distance (3.15 Å). The magnitude of the a parameter is in agreement with the sum of two Co—O bonds (*ca.* 2 Å) and the distance (7.4 Å) between terminal oxygen atoms of the terephthalate and angle β of 96°. The idealized structures based on the structures of $Mg(OH)_2$ and $Zn_5(OH)_8(H_2O)_2(NO_3)_2$ for the nitrate and terephthalate compounds are shown in figure 2.

Table 1. *Energies and intensities of bands observed in the infrared spectra of* **1**, **2**, **3** *and* **4**

compound	observed band energy (cm^{-1})
1 $Co_5(OH)_8(H_2O)_2(NO_3)_2$	511m, 659m, 832vw, 879w, 1370vs, 1476sh, 1615w, 3455s br
2 $Co_5(OH)_8(O_2CC_6H_4CO_2) \cdot 2H_2O$	513m, 656m, 749m, 815m, 1359s, 1392sh, 1501m, 1585s, 3330sh, 3470br
3 $Co_2(OH)_3(NO_3)$	500w, 633s, 689s, 726s, 804m, 1001m, 1313s, 1487s, 3580s, 3612sh
4 $Co_4(OH)_2(O_2CC_6H_4CO_2)_3 \cdot (NH_3)_{1.5}(H_2O)_{2.5}$	515m, 743s, 807s, 1015w, 1359s, 1392sh, 1498m, 1577s, 1705w, 3140br, 3500br

(c) Infrared

The energies and intensities of the observed bands are listed in table 1. The bands at low energy (less than $800 \ cm^{-1}$) are assigned to the bending mode M—O—H (Nakamoto 1986) and those at high energy (greater than $2500 \ cm^{-1}$) are assigned to the C—H and O—H stretching modes. Compounds containing the water molecules, either coordinated or as solvent, show the bending mode of water at *ca.* $1600 \ cm^{-1}$. The intermediate energy bands are those of the anions, NO_3 and terephthalate (Stählin & Oswald 1971; Zotov *et al.* 1990).

The free nitrate ion adopts the D_{3h} point group and on coordination its symmetry is lowered. Consequently, the totally symmetric mode at $1050 \ cm^{-1}$ becomes IR active. Furthermore, the position of this band changes according to its mode of coordination; for unidentate complexes (local point group C_{2v}) it is expected at *ca.* $1000 \ cm^{-1}$ and for bidentate chelation it moves to *ca.* $1025 \ cm^{-1}$. The observation of this band at $1001 \ cm^{-1}$ in **3** suggests that the nitrate has a unidentate coordination to the cobalt, in agreement with the proposed crystal structure (figure 2). In **1**, on the other hand, the weakness of this band and the observation of a strong central band at $1370 \ cm^{-1}$ suggest that the nitrate ion is not coordinated. In this case the water molecules must bind to the metal to satisfy the coordination number and charge. A similar structure is observed for $Zn_5(OH)_8(H_2O)_2(NO_3)_2$, where the water molecules make the coordination (Stählin & Oswald 1970).

The two terephthalate compounds show two strong bands at *ca.* 1360 and *ca.* $1580 \ cm^{-1}$, which are assigned to the antisymmetric and symmetric stretching modes of the carboxylate group, respectively. The energy difference between these two bands suggests that the carboxylate groups are unidentate (Nakamoto 1986). Observation of only one band of each suggests that the two carboxylate ends of the terephthalate are equivalent, confirming that the anion bridges the cobalt atoms of adjacent layers in both compounds.

(d) UV–vis

The UV–vis spectra of the compounds are shown in figure 3 and energies and assignments (Lever 1986) of the observed bands are given in table 2. The two types of

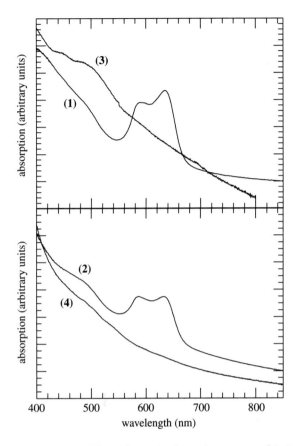

Figure 3. Transmission UV–vis electronic absorption spectra of **1**, **2**, **3** and **4**.

Table 2. *Energies and assignments of bands observed in the UV–vis spectra of* **1**, **2**, **3** *and* **4**

compound	transition energy (cm^{-1})	assignment
1 $Co_5(OH)_8(H_2O)_2(NO_3)_2$	15 800	$^4A_2 \rightarrow {}^4T_1(P)$
	17 000	$^4T_{1g}(F) \rightarrow {}^4A_{2g}(P)$
	20 300	$^4T_{1g}(F) \rightarrow {}^4T_{1g}(P)$
2 $Co_5(OH)_8(O_2CC_6H_4CO_2) \cdot 2H_2O$	15 800	$^4A_2 \rightarrow {}^4T_1(P)$
	17 000	$^4T_{1g}(F) \rightarrow {}^4A_{2g}(P)$
	21 800	$^4T_{1g}(F) \rightarrow {}^4T_{1g}(P)$
3 $Co_2(OH)_3(NO_3)$	19 800	$^4T_{1g}(F) \rightarrow {}^4A_{2g}(P)$
	22 500	$^4T_{1g}(F) \rightarrow {}^4T_{1g}(P)$
4 $Co_4(OH)_2(O_2CC_6H_4CO_2)_3 \cdot (NH_3)_{1.5}(H_2O)_{2.5}$	20 000 weak and broad	$\begin{cases} {}^4T_{1g}(F) \rightarrow {}^4A_{2g}(P) \\ {}^4T_{1g}(F) \rightarrow {}^4T_{1g}(P) \end{cases}$

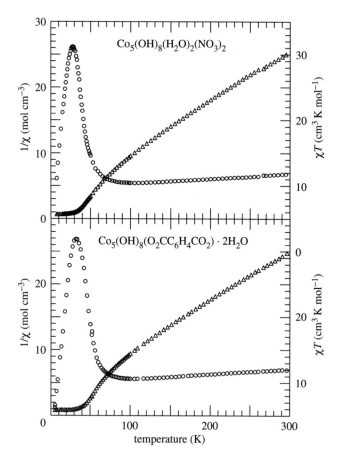

Figure 4. Temperature dependence of the inverse DC susceptibilities (triangles) and the products susceptibility and temperature (circles) of **1** and **2** in an applied field of 13 000 Oe.

compounds are quite distinct; the pink compounds show weak d–d absorption corresponding to Co(II) in octahedral coordination of a weak ligand and the green compounds show two additional sharp peaks at lower energies corresponding to Co(II) in tetrahedral coordination geometry.

(e) Magnetic properties

For **1** and **2** The temperature dependence of the inverse magnetic susceptibilities and the products of susceptibility and temperature $[\chi(T - \Theta) = C]$ of the two compounds are shown in figure 4. The magnetic behaviour of the two compounds is identical. The high-temperature ($T > 150$ K) data fit the Curie–Weiss law with Curie constants of 13.42(5) and 13.3(1) cm^3 K mol^{-1}, and Weiss temperatures of $-37(1)$ and $-40(2)$ K, respectively. The Curie values are consistent with that expected for the sum of three octahedral and two tetrahedral divalent high-spin cobalt atoms. The effective moment shows a minimum at 110 K, suggesting ferrimagnetic behaviour and a maximum at *ca.* 30 K. The latter is due to saturation since the measurements were made in 13 kOe. We should emphasize that the value of the moment at the maximum

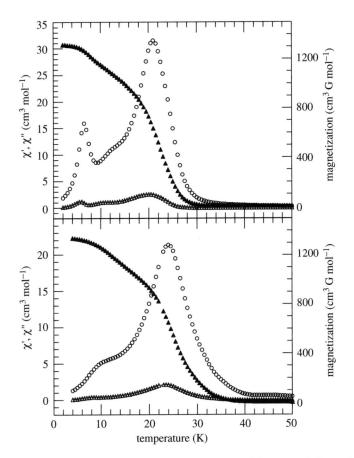

Figure 5. Temperature dependence of the DC magnetization (filled triangles) in a field of 1 Oe and the real (open circles) and imaginary (open triangles) AC susceptibilities in a field of 1 Oe for **1** (top) and **2** (bottom).

in any given field has no real information (Laget *et al.* 1998) because it is dependent on the applied field.

The Curie temperatures were obtained by measuring the magnetization on cooling in a small applied field of 1 Oe and by the onset of the imaginary component of the AC susceptibility. The transitions are quite broad (figure 5), possibly due to the two-dimensional nature of these compounds (De Jongh 1990). The Curie temperatures are 30(3) and 40(3) K for **1** and **2**, respectively. In contrast to the smooth temperature dependence of the magnetization data in the high field, the low-field DC data and, in particular, the AC data exhibit several features. These may be due to a minor impurity phase in the samples. For the nitrate salt, the sharp peak in χ' and its associated χ'' suggest that the minor phase is also magnetic. A closer look at the XRD does not reveal any other weak features, whereas the infrared data contains a weak shoulder at 1476 cm^{-1} which may be interpreted as arising from a coordinated nitrate in the minor phase. We should point out that in the zinc analogues, an example with coordinated ammonia $Zn_5(OH)_8(NH_3)_2(NO_3)_2$ is known (Louer *et al.* 1973; Benard

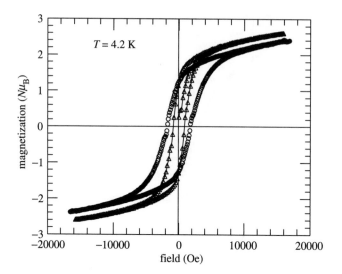

Figure 6. Isothermal magnetization at 4.2 K for **1** (triangles) and **2** (circles).

et al. 1994). Chemical analyses and the infrared data of our compounds reveal the absence of ammonia.

The isothermal magnetization of **1** and **2** at 4.2 K is shown in figure 6. Both compounds exhibit hysteresis loops with the coercive field increasing with lowering temperature and reaching 850 and 1750 Oe at 4.2 K, respectively. The remanent magnetization, after applying a field of 18 000 Oe, has almost the same temperature dependence as the magnetization in a small field of 1 Oe, reaching zero at the Curie temperatures. The values of the magnetization at 4.2 K in a field of 18 000 Oe are marginally lower than $3\mu_B$, the value expected for a ferrimagnet with one sublattice consisting of three cobalt(II) and another of two cobalt(II).

For **3** The temperature dependence of the magnetic susceptibility in different applied magnetic fields is shown in figure 7. The susceptibility increases as the temperature of the sample is lowered according to a Curie–Weiss law with Curie and Weiss constants of 6.11(1) $cm^3\,K\,mol^{-1}$ and +8.6(3) K, respectively, suggesting ferromagnetic short-range interaction between nearest cobalt neighbours. The Curie constant is less than that reported previously (Rabu *et al.* 1993) and is within the range expected for octahedral cobaltous ion (Figgis 1966). Below 20 K nonlinear susceptibility behaviour is observed. In a small applied field (less than 1000 Oe), the susceptibility decreases below 10 K and tends toward half the value at the maximum, a behaviour consistent with antiferromagnetic long-range ordering. In an applied field of 3200 Oe, the susceptibility increases continuously. In applied fields in excess of 5000 Oe the susceptibility is lowered due to saturation effects. The isothermal magnetization at 4.5 K is shown in figure 9. The magnetization increases slowly in the low field and above a critical field of *ca.* 1000 Oe it increases more rapidly before reaching a plateau at $3.2\mu_B$ in a field of 16 000 Oe. There is a very slight hysteresis of 100 ± 50 Oe around the critical field.

For **4** the magnetic susceptibility is field dependent as in the case of **3**. In a small applied field of 10 Oe it increases to a maximum at T_N (38 K) and in a field 5000 Oe, larger than the critical field, it increases to a saturation plateau (fig-

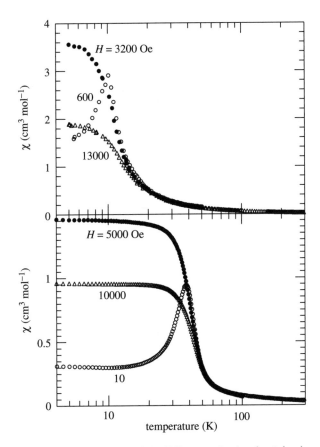

Figure 7. Temperature dependence of the DC magnetization for **3** (top) and **4** (bottom).

ure 7). The Curie constant derived from a Curie–Weiss fit of the inverse susceptibility versus temperature for the low-field data and for $T > 150$ K gives $C = 11.86(3)$ cm^3 K mol^{-1} and $\Theta = -67(1)$ K. The Curie constant, 2.97 cm^3 K per mole cobalt ($\mu_{\text{eff}} = 4.87\mu_{\text{B}}$/cobalt), is in good agreement to that expected for octahedral cobalt(II). The product of susceptibility and temperature, which is the square of the effective moment, decreases on lowering temperature to a minimum at 90 K and peaks at T_N at a value three times that at room temperature. The decrease from high temperature to 90 K is due to the effect of spin-orbit coupling (Mabbs & Machin 1973) resulting from an increasing population of the $s = \frac{1}{2}$ state at low temperatures, followed by an increase due to short-range ferromagnetic interactions between the cobalt atoms within the layer becoming more important than spin-orbit coupling. At 38 K, long-range ordering sets in due to antiferromagnetic interactions between moments in adjacent layers (Nagamiya *et al.* 1955). In a field higher than the anisotropy field, the moments are forced to be parallel.

The AC susceptibility in an applied AC field of 1 Oe is shown in figure 8. The real part shows a peak of magnitude similar to that observed for a low DC field and no anomalies are observed for the imaginary component, as expected for antiferromagnetic compounds (Goodenough 1963).

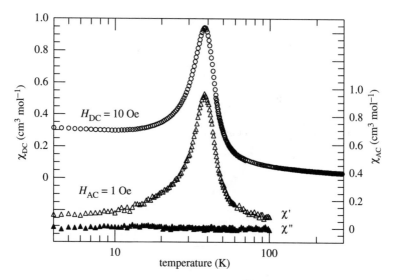

Figure 8. Temperature dependence of the DC susceptibility (open circles) and the real (open triangles) and imaginary (filled triangles) AC susceptibilities in a field of 1 Oe for **4**.

Isothermal magnetization at several temperatures is shown in figures 9 and 10. At high temperatures $(T > T_N)$ the magnetization is linear and reversible as for paramagnets, while just above T_N there is a slight curvature that is a typical signature of short-range ferromagnetic interactions. Just below T_N the magnetization takes an S-shape (figure 9) and remains reversible. The critical field to align the moments parallel increases as the temperature is lowered (figure 11). Below 22.5 K the magnetization begins to exhibit hysteresis (remanent magnetization and coercive field) and the hysteresis width increases on lowering the temperature. The shape of the loops changes gradually from the S-shape to one more commonly associated with a ferromagnet. The behaviour down to a temperature of 15 K is typical of that observed for metamagnets with very strong anisotropy and with alignment of moments perpendicular to the layers. For most of the known metamagnets the hysteresis occurs around the critical field region and the magnetization passes through zero as the field is removed, implying no remanent magnetization and no coercive field. The unusual feature of **4** is that there is a large remanent magnetization and a large coercive field (attaining in excess of 5 T at 2 K) below the tricritical temperature of 22.5 K (figure 10). The only other example which may show such an effect is NpO_2 (oxalate) (Jones & Stone 1972); unfortunately only part of the hysteresis loops was recorded. Figure 10 also show the first magnetization after zero-field cooling (ZFC) and the hysteresis loops after field cooling (FC) the sample in 5 T from 100 K to the measuring temperature. Below 10 K, if the full (± 5 T) hysteresis loop is recorded after ZFC a symmetric loop around $H = 0$ Oe is observed which has the shape known as a Rayleigh loop, also called a primary loop. This indicates that we are unable to reverse the moments with the maximum available field of our SQUID and the sample is still in the irreversible part of the first magnetization. Figure 11 presents the temperature-field phase diagram, in which the critical field to reverse the moments is estimated by the intersection of two straight lines from the opposite ends. The error bars on these points are the maximum width of the hysteresis loop or coercive field

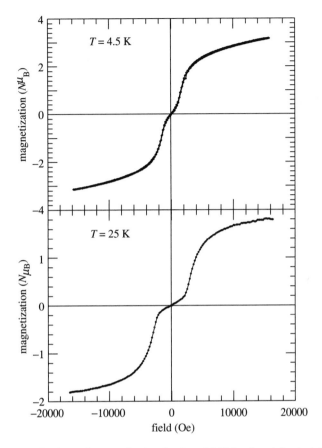

Figure 9. Isothermal magnetization for **3** at 4.5 K (top) and for **4** at 25 K.

observed. We have also displayed the thermoremanent magnetization after field cooling in 5 T from 100 to 2 K. This is a very elegant way, in this case, to identify the tricritical point at 22.5 K.

The long-range magnetic ordering mechanism in **4** is similar to that for the ferrimagnetic systems discussed above (Kurmoo *et al.* 1999; Drillon & Panissod 1998). The only difference is that in this case the interaction is antiferromagnetic, whereas it is ferromagnetic for the triple-deck compounds. The mechanism of hysteresis for a metamagnet may be viewed as follows: cooling the sample in a very small applied field establishes antiferromagnetic ordering (figure 12) below T_N. Application of a field along the easy-axis at a temperature below T_N first reverses some of the moments, usually at some defects (nucleation sites), which then polarize their surroundings. As the field is increased the created domains increase in size until all the moments are aligned with the applied field. At the critical field the sample will contain ferromagnetic clusters embedded in the antiferromagnetic bulk. This situation is similar to the process involved in mictomagnetism. The main difference here is that the size of the domains is not decided at the manufacturing stage but is temperature dependent. At a fixed temperature, the average size of these clusters will increase with field, the effective moment of the ions and the near-neighbour exchange interaction

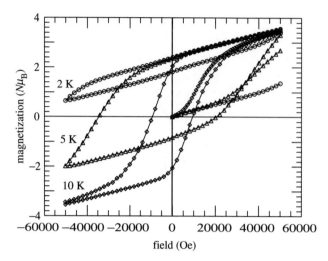

Figure 10. Isothermal first magnetization (ZFC) and hysteresis (5 T FC) loops for **4** at 2 K (circles), 5 K (triangles) and 10 K (diamonds).

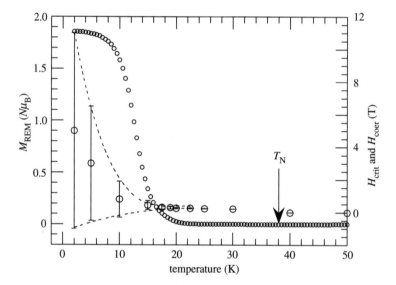

Figure 11. Field-temperature phase diagram showing the critical field (circles) and width of hysteresis (error bars) and thermoremanent magnetization for **4**. The dotted lines are guides to the eye.

(Chikazumi 1978). As the size becomes large the magnetocrystalline anisotropy energy is also increased and exceeds the exchange energy such that on removal of the applied field the compound retains its magnetic state and remanent magnetization and a coercive field are observed. Therefore, the critical and coercive fields will depend on the single-ion anisotropy and the magnitude of the exchange interactions. This is in contrast to that for ferromagnets or ferrimagnets where the coercive field depends solely on the single-ion anisotropy, assuming the shape is the same. To verify this

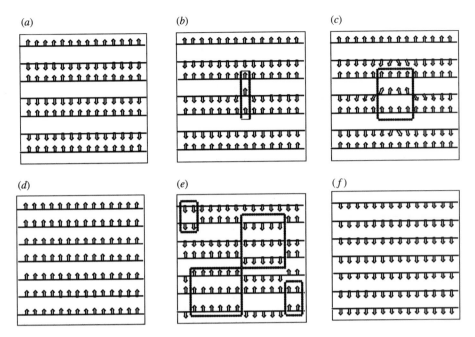

Figure 12. Schematic models of the mechanism of magnetization and demagnetization processes for **4** at a temperature below T_N after zero-field cooling. (a) $T < T_N$, $H = 0$; (b) $H = H_{crit}$; (c) $H > H_{crit}$; (d) $H \gg H_{crit}$; (e) $H = H_{coer}$; (f) $H \gg H_{coer}$.

hypothesis, we have prepared cobalt–nickel solid solutions to determine the dependence of the critical and coercive fields on the magnitude of the spin-orbit coupling. Nickel substitution has two effects: first, it reduces the average moment per ion and secondly, it reduces the effective spin-orbit coupling. For the nominal Ni_2Co_2 hydroxy-terephthalate, the Néel temperature is lowered to 22 K and the coercive field to 12 000 Oe. Furthermore, the importance of dipolar interactions was verified by replacing the terephthalate ion by a longer bridging ion, 4, 4′-biphenyl-dicarboxylate. Here, the interlayer spacing is increased to 14.1 Å, T_N is lowered only slightly to 34 K and the coercive field is 30 500 Oe at 5 K.

4. Conclusion

The present observations suggest that the Brucite-type layered compounds of cobalt having octahedral sites only are metamagnets, while those adopting the triple-deck $Zn_5(OH)_8(H_2O)_2(NO_3)_2$ structure, containing both octahedral and tetrahedral sites, are ferrimagnets. The long-range ordering in both structural types is brought about by dipolar interaction due to precursor correlation creating large effective moments in domains just above the transition temperatures. The exceptional hardness of these two structural types is due to the synergy of single-ion and crystalline shape anisotropies and, most importantly, the alignment of moments perpendicular to the layer. **4** is the first metamagnet to exhibit a coercive field and the first magnetic material which has a coercive field in excess of 5 T.

This work was funded by the CNRS-France. The technical assistance of R. Poinsot and A. Derory is much appreciated. I am grateful to Dr S. Vilminot, Dr C. J. Kepert, Dr S. Blundell and Mr B. Lovett for many fruitful discussions.

References

Allmann, R. 1968 Verfeinerung der structur des zinkhydroxidchlorids II, $Zn_5(OH)_8Cl_2 \cdot H_2O$. *Z. Kristallogr.* **126**, 417–426.

Andrä, W., Danan, H. & Mattheis, R. 1991 Theoretical aspects of perpendicular magnetic recording media. *Physica Status Solidi* **125**, 9–55.

Aruga Katori, H. & Katsumata, K. 1996 Specific-heat anomaly in the Ising antiferromagnet $FeBr_2$ in external magnetic fields. *Phys. Rev.* B **54**, R9620–R9623.

Atovmyan, L., Shilov, G. V., Lyubovskaya, R. N., Zhilyaeva, E. I., Ovaneseyan, N. S., Gusakovskaya, I. G. & Morozov, G. 1993 Crystal structure of molecular ferromagnet $NBu_4[MnCr(C_2O_4)_3]$ ($Bu = n\text{-}C_4H_9$). *JETP Lett.* **58**, 767–769.

Bellitto, C & Day, P. 1992 *J. Mater. Chem.* **2**, 265.

Bellitto, C., Federici, F. & Ibrahim, S. A. 1998 Synthesis and properties of new chromium(II) organophosphonates. *Chem. Mater.* **10**, 1076–1082.

Benard, P., Auffredic, J. P. & Louer, D. 1994 A study of the thermal decomposition of ammine zinc hydroxide nitrates. *Thermochem. Acta* **232**, 65–76.

Burrows, A. D., Mingos, D. M. P., Lawrence, S. E., White, A. J. P. & Williams, D. J. 1997 Platinum(II) phosphine complexes of dicarboxylates and ammonia: crystal structures of $[\{Pt(PPh_3)_2\}_2 \{\mu\text{-}1,3\text{-}(O_2C)_2C_6H_4\}_2]$, $[\{Pt(PPh_3)_2(NH_3)_2 \{\mu\text{-}1,4\text{-}(O2C)_2C_6H_4\}][PF_6]_2$ and *cis*-$[Pt(PPh_3)_2(NH_3)_2][NO_3]_2$. *J. Chem. Soc. Dalton Trans.*, pp. 1295–1300.

Cano, J., De Munno, G., Sanz, J. L., Ruiz, R., Faus, J., Lloret, F., Julve, M. & Caneschi, A. 1997 Ability of terephthalate (ta) to mediate exchange coupling in ta-bridged copper(II), nickel(II), cobalt(II) and manganese(II) dinuclear complexes. *J. Chem. Soc. Dalton Trans.*, pp. 1915–1923.

Carling, S. G., Day, P. & Visser, D. 1995 Crystal and magnetic structures of layer transition metal phosphate hydrates. *Inorg. Chem.* **34**, 3917–3927.

Carling, S. G., Mathonière, C., Day, P., Malik, K. M. A., Coles, S. J. & Hursthouse, M. B. 1996 Crystal structure and magnetic properties of the layer ferrimagnet $N(n\text{-}C_5H_{11})_4Mn^{II}Fe^{III}(C_2O_4)_3$. *J. Chem. Soc. Dalton Trans.*, pp. 1839–1844.

Carlino, S. 1997 The intercalation of carboxylic acids into layered double hydroxides: a critical evaluation and review of the different methods. *Solid State Ionics* **98**, 73–84.

Chikazumi, S. 1978 *Physics of magnetism*. Wiley.

Clearfield, A. (ed.) 1991 *Inorganic ion exchange materials*. Boca Raton, FL: CRC.

Clemente-Leon, M., Coronado, E., Galan-Mascaros, J.-R. & Gomez-Garcia, C. 1997 Intercalation of decamethylferrocenium cations in bimetallic oxalate-bridged two-dimensional magnets. *Chem. Commun.*, pp. 1727–1728.

Cotton, F. A., Lin, C. & Murillo, C. A. 1998 Coupling Mo_2^{n+} units via dicarboxylate bridges. *J. Chem. Soc. Dalton Trans.*, pp. 3151–3153.

Day, P. 1988 Nickel dibromide: a magnetic detective story. *Acc. Chem. Res.* **211**, 250–254.

Day, P. 1997 Coordination complexes as organic–inorganic layer magnets. *J. Chem. Soc. Dalton Trans.*, pp. 701–705.

Decurtins, S., Ferlay, S., Pellaux, R., Gross, M. & Schmalle, 1998 Examples of supramolecular coordination compounds and their supramolecular functions. In *Supramolecular engineering of synthetic metallic materials* (ed. J. Veciana, C. Rovira, C. & D. B. Amabilino), NATO ASI C518, pp. 175–196. Dordrecht: Kluwer.

De Jongh, L. J. (ed.) 1990 *Magnetic properties of layered transition metal compounds*. Dordrecht: Kluwer.

De Jongh, L. J. & Miedema, A. R. 1974 Experiments on simple magnetic model systems. *Adv. Phys.* **23**, 1–260.

De Jongh, L. J., Botterman, A. C., De Bose, F. R. & Miedema, A. R. 1969 Transition temperature of the two dimensional Heisenberg ferromagnet with $s = \frac{1}{2}$. *J. Appl. Phys.* **40**, 1363–1365.

Drillon, M. & Panissod, P. 1998 Long range ferromagnetism in hybrid compounds: the role of dipolar interactions. *J. Mag. Mag. Mater.* **188**, 93–99.

Fair, M. J., Gregson, A. K., Day, P. & Hutchings, M. T. 1977 Neutron scattering study of the magnetism of Rb_2CrCl_4, a two dimensional easy-plane ferromagnet. *Physica* B **86–88**, 657–659.

Fanucci, G. E., Nixon, C. M., Petruska, M. A., Seip, C. T., Talham, D. R., Granroth, G. E. & Meisel, M. W. 1998 Metal phosphonate Langmuir–Blodgett films: monolayers and organic/inorganic dual network assemblies. In *Supramolecular engineering of synthetic metallic materials* (ed. J. Veciana, C. Rovira & D. B. Amabilino). NATO ASI C518, pp. 465–475. Dordrecht: Kluwer.

Figgis, B. N. 1966 *Introduction to ligand fields*. London: Wiley-Interscience.

Fogg, A. M., Dunn, J. S. & O'Hare, D. 1998a Formation of second-stage intercalation reactions of the layered double hydroxide $[LiAl_2(OH)_6]Cl \cdot H_2O$ as observed by time-resolved *in situ* X-ray diffraction. *Chem. Mater.* **10**, 356–360.

Fogg, A. M., Dunn, J. S., Shyu, S.-G., Cary, D. R. & O'Hare, D. 1998b Selective ion exchange intercalation of isomeric dicarboxylate anions into the layered double hydroxide $[LiAl_2(OH)_6]Cl \cdot H_2O$. *Chem. Mater.* **10**, 351–355.

Fujita, W. & Awaga, K. 1996 Magnetic properties of $Cu_2(OH)_3$(alkanecarboxylate) compounds: drastic modification with extension of the alkyl chain. *Inorg. Chem.* **35**, 1915–1917.

Goodenough, J. B. 1963 *Magnetism and the chemical bond*. Wiley.

Heinrich, B. & Bland, J. A. C. 1994 *Ultrathin magnetic structures*. Springer.

Herpin, A. 1968 *Théorie du Magnétisme*. Paris: Presse Universitaire de France.

Jacobsen, A. J. 1992 Intercalation reactions of layered compounds. In *Solid state chemistry: compounds* (ed. P. Day & A. Cheetham). Oxford University Press.

Jones, E. R. & Stone, J. A. 1972 Metamagnetism in neptunium(V) oxalate. *J. Chem. Phys.* **56**, 1343–1347.

Kahn, O. 1993 *Molecular magnetism*. New York: VCH.

Kahn, O. (ed.) 1996 *Magnetism: a supramolecular function*. NATO ASI series C484. Dordrecht: Kluwer.

Kurmoo, M. & Kepert, C. J. 1998 Hard magnets based on transition metal complexes with the dicyanamide anion, $\{N(CN)_2\}^-$. *New J. Chem.*, pp. 1515–1524.

Kurmoo, M., Graham, A. W., Day, P., Coles, S. J., Hursthouse, M. B., Caulfield, J. L., Singleton, J., Pratt, F. L., Hayes, W., Ducasse, L. & Guionneau, P. 1995 Superconducting and semiconducting magnetic charge transfer salts: $(BEDT-TTF)_4AFe(C_2O_4)_3 \cdot C_6H_5CN$ $(A = H_2O, K, NH_4)$. *J. Am. Chem. Soc.* **117**, 12 209–12 217.

Kurmoo, M., Day, P., Derory, A., Estournès, C., Poinsot, R., Stead, M. J. & Kepert, C. J. 1999 3D-long range magnetic ordering in layered metal-hydroxide triangular lattices 25 Å apart. *J. Solid State Chem.* **145**, 452–459.

Laget, V., Rouba, S., Rabu, P., Hornick, C. & Drillon, M. 1996 Long range ferromagnetism in tunable cobalt(II) layered compounds up to 25 Å apart. *J. Mag. Mag. Mater.* **154**, L7–L11.

Laget, V., Hornick, C., Rabu, P., Drillon, M. & Ziessel, R. 1998 Molecular magnets hybrid organic-inorganic layered compounds with very long-range ferromagnetism. *Coord. Chem. Rev.* **178–180**, 1533–1553.

Lever, A. P. B. 1986 *Inorganic electronic spectroscopy*. Elsevier.

Li, H., Eddouadi, M., Groy, T. L. & Yaghi, O. M. 1998 Establishing microporosity in open metal–organic frameworks: gas sorption isotherms for Zn(BDC) (BDC = 1,4-benzenedicarboxylate). *J. Am. Chem. Soc.* **120**, 8571–8572.

Louer, M., Louer, D. & Grandjean, D. 1973 Etude structurale des hydroxynitrates de nickel et de zinc. I. Classification structurale. *Acta Crystallogr.* B **29**, 1696–1710.

Mabbs, F. E. & Machin, D. J. 1973 *Magnetism and transition metal complexes.* London: Chapman & Hall.

Mathonière, C., Nuttall, C. J., Carling, S. G. & Day, P. 1996 Ferrimagnetic mixed-valency and mixed-metal tris(oxalato)iron(III) compounds: synthesis, structure and magnetism. *Inorg. Chem.* **35**, 1201–1206.

Meyn, M., Beneke, K. & Lagaly, G. 1993 Anion exchange reactions of hydroxy double salts. *Inorg. Chem.* **32**, 1209–1215.

Miedema, A. R., van Kempen, H. & Huiskamp, W. J. 1963 Experimental study of the simple ferromagnetism in $CuK_2Cl_4 \cdot 2H_2O$ and $Cu(NH_4)_2Cl_4 \cdot 2H_2O$. *Physica* **29**, 1266–1280.

Mitchell, I. V. 1990 *Pillared layered structures: current trends and applications.* Elsevier.

Nagamiya, T., Yosida, K. & Kubo, R. 1955 Antiferromagnetism. *Adv. Phys.* **4**, 1–112.

Nakamoto, K. 1986 *Infrared and Raman spectra of inorganic and coordination compounds,* p. 230. Wiley.

Newman, S. P. & Jones, W. 1998 Synthesis, characterization and applications of layered double hydroxides containing organic guests. *New J. Chem.* **22**, 105–115.

Nicoud, J.-F. 1994 Towards new multi-property materials. *Science* **63**, 636–637.

Nuttall, C. J. & Day, P. 1998 Magnetization of the layer compounds $AFe^{II}Fe^{III}(C_2O_4)_3$ (A = organic cation), in low and high magnetic fields: manifestation of Néel N- and Q-type ferrimagnetism in a molecular lattice. *Chem. Mater.* **10**, 3050–3057.

O'Hare, D. 1993 Inorganic intercalation chemistry. In *Inorganic materials* (ed. D. W. Bruce & D. O'Hare). London: Wiley.

Rabu, P., Angelov, S., Legoll, P., Belaiche, M. & Drillon, M. 1993 Ferromagnetism in triangular cobalt(II) layers: comparison of $Co(OH)_2$ and $Co_2(NO_3)(OH)_3$. *Inorg. Chem.* **32**, 2463–2468.

Rodriguez-Carvajal, J. 1997 *FullProf: short reference guide to the program,* edn 3.2. Paris: CEA-CNRS.

Rouba, S. 1996 Corrélations structures-propriétés magnetiques dans une série d'hydroxynitrate de métaux de transition 1d et 2d. Thèse doctorat, Université Louis Pasteur, Strasbourg, France.

Schollhorn, R. 1984 *Intercalation compounds.* London: Academic.

Schieber, M. M. 1967 *Experimental magnetochemistry, selected topics in solid state physics* (ed. E. P. Wohlfarth), vol. VIII. North-Holland.

Stählin, W. & Oswald, H. R. 1970 The crystal structure of zinc hydroxide nitrate, $Zn_5(OH)_8$ $(NO_3)_2(H_2O)_2$. *Acta Crystallogr.* B **26**, 860–863.

Stählin, W. & Oswald, H. R. 1971 The infrared spectrum and thermal analysis of zinc hydroxide nitrate. *J. Solid State Chem.* **2**, 252–255.

Stryjewski, E. & Giordano, N. 1977 Metamagnetism. *Adv. Phys.* **26**, 487–650.

Takada, T., Bando, Y., Kiyama, M. & Mitamoto, K. 1966a The magnetic property of β-$Co(OH)_2$. *J. Phys. Soc. Jap.* **21**, 2726.

Takada, T., Bando, Y., Kiyama, M., Mitamoto, K. & Sato, T. 1966b The magnetic property of $Ni(OH)_2$. *J. Phys. Soc. Jap.* **21**, 2745.

Tamaki, H., Mitsumi, M., Nakamura, K., Matsumoto, N., Kida, S., Okawa, H. & Iijima, S. 1992a Metal-complex ferrimagnets with the formula $\{NBu_4[M(II)Fe(III)(ox)_3]\}_{3\infty}$ (NBu_4^+ = tetra(n-butyl)ammonium ion, ox^{2-} = oxalate ion, M = Fe^{2+}, Ni^{2+}). *Chem. Lett.*, pp. 1975–1978.

Tamaki, H., Zhong, Z. J., Matsumoto, N., Kida, S., Koikawa, M., Achiwa, N., Hashimoto, Y. & Okawa, H. 1992b *J. Am. Chem. Soc.* **114**, 6974.

Wells, A. F. 1984 *Structural inorganic chemistry*. Oxford University Press.

Werner, P. E., Eriksson, L. & Westdahl, M. 1985 A semi-exhaustive trial-and-error powder indexing program for all symmetries. *J. Appl. Crystallogr.* **18**, 367–370.

Zotov, N., Petrov, K. & Dimitova-Pankova, M. 1990 Infrared spectra of Cu(II)–Co(II) mixed hydroxide nitrates. *J. Phys. Chem. Solids* **51**, 1199–1205.

Discussion

B. J. BUSHBY (*School of Chemistry, University of Leeds, UK*). Some of Dr Kurmoo's data on finely divided crystals suggest that we are approaching a mono-domain structure. Given the size of the crystals, what does this imply regarding the number of spins in a domain? Is this a reasonable figure, and, indeed, what sizes do we expect for domains in molecular systems?

M. KURMOO. Domain size is an important parameter, among others including exchange and anisotropy energies and form of the samples, when quoting coercive field. The samples presented, showing magnetic behaviour approaching single domain, were elongated plates of maximum dimension of *ca.* 100 nm. This translates to *ca.* 10^6 magnetic centres. This is larger than those of metallic particles, which are of the order of 25 nm, and is close to those of oxides. In my opinion, the domain size will increase when the exchange energy decreases and will decrease when the anisotropy energy increases. Therefore, it depends on the particular system and on the single-ion anisotropy and shape of the particles.

M. VERDAGUER (*Laboratoire de Chimie Inorganique et Matériaux Moléculaires, Université Pierre et Marie Curie, Paris, France*). The dicyanamide ligand has two different coordination sites. Did Dr Kurmoo try (and did he succeed) to get structurally ordered bimetallic systems such as

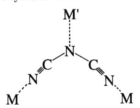

M. KURMOO. The dicyanamide contains three nitrogen atoms and two different coordination sites (two nitrile and one amide). Yes, it would be of interest to prepare structurally ordered bimetallic systems, but we have not yet tried this. The method of preparation of the binary compounds, $M(N(CN)_2)_2$, is a one-pot procedure.

Towards magnetic liquid crystals

By Koen Binnemans[1], Duncan W. Bruce[2], Simon R. Collinson[2],
Rik Van Deun[1], Yury G. Galyametdinov[3]
and Françoise Martin[1]

[1]K. U. Leuven, Department of Chemistry, Coordination Chemistry Division,
Celestijnenlaan 200F, B-3001 Heverlee (Leuven), Belgium
[2]School of Chemistry, University of Exeter, Stocker Road, Exeter EX4 4QD, UK
[3]Physical-Technical Institute, Russian Academy of Science,
Sibirsky Tract 10/7.420029, Kazan, Russia

In this paper, we present the results of studies on the synthesis and properties of a
series of liquid-crystalline lanthanide complexes of imine ligands. We describe the
liquid-crystalline behaviour as a function of the metal, ligand and anion employed and
we report on the nature of the coordination between different ligand types and the
metal centre.

Keywords: lanthanides; liquid crystals; metallomesogens; Schiff bases

1. Introduction

In this paper, we intend to present an introduction to the study of metal-based paramagnetic liquid crystals concentrating primarily on our own work with lanthanide-
based materials. The development of this subject goes largely hand-in-hand with the
development of the synthesis of the so-called *metallomesogens*—metal-based liquid
crystals—which has developed strongly in the last 15 years or so (Bruce 1996).

The liquid-crystal state is the fourth state of matter and exists between the solid
and liquid states for certain molecules. As such, it has properties reminiscent of both
phases and so like a liquid it has fluidity and like a solid it has order. It is perhaps
convenient to consider a liquid crystal as a rather ordered liquid which is, therefore,
an anisotropic fluid stabilized by anisotropic dispersion forces consequent on it being
composed of anisotropic molecules. If the molecules are rod-like, then nematic and
lamellar phases tend to result, while disc-like molecules tend to give rise to nematic
and columnar phases.

The most disordered of the liquid crystal mesophases is the nematic phase in which
the unique molecular axes are orientationally correlated about a director (n) in one
dimension in the absence of positional order. A schematic of a nematic phase of rods is
given in figure 1.

For rod-like molecules, lamellar or *smectic* phases are also known which differ from
the nematic phase in that they possess partial translational ordering. Thus, in the
smectic A (S_A) phase, the molecules are on average ordered in one direction while be
partly ordered into layers (lamellae). The smectic C (S_C) phase is analogous except
that the molecules are tilted within these layers. In fact, these layers are not real,
being better described by sinusoidal molecular distribution functions as described

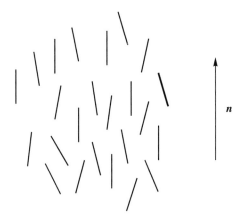

Figure 1. Schematic of a nematic mesophase.

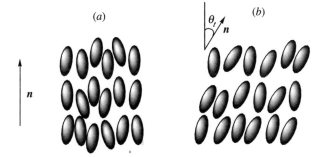

Figure 2. Schematic of a smectic A (*a*) and smectic C (*b*) phase.

well in Leadbetter (1987). The S_A and S_C phases are illustrated schematically in figure 2.

Because the phases are anisotropic, their physical properties are also anisotropic. Most commonly, we might consider the anisotropy of dielectric response ($\Delta\varepsilon$), refractive index (Δn) and diamagnetism ($\Delta\chi$), which are defined as (example for diamagnetism):

$$\Delta\chi = \chi_\parallel - \chi_\perp,$$

where $\chi_\parallel$ is the diamagnetic susceptibility *parallel* to the director and $\chi_\perp$ is the diamagnetic susceptibility *perpendicular* to the director. For $\Delta\chi > 0$, the diamagnetic anisotropy is termed positive. It is already possible to use this diamagnetism, and cooling a liquid crystalline material into its nematic phase from the normal or isotropic liquid state in the presence of a magnetic field of, say, 0.7 T will produce a macroscopically ordered sample.

However, the interaction of a paramagnetic material with an external magnetic field will be stronger than that of a diamagnetic material, which means that any alignment or *re*alignment might be realized with a much smaller field. Thus, it may in turn be possible to look at the orientational switching of a liquid-crystalline material by using an external magnetic field.

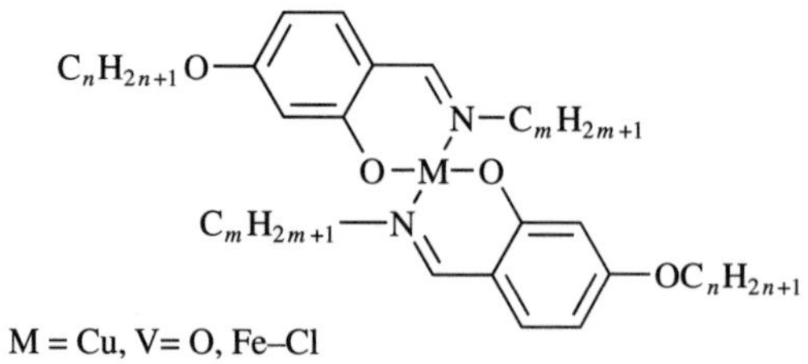

Figure 3. Early paramagnetic complexes.

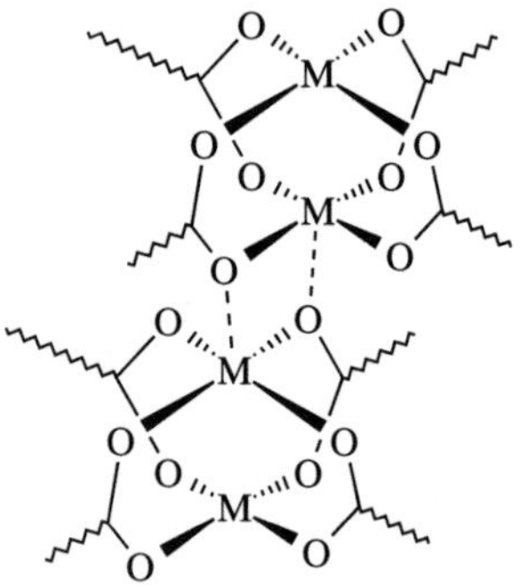

Figure 4. Mesomorphic dinuclear metal carboxylates (M = Cu, Rh, Ru, Mo).

2. Non-lanthanide liquid crystals

In terms of paramagnetic rod-like metallomesogens, the earliest examples were the Cu(II) and V(IV) complexes of salicylaldimines (figure 3) originally described by Galyametdinov and co-workers (Ovchinnikov et al. 1984). A large number of groups then became involved in the study of salicylaldimine-based mesogens, and Fe(III) complexes were added to the collection of paramagnetic systems available (Hoshino 1998). A wider discussion of other paramagnetic metallomesogens is found in Alonso (1996).

Metallomesogens forming columnar phases are also known, and probably the best-studied examples are the dinuclear dicoppertetra(alkanoates) studied by Marchon, Giroud-Godquin, Guillon and Skoulios in the early 1980s (figure 4). The dinuclear Cu(II) structure allows for antiferromagnetic exchange between the two copper centres and, consequently, the complex is diamagnetic, although SQUID studies did show an abrupt change in $\chi(T)$ at the transition from the solid to columnar phase. Later, the dinuclear Ru(II)–Ru(II) and Ru(II)–Ru(III) analogues were investigated and it was possible to evaluate the magnitude of the zero-field splitting energy between the $M_S = 0$ ground state and $M_S = \pm 1$ excited state in the Ru(II)–Ru(III) system ($D \approx 300 \text{ cm}^{-1}$) (Giroud-Godquin 1998).

3. Lanthanide liquid crystals

One of the great attractions for some groups in the field has been to incorporate lanthanide elements into liquid-crystalline systems. The reasons for this relate not only to the magnetic and photophysical properties which will result, but also to the

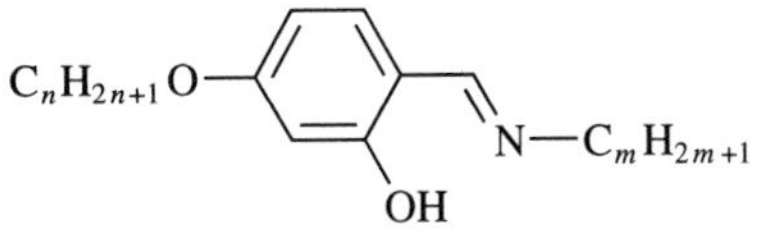

Figure 5. Salicylaldimine ligands.

synthetic challenge of building a metal with a high coordination requirement into a highly anisotropic molecule (Bruce 1994).

The first example of a lanthanide-based liquid crystal was a lutetium phthalocyanine sandwich described by Simon (Piechocki *et al.* 1985), which showed an anisotropy of conductivity in the columnar phase of the order of 10^7 and an absolute conductivity of 3.9×10^{-5} S cm^{-1} at 10 GHz.

There was later a report of mesomorphic salicylaldimine complexes of some lanthanides by Galyametdinov. These complexes were remarkable in that they seemed to take very straightforward ligands (figure 5) which generated, in combination with the metal, a species which was sufficiently anisotropic to show a smectic A phase (Galyametdinov *et al.* 1991). The formula of the complexes was given as $[Ln(L-H)(L)_2][X]_2$, where L is the salicylaldimine ligand, $L-H$ its deprotonated form and X a nitrate or chloride counter anion. It was assumed that the ligands were arranged around the metal in a trigonal prismatic geometry to give a toblerone-shaped mesogen. Indeed, some magnetic data obtained for related complexes were not inconsistent with such a hypothesis (Bikchantaev *et al.* 1996).

Through collaborations supported under INTAS, the team from Exeter began to work with Galyametdinov's team and also with Binnemans in Leuven (supported by the FWO-Flanders and the British Council); this paper gives an overview of some of what we have achieved in collaboration; however, this is very much an ongoing effort and so much of what is here is in the form of preliminary results. A primary aim was to extend the number and type of the complexes and to see how changing the ligand, metal and any anion would affect the properties. These factors are treated separately below, following a brief discussion of the synthesis.

4. Synthesis and structure

One fact that quickly came to light during these studies was that while the initial synthetic work had been accomplished using ethanol as a solvent, in many cases the results were unsatisfactory resulting in low yields or no yield at all. We eventually settled on THF as a solvent although even so, in some systems yields and purity are not always reliable and crystallization is not always a readily available option as the purity of the complex often decreases thereafter. The reason for this irreproducibility is certainly related to the ability of lanthanide ions to act as rather effective Lewis acids which can catalyse the hydrolysis of the imine bond. Such an effect has previously been observed (Smith *et al.* 1998; Berg *et al.* 1991).

Having looked at *N*-alkyl-substituted ligands and bearing in mind the wealth of liquid-crystalline metal complex chemistry based on *N*-aryl-substituted ligands, we next looked to this latter series of ligands for complexes. However, in whichever laboratory the chemistry was carried out, we were not able to complex *N*-arylsalicylaldimines to lanthanide centres and the reason for this is not entirely clear. However, one clue comes from a single crystal X-ray structure obtained on a

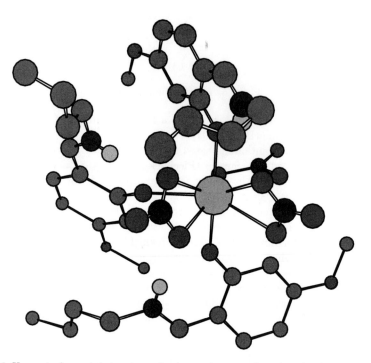

Figure 6. X-ray single-crystal structure of a dysprosium complex of a salicylaldimine with all except the iminium hydrogens removed for clarity.

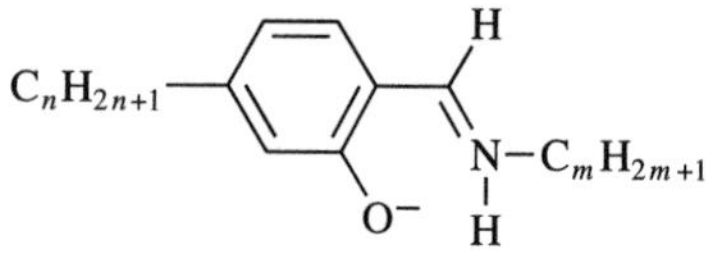

$$C_nH_{2n+1} \text{—} \underset{\underset{O^-}{\quad}}{\overset{\overset{H}{\quad}}{\bigcirc}} \text{—} \overset{H}{\underset{H}{\overset{|}{N}}} \text{—} C_mH_{2m+1}$$

Figure 7. Zwitterionic form of a salicylaldimine.

dysprosium complex of the salicylaldimine shown in figure 6 (Polishuk & Galyamet-dinov, unpublished work).

The structure seemed to show that, as schematized in figure 7, the phenolic proton of the ligand transferred to the imine nitrogen to give a zwitterion and that the binding of the ligand was entirely through a phenate oxygen. Subsequent re-examination of the ^{1}H NMR spectra of some diamagnetic La derivatives revealed that the imine proton was often broad and, one on occasion, was found to be split into a doublet. What we had presumed to be the phenolic proton was found just above $\delta 12$. If in the complex the zwitterionic form were present, then we would expect a coupling between the N—H and imine protons. We found that selective irradiation of the signal at about $\delta 12$ removed the broadening in the imine signal, showing conclusively that the protons are coupled and, therefore, that in many cases it is the zwitterionic form which coordinates (figure 8). Thus, it may be that the difference in basicity between the nitrogens of N-alkyl and N-aryl imines is sufficient to tip the balance, allowing the zwitterion (and therefore complexes) to form in one case and not the other.

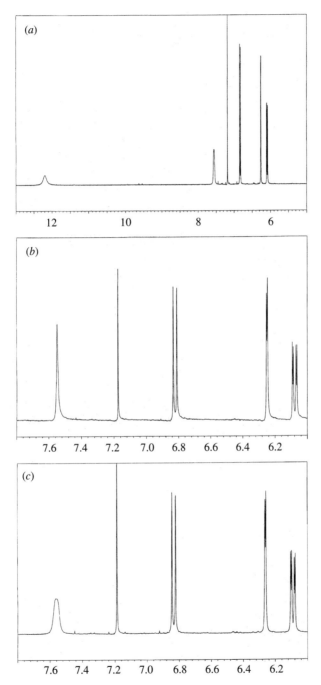

Figure 8. Lower field portion of the ^{1}H NMR spectrum of $[La(L)_2(L—H)][NO_3]_2$ (main spectrum (a)); with enlargement of area of imine signal with (b) and without (c) decoupling of the N—H$^+$ proton at $\delta 12.3$.

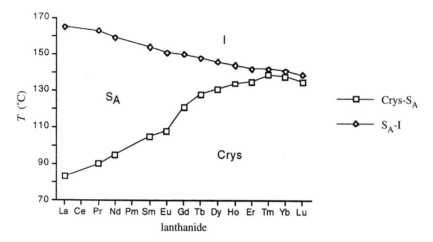

Figure 9. Plot of transition temperature as a function of lanthanide.

However, another major feature shown by the structure determination is that the initially proposed formula of $[Ln(L)_2(L—H)][X]_2$ was in error as the complex clearly contained three nitrate groups, bound in a bidentate fashion to the central dysprosium. Thus, the formula for these complexes is clearly more appropriately represented as $[Ln(L^*)_3][X]_3$, where L^* implies a rearranged (i.e. the zwitterionic) ligand. We now suppose that for simple ligands of the type shown in figure 5 complexed to lanthanide nitrates, this formula is correct and appropriate. However, we would note at this stage that preliminary data from other related complexes made in Exeter show that this situation is not necessarily common to all salicylaldimine complexes of lanthanides. A full investigation of the structure of these complexes is currently underway.

5. The effect of metal

In order to examine systematically the effect of the metal on the mesomorphism, a complex of each lanthanide (with the exceptions of cerium and promethium) was made using the ligand shown in figure 5 ($n = 8$, $m = 18$), starting from the lanthanide nitrate (Binnemans et al. 1999).

The complexes, all of which had the formula $[Ln(L)_3][NO_3]_3$, each showed only a smectic A mesophase which was readily characterized by the appearance of a focal conic fan texture and the presence of homeotropic areas consistent with a uniaxial smectic phase.

However, when the transition temperatures were plotted (figure 9) it was readily evident that the metal was having a profound effect on the mesophase range and that while the melting point was increasing with increasing atomic number, the clearing point was decreasing, reducing the mesomorphic range. The explanation for this is not entirely clear. Of course, as the lanthanide series is crossed, the ionic radius decreases (the lanthanide contraction) and so it may be possible to explain the increase in crystal-phase stability in terms of the smaller metal ion leading to a more compact arrangement of the ligands around the metal leading to a more efficient packing. This point will be difficult to delineate as, from the crystal structure shown

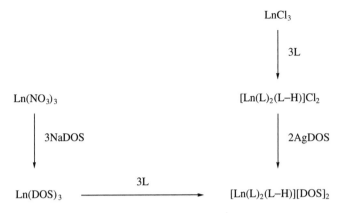

Figure 10. Preparation of the DOS salts.

above in figure 7, it is apparent that the ligand is bound to the metal only via a single atom (the phenate oxygen) and so the orientation of the ligand will be determined by the alkyl chains on the ligand and how they are required to pack under the control of lattice energy forces. Thus, crystal structures of model complexes with shorter alkyl chains will not necessarily provide a guide to what is happening in these systems.

If the arrangement of the ligands were to become more compact about the metal as its radius decreased, then the more compact anisotropic arrangement would be expected to lead to a higher clearing point which is the opposite of what is observed. However, it is possible to argue that with the increasing charge density on the metal as the lanthanide series in crossed, the polarizability due to the metal will decrease due to the higher charge density, which would lead to a destabilization of the smectic A phase. We have previously shown (Bertram *et al.* 1991) that the metal has a major contribution to make to the polarizability of metallomesogens and so we feel that such an effect could well contribute to the observed behaviour.

6. Effect of the anion

In the initial studies of these metallomesogens, Galyametdinov *et al.* (1991) had synthesized nitrate and chloride salts and had typically found transition temperatures ranging from 120 to 180 °C. While physical studies of materials at such temperatures are possible, they become easier if the temperature can be moderated somewhat. In previous studies of complexes of silver(I), we had found previously that the replacement of a small anion such as nitrate or triflate by the long-chain dodecylsulphate (DOS) led to greatly reduced transition temperatures as well as to interesting mesomorphism (Bruce *et al.* 1992; Donnio *et al.* 1997). With this in mind, we set about the introduction of the dodecylsulphate anion into these complexes.

Two approaches were employed. In the first, the chloride complex was reacted with silver dodecylsulphate, precipitating silver chloride and leading smoothly to the complex (Binnemans *et al.* 1998). In the second (Galyametdinov *et al.* 1999), $Ln(DOS)_3$ was prepared in advance from $Ln(NO_3)_3$ and NaDOS in water and then reacted with the ligand (figure 10).

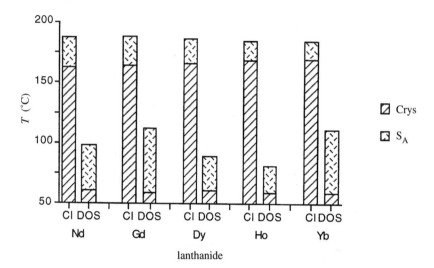

Figure 11. Graphical representation of the thermal behaviour of chloride
and DOS salts of some lanthanide complexes.

The result was as we had predicted and the transition temperatures of the complexes were significantly reduced. This is shown graphically in figure 11.

The transition temperatures for these complexes show the starkest difference with melting points reducing from *ca.* 160 to 60 °C and clearing points from *ca.* 180 °C to *ca.* 100 °C. However, it is interesting to note that the transition temperatures for the chloride salts are rather insensitive to the metal, while those for the DOS salts do show a metal dependence. The high temperatures and metal insensitivity lead us to suggest that these former complexes may have a significant ionic character (strong interaction between the hard lanthanide acids and the borderline chloride base would not be expected). For the purposes of comparison, the transition temperatures of the related nitrate complexes are somewhat intermediate between those of the chlorides and dodecylsulphates. Another possibility would be a chlorobridged, dimeric structure (*vide infra*).

7. Effect of the ligand

As it appeared that the use of *N*-arylamines was out of the question, it was therefore necessary to extend the salicylaldimines at the other end of the molecule. Thus, in another strategy aimed at reducing transition temperatures, we introduced a 3,4-dialkoxybenzoate group (figure 12) as the alkoxy chain in the 3-position of the terminal ring would lower the anisotropy of the complex.

With this ligand, a systematic approach to the structure was adopted. Thus, initially n and m were systematically varied using lanthanum nitrate as the lanthanide source in order that NMR data could be readily obtained. Some selected data are recorded in table 1.

Several things are of note from these initial results. First, it does not appear as though there is a ready correlation between the two chain lengths, n and m, and the transition temperatures; a fuller picture will emerge as more derivatives become

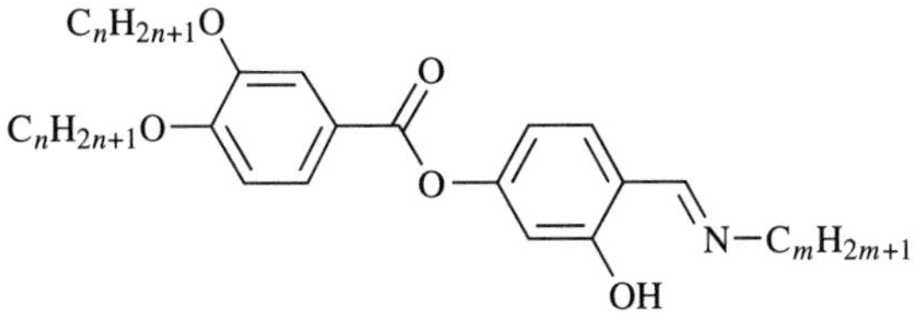

Figure 12. Structure of the 3,4-dialkoxysalicylaldimines.

Table 1. *Transition temperatures for some lanthanum nitrate salts*

n	m	complex formula	mesomorphism
4	4	$[LaL_2][NO_3]_3$	Crys · 84 · S_A · 124 · I
4	18	$[LaL_2][NO_3]_3$	Crys · 103 · S_A · 191 · I
8	4	$[LaL_2][NO_3]_3 \cdot H_2O$	Crys · 100 · S_A · 200 · I
8	12	$[LaL_2(L—H)][NO_3]_2 \cdot 2H_2O$	Crys · 83 · S_A · 123 · I

Table 2. *Selected transition temperatures for lanthanide complexes of the ligand with* $n = m = 12$

complex formula	mesomorphism
$[Dy(L)_2(L—H)][NO_3]_2$	Crys · 103 · S_A · 144 · I
$[Dy(L)_2(L—H)][OTf]_2$	Crys · 51 · S_A · 84 · I
$[Nd(L)_2(L—H)][NO_3]_2 \cdot H_2O$	Crys · 84 · S_A · 112 · I
$[Nd(L)_2(L—H)][OTf]_2$	Crys · 57 · S_A · 87 · I
$[Ho(L)_2][NO_3]_3$	Crys · 85 · S_A · 155 · I
$[Er(L)_2(L—H)][OTf]_2$	Crys · 51 · S_A · 81 · I

available. Second, note that the first three complexes in table 1 possess only two ligands. If we assume that each nitrate is bound in a bidentate fashion to the lanthanum taking up a total of six coordination sites, and that the ligands are bound in the zwitterionic form, then the metal is eight-coordinate (we are assuming for now that in the third entry, the water is in the lattice). To our knowledge, this stoichiometry has not previously been observed with salicylaldimine ligands bound to lanthanide centres. However, in the final table entry, a more familiar stoichiometry appears in which the metal appears bound both to the neutral (zwitterionic) ligand and to the ligand anion.

Next, we used a particular ligand ($n = m = 12$) in complexes with a range of lanthanides using both nitrate and triflate anions. Two patterns emerged. First, the stoichiometry of the triflate complexes was reliable, being $[Ln(L)_2(L—H)][OTf]_2$, occasionally with a water of crystallization, while the stoichiometry of the nitrate complexes varied between $[Ln(L)_2(L—H)][NO_3]_2$ and $[Ln(L)_2][NO_3]_3$ with no apparent pattern. Second, the transition temperatures of the triflate complexes were always lower than those of the corresponding nitrate complex, almost approaching room temperature. Some examples are shown in table 2.

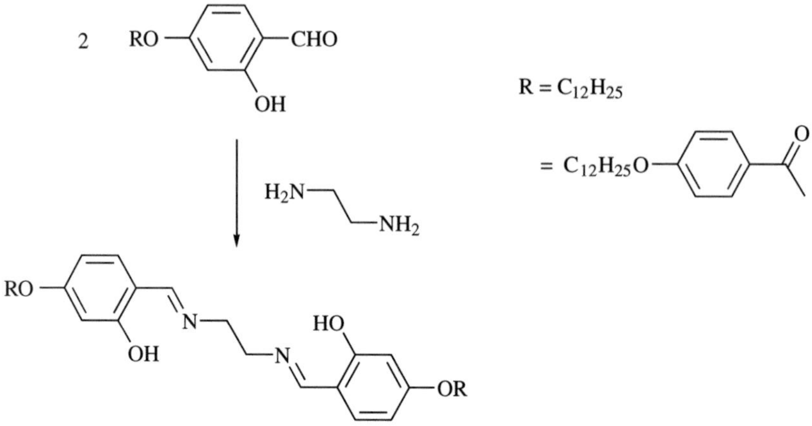

Figure 13. Structure of the dimeric ligands.

Table 3. *Mesomorphism of lanthanide complexes of dimeric ligands*

R	complex formula	mesomorphism
$C_{12}H_{25}$	$[Nd(L-2H)(L-H)]$	Crys · 150 · S · 190 · I
$C_{12}H_{25}$	$[Er(L-2H)(L-H)] \cdot 3H_2O$	Crys · 165 · S · 230 · I
$C_{12}H_{25}$	$[Eu(L-2H)(L-H)] \cdot 2H_2O$	Crys · 150 · S · 190 · I
$C_{12}H_{25}OC_6H_4CO$	$[Nd(L-2H)(L-H)]$	Crys · 140 · S · 200 · dec

8. Complexes of dimeric ligands

Previously, Paschke had synthesized (Paschke *et al.* 1990) dimeric salicylaldimine ligands which he had used to form complexes of nickel(II) and copper(II). We then made modifications of such ligands (figure 13) and used them to form complexes of various lanthanides.

The ligands themselves were mesomorphic (Cerrada *et al.* 1989) and with $R = C_{12}H_{25}$, a smectic C phase was found between 85 and 107 °C, while for $R = C_{12}H_{25}O$ C_6H_4CO, a smectic A phase was seen from 140 to 230 °C.

With these ligands, we were able to form neutral complexes in which the metal was bound to one monoanionic ligand and one dianionic ligand, the base used in the synthesis being piperidine. The mesomorphism of these complexes is shown in table 3.

For these complexes, we have not yet been able to identify the mesophase, as it is somewhat viscous, preventing the establishment of a well-defined texture; X-ray diffraction measurements are in progress. The mesophase is seen at slightly elevated temperatures, which is perhaps a little surprising since we might have expected that a totally neutral complex might be mesomorphic at lower temperatures. However, a possible explanation is found in the next section, on tripodal ligands.

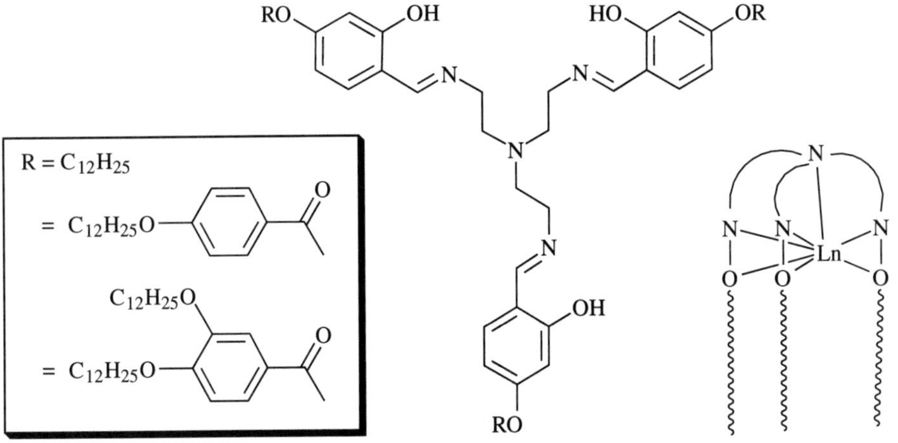

Figure 14. Structure of the tripodal ligands and a schematic of how they might coordinate.

9. Complexes of tripodal ligands

We then took the ligand design a stage further and thought about how we might encapsulate the lanthanide within a ligand system. We felt that this might be possible using a tripodal ligand to give a coordination geometry as shown in figure 14; the ligand used is shown in the same figure. The same ligand has also been used by the Zaragoza group for the synthesis of metallomesogens (Serrano, personal communication).

In one case (with R = Me), we were able to grow single crystals and the results of the structure determination are shown in figure 15 (Slawin, unpublished work). Thus, figure 15a shows what turned out to be half of the molecule, and reveals that while a tripodal geometry is adopted, the ligand appears insufficiently large to encapsulate the metal entirely and there is an open face with a phenolic oxygen protruding. Note that in this case, the ligand coordinates through the imine (and apical) nitrogen and a phenate anion, i.e. the zwitterion does not form. The presence of the open face suggests the full structure of a dimer which is held together through two bridging oxygen atoms and which is shown in figure 15b.

While we have not yet been able to determine whether the ligands are mesomorphic, their complexes, obtained as neutral materials of the general empirical formula [Ln(L—3H)], showed liquid crystal phases, although again here we do not yet have an unequivocal identification of the nature of what we believe to be the smectic phase. Representative data are collected in table 4.

In this case, the transition temperatures are rather higher than we might have anticipated, but the result of the crystal structure determination suggests a possible explanation for this, both for the tripodal ligands and for the dimeric ligands above. In the former case, we see clearly the dimeric arrangement of the complex and we suspect that such a large species would indeed have rather high transition temperatures. Of course, the dimeric ligands are closely related to the tripodal ligands and so it may well be that the neutral complexes of the dimer ligands also give rise to dimeric complexes which, being similarly large, lead to high melting systems.

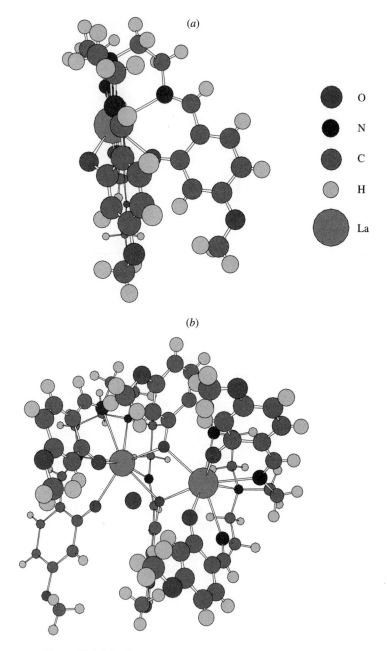

Figure 15. Molecular structure of a tripodal lanthanum complex.

10. Conclusion

In this paper, we have shown that there are several ways in which liquid-crystalline complexes of lanthanide elements can be realized, although a full development and

Table 4. *Mesomorphism of lanthanide complexes of trimeric ligands*

R	complex formula	lcmesomorphism
$C_{12}H_{25}$	$[La(L—3H)] \cdot H_2O$	Crys · 209 · S · 221 · I
$C_{12}H_{25}$	$[Gd(L—3H)] \cdot H_2O$	Crys · 205 · S · 278 · I
$C_{12}H_{25}$	$[Er(L—3H)] \cdot 2H_2O$	Crys · 230 · S · 270 · I
$C_{12}H_{25}OC_6H_4CO$	$[La(L—3H)] \cdot 4H_2O$	Crys · 216 · S · 230 · I
$3,4-(C_{12}H_{25}O)_2C_6H_3CO$	$[La(L—3H)] \cdot H_2O$	Crys · 122 · S · 180 · I

appreciation of the subject will need to go hand-in-hand with some fundamental lanthanide coordination chemistry. At the time of writing, we await with anticipation the results of the first physical measurements made on our new complexes and look forward to exploiting the phenomenon of magnetism in the anisotropic environment that is a liquid crystal mesophase.

The authors thank the following agencies for their support: the Leverhulme Trust, the EPSRC, FWO Flanders, IWT (Flanders), the British Council and INTAS (contract: INTAS 96-1198). Crystal structure data from Professor A. Polishuk (Kiev University) and Dr A. Slawin (Loughborough University) are gratefully acknowledged, as is expert NMR assistance from Dr V. Sik (Exeter).

References

Alonso, P. J. 1996 In *Metallomesogens: synthesis, properties and applications* (ed. J. L. Serrano), ch. 10, pp. 387–418. Weinheim: Wiley-VCH.

Berg, D. J., Rettig, S. J. & Orvig, C. 1991 *J. Am. Chem. Soc.* **113**, 2528–2532.

Bertram, C., Bruce, D. W., Dunmur, D. A., Hunt, S. E., Maitlis, P. M. & McCann, M. 1991 *J. Chem. Soc. Chem. Commun.*, pp. 69–70.

Bikchantaev, I., Galyametdinov, Yu. G., Kharitonova, O., Ovchinnikov, I., Bruce, D. W., Dunmur, D. A., Guillon, D. & Heinrich, B. 1996 *Liq. Cryst.* **20**, 489–492.

Binnemans, K., Galyametdinov, Yu. G., Collinson, S. R. & Bruce, D. W. 1998 *J. Mater. Chem.* **8**, 1551–1553.

Binnemans, K., Van Deun, R., Bruce, D. W. & Galyametdinov, Yu. G. 1999 *Chem. Phys. Lett.* **300**, 509–514.

Bruce, D. W. 1994 *Adv. Mater.* **6**, 699–701.

Bruce, D. W. 1996 In *Inorganic materials* (ed. D. W. Bruce & D. O'Hare), 2nd edn, ch. 8, pp. 429–522. Chichester: Wiley.

Bruce, D. W. (and 11 others) 1992 *Mol. Cryst. Liq. Cryst.* **206**, 79–92.

Cerrada, P., Marcos, M. & Serrano, J. L. 1989 *Mol. Cryst. Liq. Cryst.* **170**, 79–87.

Donnio, B., Bruce, D. W., Heinrich, B., Guillon, D., Delacroix, H. & Gulik-Krzywicki, T. 1997 *Chem. Mater.* **9**, 2951–2961.

Galyametdinov, Yu. G., Ivanova, G. I. & Ovchinnikov, I. V. 1991 *Bull. Acad. Sci. USSR Div. Chem. Sci.* **40**, 1109.

Galyametdinov, Yu. G., Ivanova, G. I., Ovchinnikov, I. V., Binnemans, K. & Bruce, D. W. 1999 *Russ. Chem. Bull.* **48**, 387–389.

Giroud-Godquin, A. M. 1998 *Coord. Chem. Rev.* **178–180**, 1485–1499.

Hoshino, N. 1998 *Coord. Chem. Rev.* **174**, 77–108.

Leadbetter, A. J. 1987 In *Thermotropic liquid crystals* (ed. G. W. Gray), ch. 1, pp. 1–27. Wiley.

Ovchinnikov, I. V., Galyametdinov, Yu. G., Ivanova, G. I. & Yagfarova, L. M. 1984 *Dokl. Akad. Nauk. SSSR* **276**, 126–127.

Paschke, R., Balkow, D., Baumeister, U., Hartung, H., Chipperfield, J. R., Blake, A. B., Nelson, P. G. & Gray, G. W. 1990 *Mol. Cryst. Liq. Cryst. Lett.* **188**, 105–118.

Piechocki, C., Simon, J., André, J.-J., Guillon, D., Petit, P., Skoulios, A. & Weber, P. 1985 *Chem. Phys. Lett.* **122**, 124–128.

Smith, A., Rettig, S. J. & Orvig, C. 1988 *Inorg. Chem.* **27**, 3929–3934.

Quantum size effects in molecular magnets

By Dante Gatteschi

Department of Chemistry, University of Florence,
via Maragliano 77, I-50144 Florence, Italy

The unique role of molecular clusters in the investigation of quantum size effects in magnetism is reviewed. The types of quantum effects to be expected in mesoscopic clusters, which can be obtained by a quasi-classical approach, are compared with those actually observed in molecular clusters. In particular, the focus of attention is put on antiferromagnetic rings, which show stepped hysteresis and anomalies in the proton relaxation behaviour. Ferrimagnetic clusters have been shown to undergo quantum tunnelling of magnetization, and have recently provided the first experimental evidence of the so-called Berry phase in magnets.

Keywords: molecular magnetism; nanomagnet; cluster; quantum tunnelling;
quantum size effects; Berry phase

1. Introduction

Particles whose sizes are of the order of some nanometres in at least one direction are of great theoretical interest at the moment because they are the ideal systems for observing phenomena associated with the coexistence of quantum and classical effects (Jain 1989; Halperin *et al.* 1993). This point is well illustrated for conductors by so-called quantum layers, quantum wires and quantum dots, which correspond to nanometric in one, two and three directions, respectively. These materials are actively investigated for basic science, but they already provide many possibilities for application (Bastard & Brum 1986).

The same theoretical interest is also being developed for magnets, in which there is a strong expectation to be able to observe quantum effects in macroscopic particles (Caldeira & Leggett 1981; Gunther & Barbara 1995; Chudnovsky & Tejada 1998). The phenomena that are actively looked for are macroscopic quantum tunnelling of the magnetization, and macroscopic quantum coherence, where the magnetization, or the Néel vector, tunnel coherently between classically degenerate directions over many periods. These phenomena are interesting both from the fundamental point of view and from the perspective of possible applications, of which quantum computing is certainly one of the most exciting (Deutsch & Jozsa 1985). They can be collected under the label 'macroscopic quantum phenomena' (MQP).

Compared with the observation of quantum phenomena in conductors, the observation of macroscopic quantum effects in magnetic particles leads to some additional difficulties. In fact, tunnelling effects are easily destroyed, or masked, if the particles are not all identical, if they are not all iso-orientated and if the interactions between them are not suitably reduced to a minimum. Several attempts have been made to prepare nanosized magnetic particles that obey all these requisites, but so far

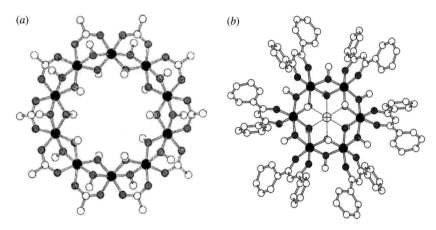

Figure 1. Structure of Fe10 and Fe6Li (after Taft & Lippard (1990)
and Caneschi *et al.* (1996), respectively).

no unambiguous success has been achieved. For instance, nanosized particles obtained
by thermal decomposition of iron salts have been investigated by microSQUID
techniques, but no clear evidence of quantum tunnelling was achieved. Analogous
doubts were left by experiments performed both on the natural form of ferritin and on
an artificially magnetite-loaded derivative (Awschalom *et al.* 1993; Gider *et al.* 1995,
1997; Garg 1995; Tejada 1995).

The situation was becoming static, when it was discovered that molecular magnetic
clusters may provide unique possibilities for the direct measurement of all kinds of
quantum size effects (Gatteschi *et al.* 1994). Two typical examples of molecular
clusters are shown in figure 1. These are molecular objects, which can be synthesized
with the techniques of molecular chemistry. They are all identical, their structures can
be exactly known, in general through X-ray crystal structure determinations and in
some cases also through neutron diffraction experiments. Furthermore, they can be
dissolved in suitable solvents, in order to reduce to a minimum the magnetic
interactions between them, or in polymer films, or even in Langmuir–Blodgett films.
Finally, they can be produced in single crystals that contain a macroscopic number of
identical magnetic subunits that are weakly coupled to each other as well as to their
surroundings. Therefore, it is possible to investigate a macroscopic ensemble, and still
measure the individual properties of the clusters.

These advantages prompted 14 to write:

> Now the chemists have come to the rescue, working up from the atomic
> scale. The promise is of many new magnetic macromolecules, behaving
> like giant spins. That would truly prop open the door to the fledgling field
> of quantum nanomagnetism, perhaps leading to the nanomolecular
> engineering of magnetic memory elements that have inherently quantum
> properties.

However, there are also some obvious disadvantages, the largest of which is the fact
that nobody has so far been able to prepare molecular clusters containing more than
about 20 magnetic centres. Under these conditions, it is certainly difficult to call the
clusters 'macroscopic' or even 'mesoscopic'. However, this paper aims to show that the

quantum effects that can be observed in these rather complex objects already have the flavour of what is expected for MQP, or, better still, that they provide unique testing grounds for theories developed at the semi-classical level. In fact, molecular clusters provide evidence of the fact that MQP are a natural extrapolation of the properties observed in quantum sized objects. This is not a totally new observation. In fact, it was well known that the thermodynamic properties of one-dimensional materials may be rather accurately obtained by extrapolation of the properties of relatively small magnetic rings (De Jongh & Miedema 1974). Now, however, quantum effects observed in clusters shed light on the possible quantum phenomena to be observed in mesoscopic objects.

As is always the case in fields of multidisciplinary research, it is necessary first of all to adopt a common language in order to achieve the best results. Therefore, I will try first of all to clarify what is commonly understood for quantum tunnelling and quantum coherence in magnetic phenomena. Second, I want to report a few results obtained in the observation of MQP in molecular clusters, starting from rings and then passing to more complex structures.

2. Quantum tunnelling and quantum coherence

Macroscopic quantum tunnelling of the magnetization must be a rare event, because a macroscopic particle is, by definition, a system that is large enough to behave classically during most of the time it is being observed. There are several theories providing information on the frequency of the tunnelling, the most popular of which is that of the so-called instantons (Langer 1967). This theory is a general one that describes the tunnelling of the particle from the metastable state in which it is prepared, using imaginary time. For this reason, motion does not take any real time, occurring in imaginary time (Chudnovsky & Tejada 1998). The tunnelling rate is expressed in the form

$$\Gamma = A(T)\exp(-U_0/T_{esc}(T)), \qquad (2.1)$$

where U_0 is the height of the barrier and T_{esc} is the characteristic escape temperature. The pre-exponential factor is called the attempt frequency and it corresponds to the small oscillations near the bottom of the potential well. Typically, at high temperature $T_{esc} = T$, i.e. the decay of the metastable state occurs through a thermal over-barrier transition. In the tunnelling regime, T_{esc} tends to a non-zero constant and the relaxation time becomes temperature independent.

The possibility of observing MQP is bound by the following conditions: (i) a barrier is present; and (ii) a suitable matrix element admixes states on the two sides of the barrier. Specializing to magnetic particles suggests a treatment that is an extension of the classical one for superparamagnetic particles. The magnetization of the particle can be either up or down, as shown in figure 2, and the barrier for reorientation is given by the axial magnetic anisotropy. The origin of the axial anisotropy is, in general, associated with magneto-crystalline anisotropy, i.e. with the contribution to the energy of the particle from the individual magnetic centres. For instance, rare earth ions, cobalt, manganese(III), are individual magnetic centres that tend to contribute a large magnetic anisotropy to the particles.

In the absence of suitable matrix elements connecting the two sides of the barrier,

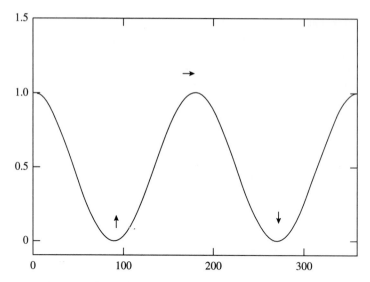

Figure 2. Two-well representation of the equilibrium states of
the magnetization of a superparamagnetic particle.

the magnetic particles can only reorient with a thermally activated mechanism. This
gives rise to a relaxation time (Morrish 1966):

$$\tau = \tau_0 \exp(\Delta/kT). \qquad (2.2)$$

However, if a suitable transverse field is present coupling the two states on the two
sides of the barrier, the tunnelling mechanism becomes possible. The transverse fields
can be provided by an external magnetic field, by magneto-crystalline contributions,
by the dipolar fields within the particles, by the dipolar fields of the other particles
that are unavoidably present in the investigated systems, or by the magnetic fields of
the nuclei that are coupled to the electrons via the hyperfine interaction (Prokof'ev &
Stamp 1996). It is important to recognize that in particles containing an odd number
of unpaired electrons the only transverse field breaking Kramer's degeneracy is an
external magnetic field. Therefore, it should not be possible to observe quantum
tunnelling effects in Kramers species in the absence of an applied transverse field.

The existence of tunnelling is, therefore, associated with the removal of the
degeneracy of a pair of magnetic levels by the action of a suitable transverse field. The
two levels are separated by the energy

$$\Delta = h\Gamma/2\pi, \qquad (2.3)$$

where Δ is the tunnel splitting. For macroscopic quantum tunnelling, the splitting Δ
is always small compared with the attempt frequency. Therefore, the wave functions
of the two low-lying states can be approximated by linear combinations of the
degenerate states in the absence of tunnelling. If a single domain particle is in the true
ground state, the probability of its magnetization having a certain orientation at a
moment of time t oscillates with time:

$$\langle \mathbf{M}(t)\mathbf{M}(0)\rangle M_0^2 \cos(\Gamma t). \qquad (2.4)$$

The system placed in a field of frequency Γ should show a resonance in the power absorption. This effect is called macroscopic quantum coherence (MQC). The observation of MQC is, however, hampered by the fact that interactions with the environment of order Δ may make its observation impossible. This is much more stringent than the condition for the observation of QTM, where dissipation towards other degrees of freedom of the system must be small compared with the attempt frequency, which is much larger than Δ.

Magnetic tunnelling has been investigated both in monodomain ferromagnetic particles, and in the Néel vector of antiferromagnetic particles. In particular, the phenomena that have been investigated are the quantum nucleation of magnetic domains, the quantum depinning of domain walls, and the relaxation of the magnetization. The investigated systems range from γ-Fe$_2$O$_3$ particles to CoFe$_2$O$_4$ particles to ferritin (Chudnovsky & Tejada 1998).

3. Magnetic rings

Magnetic rings have long been theoretically investigated for calculating the thermodynamic properties of infinite chains (De Jongh & Miedema 1974). Recently, a large number of rings, comprising metal ions with quantum ($S = 1/2$) spins (Rentschler *et al.* 1996) or classical ($S = 5/2$) spins (Taft & Lippard 1990; Caneschi *et al.* 1996; Watton *et al.* 1998), and mixed quantum–classical spins, (alternating $S = 1/2$ and $S = 5/2$ spins) (Caneschi *et al.* 1988), has been reported. Also, the coupling in the rings can be varied at will, with examples of antiferromagnetic (Taft & Lippard 1990; Caneschi *et al.* 1988, 1996; Watton *et al.* 1998), ferromagnetic (Rentschler *et al.* 1996; Abbati *et al.* 1998; Blake *et al.* 1994) and ferrimagnetic (Caneschi *et al.* 1988) rings. The ground state of the rings varies from $S = 0$ for antiferromagnetic rings to $S = 12$ for some ferromagnetic or ferrimagnetic rings. Before moving on to describe the properties of the rings relevant to MQP, we will note, on passing, that so far essentially only rings with an even number of spins have been reported. This is unfortunate because it would be of extreme interest to have some example of antiferromagnetic rings with an odd number of unpaired spins, because, in this case, spin frustration effects might be observed (McCusker *et al.* 1991). These give rise to a highly degenerate ground state, which is unstable to many perturbations.

Rather surprisingly, the most interesting results so far have not come from the rings with a large ground spin state, namely the ferromagnetic or ferrimagnetic rings, but rather from the antiferromagnetic rings, which have an $S = 0$ ground state. In fact, these systems can be used as models for studying the macroscopic quantum coherence of the Néel vector, which tunnels between classical degenerate directions over many periods. This point was realized by 27, who calculated the quantum dynamics of the Néel vector by instanton methods. This semiclassical method, valid for large spins and in the continuum limit, allows the calculation of the tunnel splitting of the Néel vector modulated by an external magnetic field applied in a direction parallel to the ring. The presence of the external magnetic field tunes the tunnel barriers by introducing an effective anisotropy. If we pass from this semiclassical approach to the quantum approach appropriate for relatively small clusters, the presence of an external magnetic field in antiferromagnetic rings induces a change in the energies of the excited S states, as sketched in figure 3. The energy of the ground level is taken as zero, and the field is taken parallel to z. It is apparent

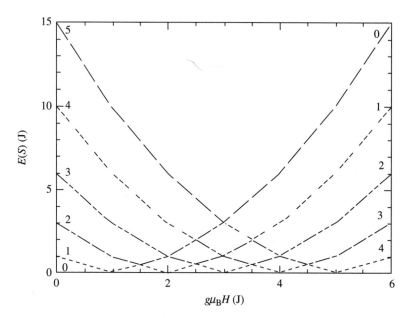

Figure 3. Field dependence of the energies of the low-lying S levels in antiferromagnetic rings.
The energy of the ground level is taken as zero.

that at a given value of the field, the $S = 1$ state, which decreases its energy, crosses
the $S = 0$ state, which is essentially field independent (Taft *et al.* 1994; Cornia *et al.*
1999*a*). When the two levels cross, the tunnel splitting is effectively quenched.
However, the process is not limited to this crossover, because the level with $S = 2$
decreases its energy faster than that of $S = 1$, and eventually a second crossover
occurs. In fact, the process continues on increasing field until the S_{max} level becomes
the ground state. S_{max} is the spin state originated by putting all the individual spins
parallel to each other. For a ring of N spins S_i, $S_{max} = NS_i$. It is apparent that for
such an antiferromagnetic ring there is the possibility of observing up to NS_i field-
determined crossovers in the ground state.

The so-called ferric wheel (Taft & Lippard 1990), the structure of which is sketched
in figure 1*a*, is the system that has been first investigated from this point of view.
Pulsed magnetization measurements at 0.7 K provided the crossover up to $S = 9$
(Taft *et al.* 1994). The crossover energies can be quantitatively calculated for
relatively small rings. However, a formula that gives the energies of the S states to a
good approximation (Caneschi *et al.* 1996) is

$$E(S) = \frac{2J}{N} S(S + 1), \qquad (3.1)$$

where N is the number of spins in the ring. Equation (3.1) is most appropriate for
rings in which the individual spins S_I are large. Therefore, it works well for quasi-
classical spins, like $S = 5/2$, and is not useful for quantum spins like $S = 1/2$.
From (3.1) we learn that the first crossover will occur when the Zeeman energy,
$g\mu_B H$ equals the energy of the lowest excited triplet, namely when $g\mu_B H = 4NJ$.
This is, in fact, the lower limit indicated by Chiolero & Loss (1998) for observing

tunnelling. It is also interesting that the following field values for tunnelling, or, in other words, the crossover fields, all appear at regular intervals, as expected on the basis of equation (3.1).

Stepped magnetization has been observed also in smaller iron(III) rings (Caneschi *et al.* 1996; Cornia *et al.* 1999*a*, *b*), like those depicted in figure 1*b*. In the centre of the ring, an alkali ion, Li^+ or Na^+, is present. In this case, it was also possible to investigate very small single crystals by new torque magnetometers using a sensitive cantilever to measure the magnetization. By measuring the magnetization in different crystal orientations, it was possible to directly measure the magnetic anisotropy of the excited S states. These investigations provided surprising results. In fact, it has been observed that the central alkali ion strongly influences both the isotropic exchange interaction, which passes from $14 \, cm^{-1}$ to $20 \, cm^{-1}$ for $M = Li^+$ and $M = Na^+$, respectively, and the magnetic anisotropy. In fact, the zero-field splitting of the first excited $S = 1$ level is $D = 1.16 \, cm^{-1}$ for the lithium and $D = 4.32 \, cm^{-1}$ for the sodium derivative.

All the measurements reported above are static measurements, which do not provide any information on the dynamics of the magnetization or on possible tunnelling phenomena. Dynamic measurements can, in principle, be made with a variety of techniques, ranging from AC susceptometry to magnetic resonance. In this respect, NMR is particularly appealing, because it can monitor the variations in the relaxation rate of the nuclear moments under the influence of the tunnelling of the electron magnetization. In fact, it can be expected that when two levels cross, the electron relaxation at low temperature is drastically influenced. Therefore, the field dependence of the nuclear relaxation rate should show anomalies corresponding to the crossover fields. Preliminary measurements performed on the ferric wheel in the temperature range 1.2–4.2 K show the presence of distinct maxima in the nuclear relaxation rates at the crossover fields (Julien *et al.* 1999). This may be experimental evidence of coherent tunnelling of the Néel vector in magnetic clusters.

4. Quantum effects in molecular ferrimagnets

The application of a strong external field may give rise to stepped variations in the magnetization also in ferrimagnets. Recently, a technique generating ultra-high magnetic fields up to 850 T by explosive compression of initial magnetic flux was implemented (Bykov *et al.* 1996). It was found that bulk ferrimagnets undergo a second-order (continuous) phase transition from the ferrimagnetic to a canted phase and then from the canted to ferromagnetic states (Zvezdin 1995). The same technique has recently been employed on the molecular cluster $[Mn_{12}O_{12}(CH_3COO)_{16}(H_2O)_4]$, Mn12Ac, the structure of which (Lis 1980)is sketched in figure 4, which has been investigated with a number of different techniques for its unique low-temperature magnetic behaviour, and for the first time some knowledge of the excited spin states has been achieved (Lubashevsky *et al.* 1999).

The ground state of Mn12Ac is now well established to be $S = 10$, as a result of non-compensation of the magnetic moments of eight manganese(III) ions, each with $S_i = 2$ and all parallel to each other, and four manganese(IV) ions, each with $S_j = 3/2$ and all antiparallel to the manganese(III) spins (Sessoli *et al.* 1993*b*). It is apparent that it is possible to increase the spin of the ground state by applying an external field that increases S. The available excited states with S larger than 10 are

228

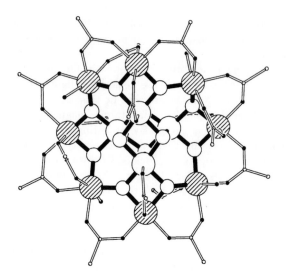

Figure 4. Sketch of the structure of Mn12Ac (after Lis (1980)).

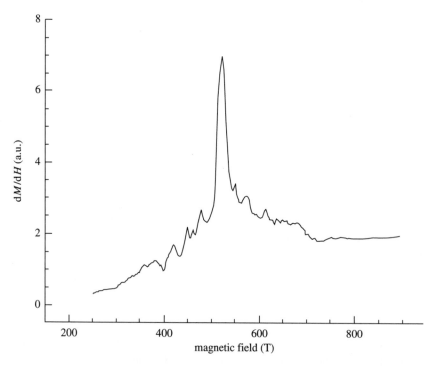

Figure 5. Differential magnetization of a single crystal of Mn12Ac in pulsed fields at 1.2 K. The duration of the pulses is of the order of 10^{-6} s.

Table 1. *Size of the matrices corresponding to the total spin states S of* Mn12Ac

S	$n(S)$	S	$n(S)$	S	$n(S)$
0	190 860	8	654 476	16	7656
1	548 370	9	428 450	17	2951
2	838 126	10	333 032	18	997
3	1 029 896	11	214 996	19	286
4	1 111 696	12	129 476	20	66
5	1 090 176	13	72 456	21	11
6	986 792	14	37 472	22	1
7	831 276	15	17 776		

$S = 11, 12, \ldots, 22$. The results of the pulsed experiments are shown in figure 5 as the $\mathrm{d}M/\mathrm{d}H$ versus H plot. No transition is observed up to $H = 350$ T, indicating that the lowest excited state with $S > 10$ is at least 350 cm^{-1} above the ground state. In fact, assuming a regular pattern of excited states, the crossover energies can be expressed as

$$H_{S \to S+1} = [E(S+1) - E(S)]/g\mu_{\mathrm{B}}. \qquad (4.1)$$

Regular here means that the energies of the lowest $S + 1$ and $S + 2$ states in zero field obey the following conditions:

$$E(S+1) > E(S) \quad \text{and} \quad 2E(S+1) - E(S) < E(S+2). \qquad (4.2)$$

Above 350 T, several spikes are observed, which strongly suggest that the energies of the excited states are quantized. Four spikes can be located at 382, 416, 448 and 475 T, suggesting that these are the crossover fields for $S = 11$, 12, 13 and 14, respectively. The following higher-intensity spike is presumably due to the fact that the crossover fields for more than one level occur at very similar values. In fact, we assign the spike to the crossover fields to $S = 15$, 16, 17 and 18. The higher-field spikes are assigned to the crossover fields to $S = 19$, 20, 21 and 22. No spikes are observed above 690 T, suggesting that the energy of the highest spin state is *ca.* 6000 cm^{-1} above the ground state. It would be desirable to attempt to record Raman spectra to confirm some of these values.

The calculation of the energies of the excited states of Mn12Ac is far from being trivial. In fact, the total number of spin states is 10^8, and even if the total spin symmetry is exploited in the best possible way, the size of the matrices corresponding to the different S values are as shown in table 1. In the Bible, the expression 'myriad of myriad' (corresponding to 10^8 in modern notation) means a number large beyond any limit.

It is apparent that the problem is tractable on the high-S side, where relatively small matrices must be calculated. Unfortunately, the crossover fields corresponding to these levels are less-well experimentally determined. We used a composite approach, using both a perturbation treatment and a direct calculation of the energies of the lowest-lying levels for each S value. In fact, the latter procedure could be used for $S = 16$–22, while the lower-lying levels were calculated through the perturbation approach. This, of course, underestimates the repulsion between levels with the same S. However, we found that although the absolute energies are not well

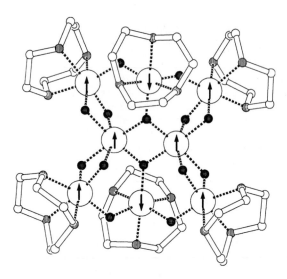

Figure 6. Sketch of the structure of Fe8 (after Wieghardt *et al.* (1984)).

calculated with this approach, the differences in crossover fields are reasonably well reproduced. The best-fit values of the parameters are: $J_1 = 122$ cm^{-1}, $J_2 = 60$ cm^{-1}, $J_3 = -11.2$ cm^{-1} and $J_4 = 30$ cm^{-1}. These results are in qualitative agreement with the values of the coupling constants usually found in the analysis of the magnetic properties of smaller clusters containing manganese ions.

Molecular ferrimagnets are the systems that have, so far, provided the best examples for quantum tunnelling of the magnetization. The reason for this is easily understood: up to now, the ferrimagnets are the systems with the highest ground spin states and with the highest magnetic anisotropy. We do not see anything particularly special in ferrimagnets, except that, by chance, they have so far provided the best systems obeying the above requirements. The systems that I will treat to some extent are Mn12Ac, introduced above, and Fe8, the structure of which (Wieghardt *et al.* 1984) is sketched in figure 6. The spins depicted on the iron(III) ions provide a justification for the ground $S = 10$ state.

Both Mn12 and Fe8 have the $S = 10$ ground state, with the former having a uniaxial anisotropy (Sessoli *et al.* 1993*b*) and the latter having a biaxial anisotropy (Delfs *et al.* 1993; Barra *et al.* 1996). In both cases the low symmetry splitting of the ground states can be described (Barra *et al.* 1997) by the spin Hamiltonian:

$$H = D[S_z^2 - S(S+1)/3] + E(S_x^2 - S_y^2) + B_4^0 O_4^0 + B_4^4 O_4^4 + B_4^2 O_4^2. \qquad (4.3)$$

The Hamiltonian assumes C_2 symmetry. The O_k^m are operator equivalents,

$$O_4^0 = 35S_z^4 - [30S(S+1) - 25]S_z^2 - 6S(S+1) + 3S^2(S+1)^2,$$
$$O_4^2 = \tfrac{1}{4}\{[7S_z^2 - S(S+1) - 5](S_+^2 + S_-^2) + (S_+^2 + S_-^2)[7S_z^2 - S(S+1) - 5]\},$$
$$O_4^4 = \tfrac{1}{2}(S_+^4 + S_-^4),$$

and the B_4^4 are parameters. Higher-order terms, in principle up to the 20th, are symmetry allowed, but they were not included for the sake of simplicity.

In Mn12Ac, the tetragonal symmetry of the cluster requires $E = 0$ and $B_4^2 = 0$. The values of the parameters, as obtained through EPR spectroscopy (Barra *et*

Table 2. *Low-symmetry parameters for* Mn12Ac *and* Fe8 (*all parameters in* K)

sample	D	E	B_4^0	B_4^4	B_4^2	ref.
Mn12Ac	−0.66	0	-3.17×10^{-5}	$+6 \times 10^{-5}$	—	Barra *et al.* 1997
Fe8	−0.2748	0.0464	—	—	—	Barra *et al.* 1996
Fe8	−0.2909	−0.0464	1.00×10^{-6}	8.54×10^{-6}	1×10^{-7}	Caciuffo *et al.* 1998

al. 1996, 1997) and inelastic neutron scattering (Caciuffo *et al.* 1998), are given in table 2.

The energies of the M states of the $S = 10$ multiplets for Mn12Ac and Fe8 are shown in figure 7. The levels are plotted according to the usual convention of the two separate potential wells, i.e the $+M$ states are plotted on one side and the $-M$ states are plotted on the other side. However, it must be stressed that this is reasonable to a good approximation for Mn12Ac, in which $E = 0$, while it is much less acceptable for Fe8. In fact, for the latter, M is no longer a good quantum number, and extensive mixing of the states occurs. Since the E term admixes states with M differing by ± 2, its effect is large on the states with small M, while it has only a small effect on the states with large M. Therefore, for the latter, the approximation of the two potential wells is still acceptable. In fact, the pairs are quasi-degenerate up to $M = \pm 5$, but the higher levels are heavily admixed. In particular, there are three reasonably well separated levels, and higher up there are two more quasi-degenerate pairs. Therefore, there is no $M = 0$ top level. In fact, the two highest quasi-degenerate levels are admixtures of various levels. One is the admixture of $M = 0$ with $M = \pm 2$ and other even M values, while the other is the admixture of $M = \pm 1$ with other odd M values. On the other hand, the levels are well behaved for Mn12. The largest splitting is observed between the $M = \pm 2$ levels, due to the B_4^4 term, which mixes states differing by 4 in M.

It is quite remarkable that a very accurate description of the split levels of the ground $S = 10$ multiplet of Fe8 could be achieved through inelastic neutron-scattering experiments (Caciuffo *et al.* 1998). This technique does not, in general, produce good-quality spectra in systems having a large number of protons, and extensive deuteration is required. We performed our experiments on two grams of non-deuterated powder, and we obtained the detailed spectra shown in figure 8. The peaks correspond to transitions between M levels of the ground $S = 10$ state. These data allowed us to also determine the fourth-order crystal field parameters that had escaped the HF-EPR analysis.

At low temperature, both clusters show slow relaxation of the magnetization (Barra *et al.* 1996; Sessoli *et al.* 1993a) with lower blocking temperature for Fe8. This is in qualitative agreement with the fact that at low temperature the pairs of levels with large M are populated. The magnetization can revert its sign if the spin value passes from $M = -S$ to $M = -S + 1$, then to $M = -S + 2$, and so on up to the top of the barrier, or to a level from which it can shortcut to the opposite side of the barrier. It is customary to take $\Delta = |D|S^2$ as a rough estimation of the height of the barrier. In this sense, it is apparent that the barrier for Mn12Ac is higher than for Fe8. Furthermore, the latter has more opportunities to shortcut given the large admixture of the levels with small M. The magnetization of Mn12Ac has been found to follow the Arrhenius law with $\tau_0 = 2 \times 10^{-7}$ s, and $\Delta/k = 62$ K, in

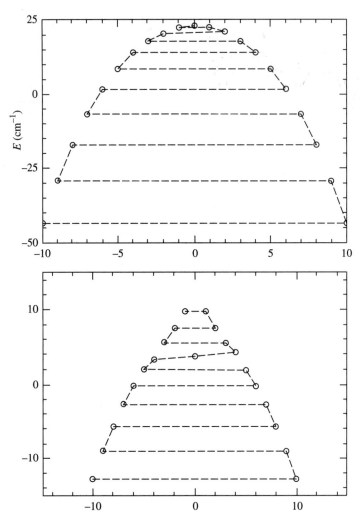

Figure 7. Energies of the M components of the $S = 10$ ground states of Mn12Ac (top), and Fe8 (bottom). The lowest components are $M = \pm 10$, while the highest components are heavily admixed for Fe8 (see text).

reasonable agreement with the value that can be calculated from the zero-field splitting parameter. Fe8 does not quite obey the Arrhenius law, except in small temperature ranges.

Both clusters show evidence of thermally assisted quantum tunnelling (Thomas *et al.* 1996; Friedman *et al.* 1997; Sangregorio *et al.* 1997). The meaning of this expression is that when a spin is in the M state, with $M < S$, it has a finite probability of tunnelling to the $-M + n$ state when the two levels have the same energy (Luis *et al.* 1998). n is an integer that depends on the value of the magnetic field applied parallel to the easy axis of the magnetization of the clusters needed in order to put the two levels at the same energy. For the simple case of axial symmetry,

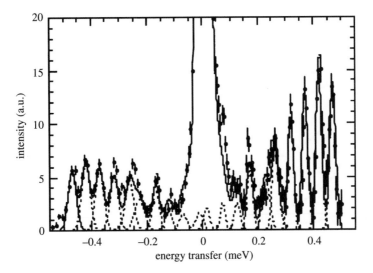

Figure 8. Inelastic neutron scattering of Fe8 at 10 K.

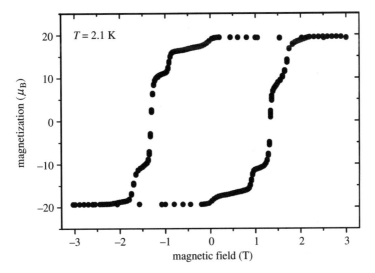

Figure 9. Stepped hysteresis of a single crystal of Mn12Ac at 2.1 K.
The field is parallel to the tetragonal axis of the cluster.

including only second-order terms

$$H_n = nD/(g\mu_\mathrm{B}).\tag{4.4}$$

The efficiency of this mechanism has been best observed in a single crystal of Mn12Ac, whose temperature-dependent hysteresis (Thomas *et al.* 1996) is shown in figure 9. The sharp steps correspond to fast relaxation, determined by quantum tunnelling between degenerate pairs of levels. At different values of the field the levels are not at the same energy, the quantum tunnelling is therefore quenched. Similar results were observed (Sangregorio *et al.* 1997) for Fe8, and the field corresponding to the strong

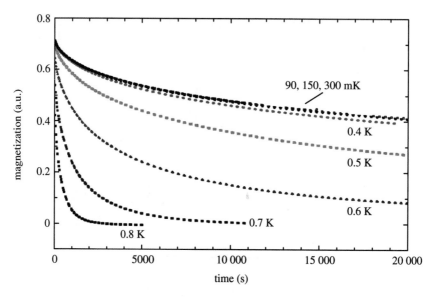

Figure 10. Time dependence of the decay of the magnetization of Fe8 below 1 K.
Below 300 mK the relaxation becomes independent of temperature.

tunnelling effects can be calculated with high accuracy, adding the Zeeman term to the spin Hamiltonian (4.1).

At very low temperature, below 2 K for Mn12Ac, the $M = \pm S$ levels are the only populated ones, but no clear evidence of tunnelling has been achieved. In fact, the transverse anisotropy is so small that the tunnelling frequency between these two levels is vanishingly small. It must be mentioned that the Arrhenius law suggests that at 1 K the relaxation time of the magnetization becomes of the order of billions of years. There is no doubt that under these conditions the Mn12Ac clusters behave as tiny magnets. It is also interesting to notice that, although the size of the clusters is so small that they are clearly within the quantum limit, quantum tunnelling is completely inefficient at low temperature.

Matters are different with Fe8, where the transverse magnetic anisotropy is large enough to establish the quantum tunnelling regime at low temperature. This is clearly shown by figure 10, where the decay of the magnetization of Fe8 becomes (Sangregorio *et al.* 1997) clearly temperature independent below 300 mK.

Recently, another unique quantum feature has been observed (Wernsdorfer & Sessoli 1999) in Fe8, which represents the first experimental evidence of the so-called Berry phase in magnets (Berry 1984). A semiclassical approach of the instanton type suggests that if a field is applied parallel to the hard axis, x, of a magnetic particle characterized by a giant spin S, the tunnelling splitting oscillates regularly as a function of the applied field (Garg 1993). The tunnel splitting is quenched whenever the following condition is met:

$$H_m = (S - m + 1/2)\Delta H, \qquad (4.5)$$

where m is an integer ranging from 1 to S. If only the second-order terms of the spin

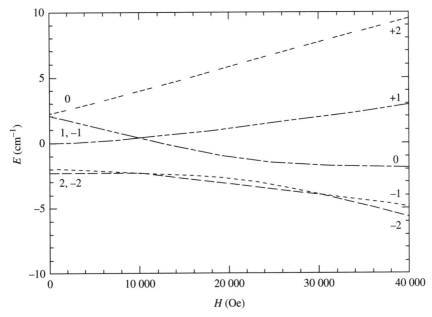

Figure 11. Field dependence of the M energy levels of an S multiplet. The field is applied parallel to the hard axis. The different labelling of the levels on the right and on the left corresponds to weak and strong field limits, respectively.

Hamiltonian (3.1) are included,

$$\Delta = [2E(E - D)]^{1/2} 2/(g\mu_{\mathrm{B}}). \qquad (4.6)$$

Since the tunnel splitting is responsible for the tunnelling relaxation it must be expected that a direct measurement of the relaxation rates should show maxima at the fields H_m given by equations (4.5) and (4.6).

Before proceeding further on this point, we want to stress that for particles with relatively small spins the conditions described by equations (4.5) and (4.6) find a simple interpretation. The results of calculation applied to an $S = 2$ state with $D = -1 \ \mathrm{cm}^{-1}$, $E/D = 1/3$ in the presence of a variable field parallel to x are shown in figure 11. The levels are all non-degenerate in zero field, and in the presence of the field they vary their energies and the admixtures. The lowest-lying levels, which in axial symmetry would be $M = \pm 2$, initially decrease their separation, then they become degenerate at $H = 1.01$ T. At the same field, the upper levels, which in axial symmetry would be $M = \pm 1$, also cross. A further increase of the field determines a second crossing of the lowest lying levels at $H = 3.03$ T. If the calculations are repeated, for instance for $S = 3$, the crossing fields will remain the same, except that there would be an additional one for the lowest-lying levels at 5.05 T. Again, what is extremely interesting is to see that the semiclassical and the quantum exact calculations give the same results.

If analogous calculations are performed for half-integer spins, the splitting is calculated as zero in zero field as expected for Kramers doublets. The transverse field removes the degeneracy and the pairs of levels to cross each other with the periodicity calculated through equations (4.5) and (4.6).

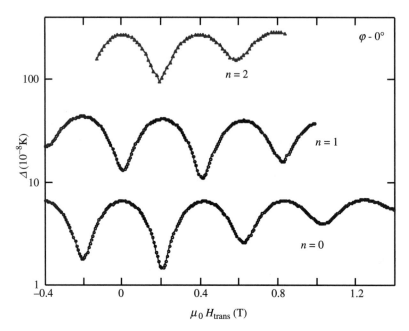

Figure 12. Experimental tunnel splitting of the lowest $M = \pm 10$ components of Fe8. The transverse field is applied parallel to the hard axis. An AC field is applied parallel to the easy axis. The n values correspond to the resonances defined in equation (4.4).

The exciting result is that for the first time Wernsdorfer & Sessoli (1999) have directly measured the splitting of the levels by measuring the tunnelling rate of Fe8 clusters in the presence of two applied fields. They used an array of microSQUIDs to measure the relaxation of the magnetization in the presence of an AC field that sweeps over the entire resonance transition, which is determined by dipolar and hyperfine fields. The AC field is applied parallel to the easy axis, while the static field is applied parallel to the hard axis. The results are shown in figure 12 for $H_z = 0$ and for H_z corresponding to the first resonance described by equation (4.5). There are two things to be noticed. The first is the periodicity of the maxima in the relaxation rate, which correspond to the fields at which the lowest levels have the maximum tunnel splitting. The second is that there is a strong parity effect. In fact, for n even, the maxima in the tunnelling rates correspond to the H_m fields of equations (4.5) and (4.6). The parity effect has a symmetry origin analogous to the Kramers degeneracy of the spin states with half-integer S. These measurements were able, for the first time, to determine very small energy splittings of the order of 10^{-8} K, which are not accessible by spectroscopic techniques.

In the semiclassical picture, the observed oscillations are due to topological quantum interference of two tunnel paths of opposite directions. The tunnelling transitions of the magnetization vector between two minima in a biaxial system can be schematized as given by two symmetrical paths in the plane perpendicular to the hard axis. In zero field, the two quantum spin paths give constructive interference when S is an integer, while a quench of the tunnelling splitting is given by destructive interference for half-integer spins (Bogacheck & Krive 1992; Loss et al.

1992; Golyshev & Popkov 1995). When the field is applied along the hard axis, a destructive–constructive interference of the quantum spin phase (known as the Berry phase (Berry 1984)), is observed and gives rise to oscillations in the tunnelling rate, similar to the Aharonov–Bohm oscillations of the conductivity in mesoscopic rings (Imry 1977).

These experiments on Fe8 have, for the first time, evidenced this phenomenon in magnetic systems and have again shown the great impact of molecular clusters for the study of magnetism at the mesoscopic scale. The Berry phase and the Haldane gap and the related parity effects in magnetic materials have been debated for a long time, but experiments on traditional nanosized particles have failed to reveal them. The Fe8 systems are the ideal candidates because of their biaxial anisotropy, their pure observable tunnelling regime at accessible temperatures, and their intrinsic quantum nature.

5. Conclusions

Molecular clusters are providing exciting new perspectives at the frontier between the quantum and the classical world. I believe that in this article I have shown how different quantum effects can be observed through the collaboration between physicists and chemists who have become able to speak to each other. I also feel that these materials are extremely productive, not only for the development of new basic science but also for new applications. If quantum effects are important for developing quantum computers, molecular clusters of increasing size and complexity provide an almost infinite number of new possibilities.

The results of the Florence laboratory have been obtained through collaboration with many groups throughout the world. They are all duly acknowledged by the references. They have been obtained thanks to the enthusiasm of my collaborators in Florence and Modena: Andrea Caneschi, Andrea Cornia, Claudio Sangregorio, Alessandro Lascialfari. The highest credit, however, should be given to Roberta Sessoli, who was the actual initiator of the field. The financial support of MURST, CNR and PFMSTAII is gratefully acknowledged.

References

Abbati, G. L., Cornia, A., Fabretti, A. C., Caneschi, A. & Gatteschi, D. 1998 *Inorg. Chem.* **37**, 1430.

Awschalom, D. D., DiVincenzo, D. P., Grinstein, G. & Loss, D. 1993 *Phys. Rev. Lett.* **71**, 4276.

Barra, A. L., Debrunner, P., Gatteschi, D., Schulz, C. & Sessoli, R. 1996 *Europhys. Lett.* **35**, 133.

Barra, A. L., Gatteschi, D. & Sessoli, R. 1997 *Phys. Rev.* B **56**, 8192.

Bastard, G. & Brum, J. A. 1986 *IEEE J. Quantum Electron.* **22**, 1625.

Berry, M. V. 1984 *Proc. R. Soc. Lond.* A **392**, 45.

Blake, A. J., Grant, C. M., Parsons, S., Rawson, J. M. & Winpenny, R. E. P. 1994 *J. Chem. Soc. Chem. Commun.*, p. 2363.

Bogachek, E. N. & Krive, I. V. 1992 *Phys. Rev.* B **46**, 14 559.

Bykov, A. I., Dolotenko, M. I., Kolokol'chikov, N. P., Pavlovskii, A. I. & Tatsenko, O. M. 1996 *Physica* B **216**, 215.

Caciuffo, R., Amoretti, G., Murani, A., Sessoli, R., Caneschi, A. & Gatteschi, D. 1998 *Phys. Rev. Lett.* **81**, 4794.

Caldeira, A. O. & Leggett, A. J. 1981 *Phys. Rev. Lett.* **46**, 211.

Caneschi, A., Gatteschi, D., Laugier, J., Rey, P., Sessoli, R. & Zanchini, C. 1988 *J. Am. Chem. Soc.* **110**, 2795.

Caneschi, A., Cornia, A., Fabretti, A. C., Foner, S., Gatteschi, D., Grandi, R. & Schenetti, L. 1996 *Chem. Eur. J.* **2**, 2329.

Chiolero, A. & Loss, D. 1998 *Phys. Rev. Lett.* **80**, 169.

Chudnovsky, E. M. & Tejada, J. 1998 *Macroscopic quantum tunnelling of the magnetic moments.* Cambridge University Press.

Cornia, A., Jansen, G. M. & Affronte, M. 1999*a Phys. Rev. B.* (In the press.)

Cornia, A., Affronte, M., Jansen, A. G. M., Abbati, G. L. & Gatteschi, D. 1999*b Angew. Chem. Int. Ed. Engl.* **38**, 2264.

De Jongh, L. J. & Miedema, A. R. 1974 *Adv. Phys.* **23**, 1.

Delfs, C., Gatteschi, D., Pardi, L., Sessoli, R. & Hanke, D. 1993 *Inorg. Chem.* **32**, 3099.

Deutsch, D. & Jozsa, R. 1985 *Proc. R. Soc. Lond.* A **239**, 553.

Friedman, J. R., Sarachik, M. P., Tejada, J. & Ziolo, R. 1997 *Phys. Rev. Lett.* **76**, 3830.

Garg, A. 1993 *Europhys. Lett.* **22**, 205.

Garg, A. 1995 *Science* **272**, 424.

Gatteschi, D., Caneschi, A., Pardi, L. & Sessoli, R. 1994 *Science* **265**, 1054.

Gider, S., Awschalom, D. D., Douglas, T., Mann, S. & Chaparala, M. 1995 *Science* **268**, 77.

Gider, S., Awschalom, D. D., DiVincenzo, D. P. & Loss, D. 1997 *Science* **272**, 425.

Golyshev, V. Yu & Popkov, A. F. 1995 *Europhys. Lett.* **29**, 327.

Gunther, L. & Barbara, B. (eds) 1995 *Quantum tunneling of magnetization—QTM 94.* Wiley.

Halperin, B. I., Lee, P. A. & Read, N. 1993 *Phys. Rev.* B **47**, 7312.

Imry, Y. 1977 *Introduction to mesoscopic physics.* Oxford University Press.

Jain, J. K. 1989 *Phys. Rev. Lett.* **63**, 199.

Julien, M. H., Jang, Z. H., Lascialfari, A., Borsa, F., Horvatic, M., Caneschi, A. & Gatteschi, D. 1999 *Phys. Rev. Lett.* **83**, 227.

Langer, J. J. 1967 *Ann. Phys.* **41**, 108.

Lis, T. 1980 *Acta Cryst.* B **36**, 2042.

Loss, D., DiVincenzo, D. P. & Grinstein, G. 1992 *Phys. Rev. Lett.* **69**, 3232.

Lubashevsky, I. A. (and 12 others) 1999 (Submitted.)

Luis, F., Bartolome, J. & Fernandez, J. F. 1998 *Phys. Rev.* B **57**, 1.

McCusker, J. K., Schmitt, E. A. & Hendrickson, D. N. 1991 In *Magnetic molecular materials* (ed. D. Gatteschi, O. Kahn, J. Miller & F. Palacio). NATO-ASI Series, vol. E198. Kluwer.

Morrish, R. 1966 *The physical principles of magnetism.* Wiley.

Prokof'ev, N. V. & Stamp, P. C. E. 1996 *J. Low Temp. Phys.* **104**, 143.

Rentschler, E., Gatteschi, D., Cornia, A., Fabretti, A. C., Barra, A.-L., Shchegolikhina, O. I. & Zhdanov, A. A. 1996 *Inorg. Chem.* **35**, 4427.

Sangregorio, C., Ohm, T., Paulsen, C., Sessoli, R. & Gatteschi, D. 1997 *Phys. Rev. Lett.* **78**, 4645.

Sessoli, R., Gatteschi, D., Caneschi, A. & Novak, M. 1993*a Nature* **356**, 141.

Sessoli, R., Tsai, H. L., Schake, A. R., Wang, S., Vincent, J. B., Folting, K., Gatteschi, D., Christou, G. & Hendrickson, D. N. 1993*b J. Am. Chem. Soc.* **115**, 1804.

Stamp, P. C. E. 1996 *Nature* **383**, 125.

Taft, K. L. & Lippard, S. J. 1990 *J. Am. Chem. Soc.* **112**, 9629.

Taft, K. L., Delfs, C. D., Papaefthymiou, G. C., Foner, S., Gatteschi, D. & Lippard, S. J. 1994 *J. Am. Chem. Soc.* **116**, 823.

Tejada, J. 1995 *Science* **272**, 425.

Thomas, L., Lionti, F., Ballou, R., Gatteschi, D., Sessoli, R. & Barbara, B. 1996 *Nature* **383**, 145.

Watton, S. P., Fuhrmann, P., Pence, L. E., Caneschi, A., Cornia, A., Abbatti, G. L. & Lippard, S. J. (eds) 1997 *Angew. Chem. Int. Ed. Engl.* **36**, 2774.

Wernsdorfer, W. & Sessoli, R. 1999 *Science* **284**, 133.

Wieghardt, K., Pohl, K., Jibril, I. & Huttner, G. 1984 *Angew. Chem. Int.* **23**, 77.

Zvezdin, A. K. 1995 In *Handbook of magnetic materials* (ed. K. H. J. Buschow), vol. 9, p. 405. Elsevier.

Large metal clusters and lattices with analogues to biology

By Daniel J. Price[1], Frederic Lionti[2], Rafik Ballou[2],
Paul T. Wood[1] and Annie K. Powell[1]

[1] School of Chemical Sciences, University of East Anglia, Norwich NR4 7TJ, UK
[2] Laboratoire Louis Néel, CNRS, 25 Avenue des Martyrs BP166,
38042 Grenoble Cedex 9, France

The magnetic properties of two types of material derived from the brucite lattice, $M(OH)_2$ are described. The structure-directing effects of simple templates on the brucite lattice parallels the processes seen in nature in formation of biominerals. The first type exemplified by Fe_{17}/Fe_{19} aggregates models the structural features of the iron storage protein ferritin. The magnetic behaviour also reveals some interesting parallels with the magnetic phenomena reported previously on ferritins. We have used a combination of experimental techniques including DC, AC and RF susceptibility measurements on powders and a microSQUID on single crystals. The second type is based on an extended-defect brucite structure with stoichiometry $M_2(OH)_2(ox)$ for $M = Fe^{2+}$, Co^{2+}. They show long-range ordering to antiferromagnetic phases and then, at much lower temperatures, undergo a phase transition to canted antiferromagnetic states. Symmetry arguments are used to predict the spin configurations in these extended materials.

Keywords: magnetism; biominerals; clusters;
hydrothermal synthesis; coordination chemistry

1. Introduction

The oxide, oxyhydroxide and hydroxide phases of paramagnetic transition-metal ions can display a variety of cooperative magnetic phenomena, such as ferromagnetism, ferrimagnetism and spin-canted antiferromagnetism, which are often useful in technological applications. The iron-containing phases prove to be particularly versatile and the chemistry and physical properties of these materials are relatively well understood (Cornell & Schwertmann 1996). The hydrolysis of iron(III) is the starting point for the formation of all the iron oxides, oxyhydroxides and hydroxides (Cornell & Schwertmann 1996; Wells 1962). In recognition of the fact that biological systems are able to manipulate the phase and form of such iron minerals (Frankel & Blakemore 1990), we have been exploring ways to achieve similar control *in vitro* by modifying the natural hydrolysis of iron(III) ions. The mineral products of iron hydrolysis adopt structures based on cubic and hexagonal lattices, but the two types can be interchanged relatively easily (Wells 1962). The initial products of both iron(III) and iron(II) hydrolysis are best described as materials adopting structures

† Present address: Institut für Anorganische Chemie, Universität Karlsruhe, Engesserstr. Geb. 30.45, D76128 Karlsruhe, Germany.

$${\{FeL_m\}^{\pm q} + xsH_2O \rightleftharpoons \{Fe_xL_n(O)_y(OH)_z(H_2O)_p\}^{\pm a} \rightleftharpoons \text{`Fe(OH)}_3\text{'}.}$$

Scheme 1.

based on hexagonal close packing. In the case of iron(II), $Fe(OH)_2$ forms initially, which has the brucite structure (typified by brucite, $Mg(OH)_2$), whereas for iron(III), the more poorly defined mineral ferrihydrite is the first identifiable phase, which can be formulated in various ways and is often denoted simply by 'Fe(OH)$_3$'. The formulation of this as $\{(Fe_2O_3)(2Fe \cdot O \cdot OH)(2.6H_2O)\}$ (Russell 1979) underlines the observation that the mineral is most likely to have a defect hexagonal close packed structure (Cornell & Schwertmann 1996; Towe & Bradley 1967) related to the better-defined hexagonal close packed oxyhydroxide, goethite, α-Fe$\cdot$O$\cdot$OH, and oxide, haematite, α-Fe$_2$O$_3$, which are the phases ferrihydrite evolves into on ageing (Cornell & Schwertmann 1996; Wells 1962). Ferrihydrite phases are often nanoparticulate, further complicating their characterization. In our research to date, we have found that modification of iron hydrolysis leads to materials based on these hexagonal close-packed (HCP) structures. It is, therefore, helpful to describe how the structures of the three well-defined phases brucite, goethite and haematite are related.

The brucite structure, $M(OH)_2$, adopts an AX_2 HCP structure related to CdI_2. It has a two-dimensional layer structure with hexagonally close packed oxygen atoms as double strips of OH^- with M(II) in octahedral holes. The hydroxide ions bridge across three metal centres. In three dimensions the layers stack up to optimize hydrogen bonding between the layers. The related structure adopted by M(III) ions, typified by goethite, is also an AX_2 structure, but, in order to preserve charge balance, there is only half the number of protons in the stoichiometry: $M \cdot O \cdot OH$. The oxygen atoms are again in an HCP arrangement with M(III) in octahedral holes, but now the protons are shared between layers in the three-dimensional structure such that the compound could also be formulated as '$M(OH_{0.5})_2$'. The oxide structure exemplified by haematite, α-Fe$_2$O$_3$, is an A_2X_3 structure, again with HCP oxygen atoms and M(III) in octahedral holes. The oxides bridge between four metal centres, the relationship to the layer structures of brucite and goethite being that a μ_4-O connects three metal ions from one layer to one in the next layer up, and so on, to give a truly three-dimensional network.

In this article we describe the structures and magnetic properties of three materials synthesized by us which can be regarded as resulting from hydrolysis reactions of transition-metal ions that have been modified by the presence of templating species and control over crystallization conditions to produce engineered metal hydroxide frameworks. The synthetic strategy is to provide a templating species, L, which enters the coordination sphere of the metal ion and in some way affects the nature of the resulting hydrolysis product, through the production of an intermediate material $\{M_xL_n(O)_y(OH)_z(H_2O)_p\}^{\pm a}$, as indicated for iron(III) in scheme 1. It is believed that this is one of the ways in which biology is able to tailor mineral species for specific functions through the process of biomineralization (Mann *et al.* 1989). We have therefore been exploring analogous interactions *in vitro* in order to produce new materials with tailored properties.

The intermediate species $M_xL_n(O)_y(OH)_z(H_2O)_p\}^{\pm a}$ may have zero values for some of the components, but must have some oxygen atoms derived from coordinated water molecules present. The compounds we describe here can be understood in terms of this intermediate material and derived from the brucite mineral structure. Firstly, we describe the situation where chelating ligands, L, can be used to encapsulate portions of the $M(OH)_2$ framework giving particles with large boundary effects and unusual magnetic properties that result from these. In general, these particles can be described in terms of $\{Fe_xL_n(O)_y(OH)_z(H_2O)_p\}^{\pm a}$, where the parameters x, n, y, z and p all have fairly small finite values, as in the case of $\{Fe_{19}L_{10}(O)_6(OH)_{14}(H_2O)_{12}\}^+$, where $L = N(CH_2COO)_2(CH_2CH_2O)^{3-}$ (see, for example, Heath & Powell 1992). Secondly, we describe the situation where the ligand, L, becomes incorporated into the brucite lattice to give a modified structure that can be regarded as a 'defect' brucite network. These extended phases can also be related to the $\{M_xL_n(O)_y(OH)_z(H_2O)_p\}^{\pm a}$ formulation, but now the overall structures are infinite and the stoichiometries are correspondingly simpler, as in the case of $\{Fe_2(L)(OH)_2\}$ where $L = C_2O_4^{2-}$ (Molinier *et al.* 1997). The lowering of the overall symmetry of the brucite lattice through these inclusions results in materials that display spin-canting phenomena.

2. Experimental

The Fe_{19} and Fe_{17} aggregates were synthesized under ambient conditions as described previously (Heath & Powell 1992; Powell *et al.* 1995). Diiron(II) dihydroxyoxalate was synthesized under hydrothermal conditions as described previously (Molinier *et al.* 1997). The sample was ground and a bar magnet was used to remove most of the ferrimagnetic magnetite impurity. Dicobalt(II) dihydroxyoxalate was also synthesized hydrothermally. A fresh cobalt hydroxide precipitate was formed by the addition of NaOH to an aqueous solution of $CoCl_2$. To this, $Na_2C_2O_4$ was added and the reaction heated to 220 °C for 2 days (Gutschke *et al.* 1999*a*).

Magnetic measurements were performed on a Quantum Design MPMS SQUID magnetometer. A few mg of each of the powdered samples were pressed into a gelatine capsule mounted onto a polyethylene straw. Field-cooled magnetization measurements were performed in longitudinally applied fields of 100 and 1000 G, over a temperature range of 2–300 K. The contribution of magnetite impurities to the iron-containing sample could not be deconvoluted, limiting the quantitative usefulness of these data. However, no such problem was encountered with the cobalt analogue. For this sample, a correction for the diamagnetic contribution to the susceptibility of -73×10^{-6} cm^3 mol^{-1} was calculated from Pascal's constants (see, for example, Kahn 1993) was applied.

The static and dynamic magnetic behaviour of the Fe_{19} and Fe_{17} aggregates were investigated in more detail at Grenoble on powdered samples using home-built AC-susceptometers and RF SQUID magnetometers equipped with dilution fridges and on a small single crystal using microSQUID devices (Wernsdorfer 1996). The real and imaginary parts of the longitudinal AC-susceptibility were measured on the powdered samples at frequencies as low as 0.005 Hz and up to 12.5 kHz in a temperature range from 50 mK up to 40 K. Some AC-susceptibility measurements were performed by superposing a static magnetic field up to 1.5 T on the weak alternating magnetic field (0.1 mT). Magnetization measurements on powdered samples could be extended to temperatures as low as 50 mK and in a range of applied magnetic fields from -5 T to 5 T. Some magnetic relaxation measurements on powdered samples were also performed. More relevant, however, were the magnetic relaxation measurements

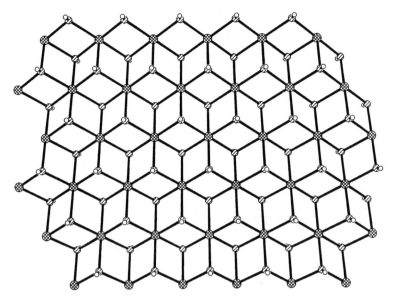

Figure 1. The layer structure of brucite showing metal ions held together in a regular triangular lattice by μ_3 bridging hydroxides.

performed on the single crystal using the microSQUID devices. For these, the applied magnetic field is limited to about 1 T but can be applied in any direction of space and, in particular, measurements can be performed under a main longitudinal field with a small transverse component. Moreover, the magnetic-field strength can be varied at a high time rate and the magnetization can be recorded with a high resolution in time, allowing clean and efficient protocols for the magnetic relaxation measurements. Working with a dilution fridge, the microSQUID data were collected within the temperature range from 30 mK to 5 K.

3. Results and discussion

(a) Materials containing trapped finite portions of the brucite lattice

Although we have been able to obtain a variety of species of general formula

$$\{M_xL_n(O)_y(OH)_z(H_2O)_p\}^{\pm a}$$

for M = Fe(III), Al(III) using the general synthetic method outlined above (Heath 1992; Womack 1999; Heath *et al.* 1995; Baissa *et al.* 1999), the most extensively studied material in terms of its magnetic behaviour is the solid formed with the ligand hydroxyethyliminodiacetic acid, $N(CH_2COOH)_2(CH_2CH_2OH) = H_3$heidi, which contains interpenetrating lattices of aggregates of 19

$$\{Fe_{19}L_{10}(O)_6(OH)_{14}(H_2O)_{12}\}^{+},$$

and 17

$$\{Fe_{17}L_8(O)_4(OH)_{16}(H_2O)_{12}\}^{3+},$$

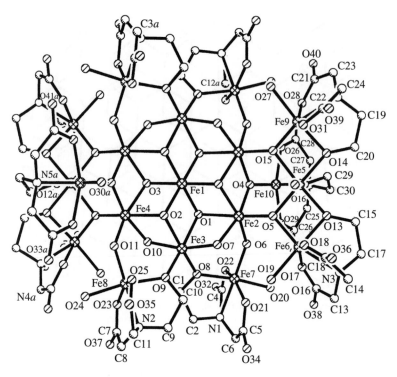

Figure 2. The molecular structure of the Fe$_{19}$ aggregate.

iron(III) centres (Heath & Powell 1992; Powell *et al.* 1995). The fact that the two aggregates co-crystallize complicates the interpretation of the bulk magnetism, but initial magnetic studies on this material indicated that the aggregates could be regarded as particles stabilizing high ground state spins with at least one of the clusters carrying a spin in the region of 33/2 (Powell *et al.* 1995). Since this material has been structurally characterized using single-crystal X-ray diffraction we were able to suggest possible pathways for the stabilization of the high ground-state spin. The aggregates both contain cores corresponding to a brucite lattice, except, of course, the metal is in the +3 oxidation state, $\{M(OH)_2\}^+$. This can readily be seen by comparing the brucite lattice (figure 1) with the structure of the Fe$_{19}$ aggregate (figure 2). The encapsulating peripheral iron oxide, hydroxide and ligand units serve to compensate for the build-up of positive charge on the core to give aggregates carrying rather small overall charges. This, in turn, creates a situation rather like that seen in the iron storage protein ferritin (figure 3), where an iron mineral akin to ferrihydrite is encapsulated in a shell of nucleating iron centres and protein ligands (Ford *et al.* 1984). We have shown previously that these smaller aggregates make excellent 'scale models' for loaded ferritin (Heath *et al.* 1996; Powell 1997). The structural parallels between loaded ferritin and the cluster models seem to carry over into the types of magnetic behaviour observed. It is reasonable to expect a finite aggregate to display strong boundary effects, particularly when the particle size falls within the nanoscale regime. These can be under-stood in terms of large surface effects giving rise to uncompensated spins (Le Brun

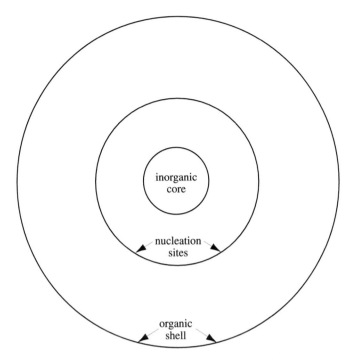

Figure 3. Representation of the structural elements of loaded ferritins.

et al. 1997, and references therein; Tejada & Zhang 1994). In the case of the Fe_{19} and Fe_{17} aggregates it is possible to propose models to explain the observed spin states. These allow for the fact that the exchange pathways amongst the iron(III) centres are different in different regions of the aggregate, also as a consequence of the boundary around the particle. In this vein, it is interesting to compare some recent ultra-low-temperature magnetic results obtained on the Fe_{19}/Fe_{17} system with some of the magnetic properties reported on both superparamagnetic molecular particles, such as $[Mn_{12}O_{12}(CH_3COO)_{16}(H_2O)_4] \cdot 2CH_3COOH \cdot 4H_2O$, commonly abbreviated to $Mn_{12}Ac$, and loaded ferritins. The former can be understood in terms of anisotropic 'single molecule magnets' of $S = 10$ displaying resonant quantum tunnelling effects (Thomas *et al.* 1996; Friedman *et al.* 1996), while the latter are examples of polydisperse systems, and, thus, the magnetic data are correspondingly harder to interpret (Awschalom *et al.* 1992). The possibility that these types of materials can be used in applications such as information storage and in quantum computing, for example, in addition to shedding light on the fundamental questions regarding the link between the molecular and the infinite, and the boundary between the quantum and macroscopic worlds, has been pointed out by several commentators (Chudvonsky 1996; Stamp 1996; Gatteschi *et al.* 1994).

The observation that $Mn_{12}Ac$ displays stepped hysteresis phenomena at temperatures where the molecules are expected to be in their ground state with a total spin per molecule of $S = 10$ has been explained in terms of resonant tunnelling between intermediate levels. Further experiments on orientated single crystals using AC susceptibility measurements (Thomas *et al.* 1996; Friedman *et al.* 1996) and also

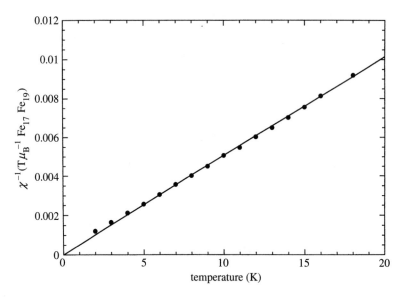

Figure 4. Plot of reciprocal susceptibility versus temperature for a powdered sample of Fe₁₇/Fe₁₉ showing Curie behaviour between 5 and 18 K.

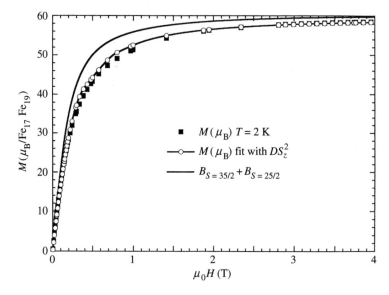

Figure 5. Magnetization isotherm at 2 K for a powdered sample of Fe_{17}/Fe_{19} (filled squares). The upper curve shows the Brillouin function for two spins of $S_1 = 35/2$ and $S_2 = 25/2$, the lower curve, with open circles, corresponds to the fit incorporating an anisotropic energy barrier.

MCD experiments on powders are in line with the bistability of this system and its superparamagnetic behaviour (Cheesman *et al.* 1997). Although there is still much debate regarding the underlying physics of these phenomena, the system itself represents a well-behaved vehicle for investigation, since it forms what can be regarded

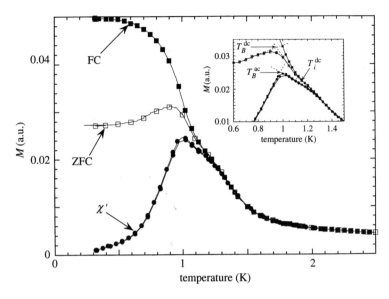

Figure 6. Thermal variation of the real part of the AC-susceptibility, zero-field-cooled and field-cooled magnetizations measured on a powdered sample of Fe_{17}/Fe_{19} showing the double-blocking behaviour.

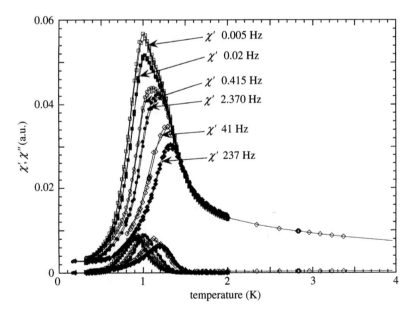

Figure 7. Thermal variation of the real and imaginary components of the AC susceptibility in the region of the blocking temperature for various driving frequencies measured on a powdered sample of Fe_{17}/Fe_{19}.

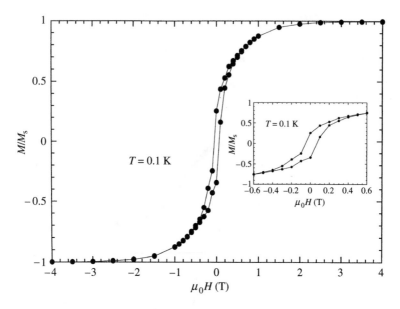

Figure 8. The magnetization isotherm of a powdered sample of Fe_{17}/Fe_{19} at 0.1 K showing a weak hysteresis.

as monodisperse highly orientated particles within a well-ordered crystal lattice with a readily identifiable easy axis of magnetization. On the other hand, materials such as loaded ferritin molecules are much harder to study. On average, a loaded ferritin molecule could contain about 4000 iron(III) centres (Ford *et al.* 1984), and the dimensions of the iron oxyhydroxide cores of these molecules correspond to spherical particles about 7 or 8 nm in diameter. Magnetic measurements such as Mössbauer spectroscopy have revealed that loaded ferritins from mammalian sources display superparamagnetism with blocking temperatures as high as 60 K. There is also evidence that such molecules display hysteresis phenomena and, from susceptibility measurements, that there could be resonant tunnelling in operation as well (Awschalom *et al.* 1992; Tejada *et al.* 1996). However, the situation here is very different from that in the $Mn_{12}Ac$ system. Although the description of both as superparamagnetic is reasonable, in the case of the loaded ferritins it becomes impossible to assign a single large spin to each molecule. Indeed, it might even be the case that, within the protein shell of a given ferritin, there are several particles occupying the 7–8 nm cavity, complicating matters even further. In any case, assuming, for simplification, that each ferritin contains a single 7–8 nm diameter particle, we can only proceed by assigning an average overall spin to the system; in other words, we are dealing with a polydisperse material.

The Fe_{17}/Fe_{19} system we describe here lies between these two examples. While the detailed molecular and crystal structure of the compound is known, as in the case of $Mn_{12}Ac$, the situation is complicated by the fact that there are two different clusters, the least-squares planes of the iron oxyhydroxide cores of which are orientated at 28° to each other. This can thus be regarded as a system in which orientation can be achieved using single crystals, but it is not as simple as in the $Mn_{12}Ac$ case since there are two sets of magnetically distinct molecules that are likely to carry different total

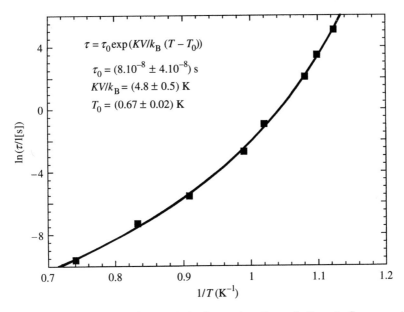

Figure 9. Logarithm of the characteristic fluctuation time of S_1 and S_2 versus inverse temperature for a powdered sample of Fe_{17}/Fe_{19}. The non-Arrhenius behaviour is indicative of a weak intercluster interaction.

spin values and for which the easy axis of magnetization is likely to lie along different vectors for each type of aggregate.

We have found that, at very low temperatures, this material displays hysteresis phenomena that bear resemblance to the reported data for loaded ferritin cores (Awschalom *et al.* 1992; Tejada *et al.* 1996). We do not yet fully understand the physical origins of the effects that we have measured, but we can propose some possible interpretations of the data, which could give some insight into the phenomena observed for loaded ferritins.

From our new data measured on both powdered samples and orientated single crystals, we find that the system can be described as containing two magnetically distinct clusters, one of spin 25/2 and the other of spin 35/2. This conclusion is drawn from the graph of reciprocal DC susceptibility versus temperature, which obeys a Curie law over the range 4–18 K fitted for a Curie constant, C, of $C = g^2[S_1(S_1+1) + S_2(S_2+1)]$, giving an excellent fit for $g = 2.00$, $S_1 = 35/2$, $S_2 = 25/2$ (figure 4). This is supported by the magnetization data with respect to B/T (B is the Brillouin function for the spin, S) between 5 and 20 K, which fits well for $B(S = 35/2) + B(S = 25/2)$. The difference in total spin for the two clusters equates to the removal or addition of two Fe(III) high-spin ions of $S = 5/2$, which fits with the difference in molecular composition between the two aggregates. Thus we can consider the system at these temperatures as comprising sets of aggregates with overall spins of 35/2 and 25/2. Below 4 K, the scaling law with respect to B/T is no longer obeyed due to the emergence of magnetocrystalline anisotropy. A fit of the curves (figure 5) gives an anisotropy barrier of 4.8 K for $S = 25/2$ and 9.4 K for $S = 35/2$. The temperature dependence of the hysteresis shows a superparamagnetic blocking of the spins around 1 K, and the thermal variation of the real and imaginary

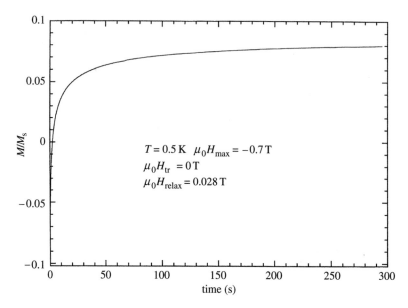

Figure 10. Temporal variation of the magnetization on a single crystal of Fe_{17}/Fe_{19} showing equilibrium magnetization, M_{eq}, is reached in about 300 s.

part of the AC susceptibility around this blocking temperature (figures 6 and 7) clearly shows a double-blocking behaviour, in agreement with the differences of the energy barriers for the two overall spins. The magnetization isotherm measured below the blocking temperature gives a weak hysteresis loop with qualitative similarities to hysteresis loops reported for loaded ferritin cores (figure 8). It is likely that there are weak intercluster interactions, as indicated by the fact that a plot of the logarithm of the fluctuation time of the spins with respect to inverse temperature does not obey an Arrhenius law (figure 9) (see, for example, Dormann *et al.* 1988). From figure 9 it can be seen that we can estimate that this interaction becomes significant at about $T = 0.67$ K.

It was also possible to measure data on a microSQUID. Since it is not possible to measure data on such an instrument on absolute scales, it is necessary to give data in terms of relative magnetization, M/M_s, where M_s is the saturation magnetization. These measurements confirmed the scaling law with respect to B/T and that this law is not obeyed below 5 K. Single-crystal studies allow relaxation phenomena to be investigated and it was found that the time to reach equilibrium of the magnetic relaxation, t_{max}, was about 300 s (figure 10). Further measurements reveal that the relaxation processes are very complex and that the measurements are difficult to interpret. The relaxation processes were investigated by measuring the variation of M_{eq}, defined as

$$M_{eq} = \tfrac{1}{2}[M(t_{max}, H_{max} \to H_{relax}) + M(t_{max}, -H_{max} \to -H_{relax})].$$

The temporal variation of $\Delta M(t)$, where $\Delta M(t) = M(t) - M_{eq}$, is more complex than is usually observed and does not follow exponential, stretched exponential or even logarithmic behaviour as shown in figure 11. A relaxation protocol was used in which the sample was initially heated from 5 K with a field of -0.7 T applied, which aligns

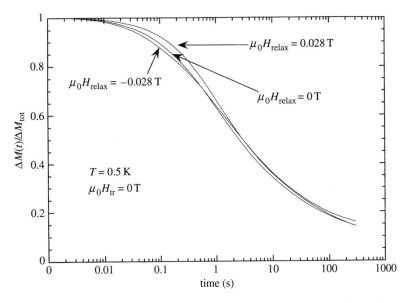

Figure 11. Temporal variation of $\Delta M(t)$ (see text) on a single crystal of Fe_{17}/Fe_{19}. The complex relaxation processes of this system are evident from the inability to fit these curves using exponential, stretched exponential or even logarithmic forms.

all the $S_1 = 35/2$ and $S_2 = 25/2$ along the field. The sample was then cooled under this field. At the given measuring temperature, the field was switched to the measuring field, H_{meas}, at a sweeping rate of $0.14\ T\ s^{-1}$. Our present analysis of the observed relaxation behaviour suggests that when H_{meas} is of the order of the molecular field, $H_{int} = -0.1\ T$, describing the antiferromagnetic interactions between the two spins, the applied field can no longer maintain the parallel alignment of these spins and $S_2 = 25/2$ tends to reverse its orientation by overcoming the anisotropy barrier of 4.8 K either thermally or by tunnelling. When H_{meas} is close to zero, the two spins are antiparallel and to reverse the orientation they have to overcome the sum of the two anisotropy barriers again either thermally or through tunnelling. Conversely, when H_{meas} is of the order of $H_{int} = 0.1\ T$, it is the spin $S_1 = 35/2$ that will tend to align parallel to the field and must overcome the associated anisotropy barrier.

Clearly, the Fe_{17}/Fe_{19} system has a much more complicated magnetic behaviour than the $Mn_{12}Ac$ system and we are still far from understanding the intricacies of the relaxation phenomena. However, comparing these results with some of those reported on ferritins, we can see that an understanding of the data we have measured could give useful insights into the physics of the polydisperse ferritin systems as well as other systems of polydisperse antiferromagnetically coupled particles.

(b) Extended-defect brucite structures

The hydrolysis of iron(III) or cobalt(II) salts followed by hydrothermal reaction conditions in the presence of the oxalate ligand leads to infinite structures incorporating the oxalate ion into a defect brucite lattice as $M_2(OH)_2(ox)$. As we have reported previously, in the case of iron(III), this reaction results in the stabilization of iron(II), probably as a result of the modification of pH and redox potential under

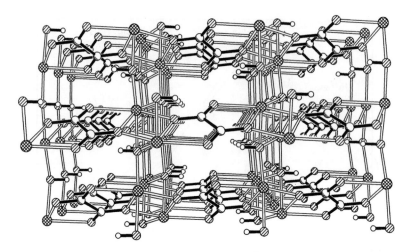

Figure 12. Packing diagram of $M_2(OH)_2(ox)$ showing the two-dimensional character of the M(OH) layers held apart by oxalato anions.

hydrothermal conditions (Molinier *et al.* 1997). It was found that using iron(II) starting materials only led to the formation of the mineral humboldtine, $Fe(ox)(H_2O)_2$, or other iron(II) minerals such as siderite, $FeCO_3$. The corresponding cobalt(II) compound, $Co_2(OH)_2(ox)$, is prepared from cobalt(II) starting materials without difficulty. $Co_2(OH)_2(ox)$ and $Fe_2(OH)_2(ox)$ are isostructural with layers of metal hydroxide connected by oxalate anions (figure 12). The metal hydroxide layers adopt a structure related to that of brucite (figure 1), except that pairs of μ_3-OH bridges are replaced by oxygen atoms from the oxalate ligands (figure 13) and the overall framework is distorted from the idealized triangular lattice of brucite (figure 14).

The magnetic behaviour of both compounds reflects this change in spin topology through the introduction of defects and distortion to the brucite structure. For the iron(II) compound, the plot of magnetization versus temperature is complicated by the presence of tiny amounts of magnetite, which are always present in the synthesis of this material. The fact that the curve is only displaced by about 11 G cm^3 mol^{-1} indicates that this impurity is at a very low level. Looking at the shape of the curve, we can identify two cusps at 120 K and 90 K and the onset of a spontaneous magnetization at 12 K (figure 15). The first and largest cusp at 120 K is due to the Vewery transition in the ferrimagnetic magnetite impurity (see, for example, Chikazumi 1964), but the second cusp at 90 K marks the onset of antiferromagnetic ordering in $Fe_2(OH)_2(ox)$. On further cooling, the rapid increase in magnetization at 12 K is indicative of spontaneous magnetization, but the absolute value attained is only a tiny fraction of what would be expected for a parallel, ferromagnetic arrangement. The comparison with the situation for the cobalt(II) compound (figure 15) shows, qualitatively, the same thermally dependent features. In this case, there is a single cusp reaching a maximum at 84 K with the onset of spontaneous magnetization at 32 K. Again, the absolute value of this magnetization is much smaller than would be expected for a parallel spin arrangement. The qualitative similarities of the magnetic behaviour of the Fe(II) and Co(II) systems allow us to suggest the following interpretation of the magnetism for both by analysing the data for the Co(II) system further.

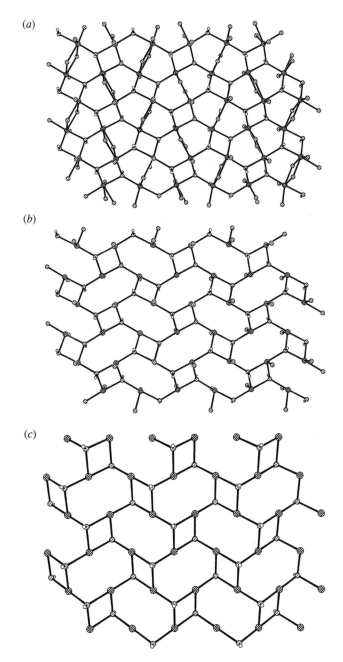

Figure 13. (*a*) Layer of $M_2(OH)_2(ox)$ showing how metal atoms are bridged by μ_3-hydroxide and μ_2 bridging oxalate oxygen atoms. (*b*) The same layer showing the hydroxy bridging only. (*c*) The equivalent structure formed by removing selected pairs of bridging hydroxides from the brucite lattice.

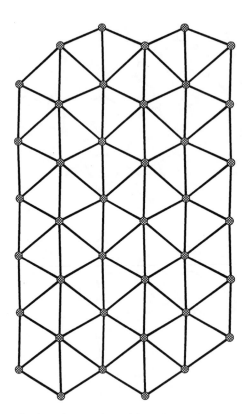

Figure 14. The distorted triangular net formed by a layer of the spin carrying metal ions in $M_2(OH)_2(ox)$. The bonds represent nearest-neighbour interactions for which at least one single bridging atom superexchange pathway exists.

The plot of reciprocal susceptibility versus temperature for $Co_2(OH)_2(ox)$ (figure 16) in the paramagnetic region (above 84 K) follows a Curie–Weiss law and can be extrapolated to give a negative intercept of $\vartheta = -117$ K, supporting the suggestion that there is a dominant antiferromagnetic interaction. This, coupled with the small absolute values for the spontaneous magnetization observed in the low temperature phase, leads to the conclusion that antiferromagnetically ordered spin canting is responsible for this. We can understand the antiferromagnetic ordering from the details of the single crystal X-ray structure analysis (Gutschke *et al.* 1999*a*), which shows that within a given layer of the structure there are six Co(II) neighbours linked through single-atom superexchange pathways at distances ranging between 3.18 and 3.69 Å. Four of these neighbours are linked by single μ_3-hydroxide oxygen atoms, a fifth is linked by two such μ_3-hydroxide oxygen atoms, and the sixth by two μ_2-oxalate oxygen atoms. The magnetic ions in each layer are connected to magnetic ions of neighbouring layers *via trans* oxalate bridges, giving a three-atom super-exchange pathway and placing the metal ions from adjoining layers at 5.759(5) Å (figure 12). Although, from a structural point of view, we might expect the material to have a two-dimensional character, it is evident from the phase transition to an ordered antiferromagnetic state that there is significant exchange between layers.

From the Néel temperature, T_N, we can estimate an average exchange interaction,

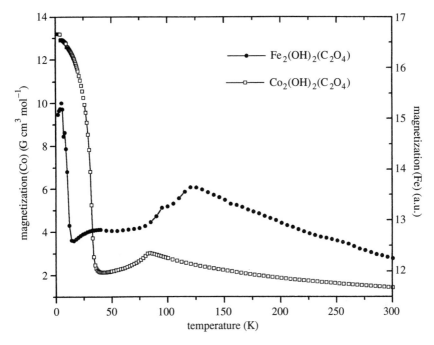

Figure 15. The temperature dependence of the magnetization curves for $Fe_2(OH)_2(ox)$ and $Co_2(OH)_2(ox)$. The displacement of the curve for $Fe_2(OH)_2(ox)$ results from the small amount of magnetite impurity (see text).

J, of -4.6 cm^{-1} using molecular field theory. This simplistic approach makes no distinction between intra- and interlayer couplings, so that, as in many other materials, the inadequacy of molecular field theory is revealed by the inequivalence of $|\vartheta|$ (117 K) and T_N (84 K).

A qualitative description of the situation can be used to help understand the magnetism of these $M_2(OH)_2(ox)$ phases. The transition from a paramagnetic state to one of long-range order for a given structure is determined principally by the strength of spin–spin interactions, namely, the coupling, J. Since T_N for the Fe(II) compound is greater than T_N for the Co(II) compound (90 K versus 84 K), the absolute value of J for the Fe(II) must be somewhat greater than that for Co(II) and also more negative as a result of stronger antiferromagnetic coupling. On the other hand, the onset of canted antiferromagnetic behaviour requires not only an exchange coupling term, but also a term favouring a non-collinear spin configuration. This can be due to non-collinear single-ion anisotropies or to an antisymmetric exchange term. Both mechanisms will contribute to an energy term, D, favouring a non-collinear spin arrangement. If D were much larger than J, then we would expect the transition from paramagnet to canted antiferromagnet to occur directly, without the appearance of the intermediate antiferromagnetic phase as the temperature is decreased (Gutschke *et al.* 1999*b*). Conversely, if D is much smaller than J, then we would expect to see the phase transition sequence on cooling follow the order observed here: paramagnet, antiferromagnet, canted antiferromagnet. Thus, we can assign a larger absolute value for J than for D. Furthermore, since the onset of the spin-canted phase occurs at a

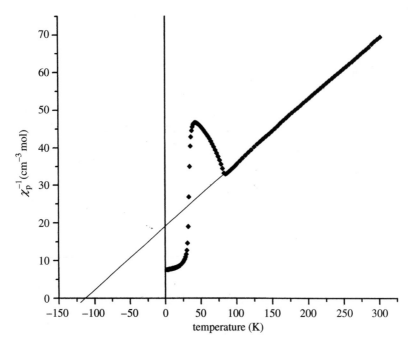

Figure 16. Reciprocal susceptibility as a function of temperature for $Co_2(OH)_2(ox)$ showing the good fit of the data in the paramagnetic region above 84 K to Curie–Weiss behaviour. The large negative intercept on the ordinate corresponds to the Weiss constant, θ, of -117 K.

higher temperature for the Co(II) system than for the Fe(II) one, we can also assign a larger absolute value of D for the Co(II) case.

Spin canting can be described phenomenologically by a term $\boldsymbol{d} \cdot \boldsymbol{S}_1 \times \boldsymbol{S}_2$ in the spin Hamiltonian. There are two distinct mechanisms by which such a term can stabilize a non-collinear spin arrangement, either through single-ion anisotropy or by antisymmetric exchange interactions (Moriya 1963; Carlin 1986). For any magnetic phase with a spontaneous magnetization, very stringent symmetry requirements must be met, namely, that the magnetic symmetry element $\bar{I}'$ must be absent from the magnetic space group of the phase. If this element is present there can be no net magnetization, thus, certain magnetic space groups are incompatible with ferromagnetic, ferrimagnetic and canted antiferromagnetic phases (Briss 1964; Joshua 1991; Opechowski & Guccione 1965). This incompatibility is easily understood, since the $\bar{I}'$ element relates two spin sites by a crystallographic inversion centre where the moments are aligned exactly antiparallel and spin magnetic moments strictly cancel. The magnetic phase transition is second order and, thus, the magnetic space group of a canted antiferromagnet and of the underlying antiferromagnetic structure must be the same. Hence, we need to find the simple antiferromagnetic structures for which $\bar{I}'$ is not a symmetry element.

$Fe_2(OH)_2ox$ and $Co_2(OH)_2ox$ are isomorphous and isostructural, crystallizing in the monoclinic space group $P2_1/c$. The four crystallographically equivalent metal ions generated by the symmetry elements of the $P2_1/c$ space group within the chemical unit cell each have seven near neighbours. However, only three independent couplings

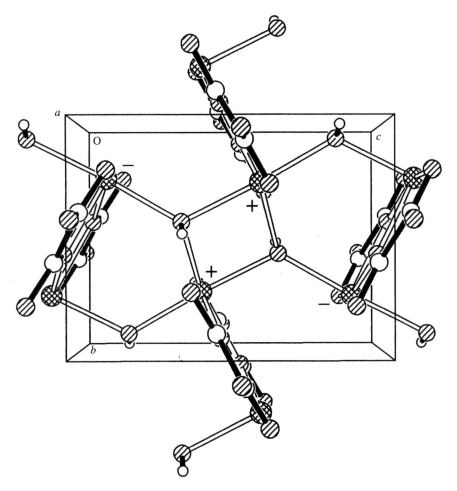

Figure 17. The proposed magnetic unit cell and spin configuration for the ordered antiferromagnetic states of $M_2(OH)_2(ox)$ (+ indicates spin-up, − indicates spin-down).

are necessary to describe the spin arrangement throughout the crystal. For the most likely arrangement of the spins according to a simple two-sublattice model, there are eight possible different spin configurations for the three principal couplings relating parallel (ferromagnetically coupled) or antiparallel (antiferromagnetically coupled) moments. Four of these correspond to magnetic unit cells that are the same size as the crystallographic chemical unit cells, with the other four corresponding to magnetic unit cells that are doubled along the *a*-axis. The magnetic space groups for each of these eight spin configurations were determined for many orientations of the spin vectors (e.g. with the sublattice magnetizations parallel to the lattice vectors **a**, **ab**, **b**, **bc**, **c**, **ac** and **abc**). Only one spin configuration (figure 17) corresponded to a set of magnetic space groups for which the $\bar{I}'$ element was absent. These were $P2_1/c$ for the sublattice magnetizations parallel to the *b*-axis, $P2_1'/c'$ if the spin vectors are confined to the *ac*-plane, and $P\bar{I}$ otherwise.

Neutron diffraction studies are planned to confirm the spin configurations and to determine the orientation of the Néel vector in the hydroxyoxalate phases. It is interesting that the very stringent requirements of canted antiferromagnetism have, in this case, enabled us to determine the probable spin configuration of the canted and antiferromagnetic phases from logical symmetry considerations. Although, to a first approximation, these materials are topologically stacked triangular lattices, the geometric distortion and consequent reduction in symmetry causes a simplification of the magnetic phase behaviour compared with the related binary divalent transition-metal hydroxides and halides (Day 1988; Aruga Katori *et al.* 1996; de Jongh 1990). We have reported elsewhere that when the hydrothermal syntheses that produce $Fe_2(OH)_2(ox)$ are allowed to proceed further or with slightly altered reaction protocols (Molinier *et al.* 1997), then the iron oxide phases of magnetite and haematite can result. Indeed, we found that the magnetite we observed displayed rhombic dodecahedral crystal morphology rather than the more usual octahedral or cubic ones. We were able to show that this was probably the result of the inclusion of the oxalate template in the $Fe_2(OH)_2(ox)$ precursor acting as a growth inhibitor. From this, we can gain insights into the way in which templating species can direct the form and function of extended phases.

4. Conclusions

The two types of system reported here represent two ways of manipulating the brucite lattice to give materials with interesting magnetic properties. The Fe_{17}/Fe_{19} aggregates, which trap small portions of the brucite lattice, prove not only to be useful structural 'scale' models for the protein ferritin, but also display magnetic phenomena that could help in the interpretation of the magnetic behaviour observed for ferritins. The second system, $M_2(OH)_2(ox)$, reveals how the introduction of defects into the extended brucite structure alters the magnetic behaviour through the distortion of the simple triangular lattice.

Thus, we have shown how the introduction of templating species can be used to manipulate a simple hydroxide structure in order to produce materials with magnetic properties significantly different from those of the parent lattice in a manner akin to that used in nature in the process of biomineralization.

We thank Wolfgang Wensdorfer for performing the microSQUID measurements and the EPSRC and Wellcome Trust for funding.

References

Aruga Katori, H., Katsumata, K. & Katori, M. 1996 *Phys. Rev.* B **54**, R9620.

Awschalom, D. D., Smyth, J. F., Grinstein, G., DiVincenzo, D. P. & Loss, D. 1992 *Phys. Rev. Lett.* **68**, 3092.

Baissa, E., Mandel, A., Anson, C. E. & Powell, A. K. 1999 Single crystal X-ray diffraction reveals supramolecular interactions between novel Al_{15} clusters to produce zeotypic lattices. (In preparation.)

Briss, R. R. 1964 *Symmetry and magnetism, selected topics in solid state physics* (ed. E. P. Wohlfarth). Amsterdam: North Holland.

Carlin, R. L. 1986 *Magnetochemistry*. Springer.

Cheesman, M. R., Oganesyan, V. S., Sessoli, R., Gatteschi, D. & Thomson, A. J. 1997 *Chem. Commun.*, p. 1677.

Chikazumi, S. 1964 *Physics of magnetism*, p. 387. Wiley.

Chudvonsky, E. M. 1996 *Science* **274**, 938.

Cornell, R. M. & Schwertmann, U. 1996 *The iron oxides.* Weinheim: VCH.

Day, P. 1988 *Acc. Chem. Res.* **21**, 250.

de Jongh, L. J. (ed.) 1990 *Magnetic properties of layered transition metal compounds.* Dordrecht: Kluwer.

Dormann, J. L., Bessais, L. & Fiorani, D. 1988 *J. Phys.* C **21**, 2015.

Ford, G. C., Harrison, P. M., Rice, D. W., Smith, J. M. A., Treffry, A., White, J. L. & Yariv, J. 1984 *Phil. Trans. R. Soc. Lond.* B **304**, 551.

Frankel, R. B. & Blakemore, R. P. (eds) 1990 *Iron biominerals.* New York: Plenum.

Friedman, J. R., Sarachik, M. P., Tejada, J. & Ziolo, R. 1996 *Phys. Rev. Lett.* **76**, 3830.

Gatteschi, D., Canneschi, A., Pardi, L. & Sessoli, R. 1994 *Science* **265**, 1054.

Gutschke, S. O. H., Price, D. J., Powell, A. K. & Wood, P. T. 1999a Synthesis, single crystal X-ray structure and properties of CO_2 $(OH)_2$ (OX). (In preparation.)

Gutschke, S. O. H., Price, D. J., Powell, A. K. & Wood, P. T. 1999b *Angew. Chem. Int. Ed. Engl.* **38**, 1088.

Heath, S. L. 1992 PhD thesis, University of East Anglia, Norwich.

Heath, S. L. & Powell, A. K. 1992 *Angew. Chem. Int.* **31**, 191.

Heath, S. L., Jordan, P. A., Johnson, I. D., Moore, G. R., Powell, A. K. & Helliwell, M. 1995 *J. Inorg. Biochem.* **59**, 785.

Heath, S. L., Charnock, J. M., Garner, C. D. & Powell, A. K. 1996 *Chem. Eur. J.* **2**, 634.

Joshua, S. J. 1991 *Symmetry principles and magnetic symmetry in solid state physics.* Bristol: Adam Hilger.

Kahn, O. 1993 *Molecular magnetism.* Weinheim: VCH.

Le Brun, N. E., Thomson, A. J. & Moore, G. R. 1997 *Struct. Bonding* **88**, 103.

Mann, S., Webb, J. & Williams, R. J. P. (eds) 1989 *Biomineralization.* Weinheim: VCH.

Molinier, M., Price, D. J., Wood, P. T. & Powell, A. K. 1997 *J. Chem. Soc. Dalton Trans.*, p. 4061.

Moriya, T. 1963 In *Magnetism* (ed. G. T. Rado & H. Suhl), vol. I, ch. 3, p. 85. Academic.

Opechowski, W. & Guccione, R. 1965 In *Magnetism* (ed. G. T. Rado & H. Suhl), vol. IIa, ch. 3, p. 105. Academic.

Powell, A. K. 1997 *Struct. Bonding* **88**, 1.

Powell, A. K., Heath, S. L., Gatteschi, D., Pardi, L., Sessoli, R., Spina, G., Del Giallo, F. & Pieralli, F. 1995 *J. Am. Chem. Soc.* **117**, 2491.

Russell, J. D. 1979 *Clay Miner.* **14**, 109.

Stamp, P. C. E. 1996 *Nature* **383**, 125.

Tejada, J. & Zhang, X. X. 1994 *J. Phys. Condens. Matter* **6**, 263.

Tejada, J., Ziolo, R. F. & Zhang, X. X. 1996 *Chem. Mater.* **8**, 1784.

Thomas, L., Lionti, F., Ballou, R., Gatteschi, D., Sessoli, R. & Barbara, B. 1996 *Nature* **383**, 145.

Towe, K. M. & Bradley, W. F. 1967 *J. Colloid Interface Sci.* **24**, 384.

Wells, A. F. 1962 *Structural inorganic chemistry*, 3rd edn. Oxford University Press.

Wernsdorfer, W. 1996 PhD thesis, Université Joseph Fourier, Grenoble, France.

Womack, T. G. 1999 PhD thesis, University of East Anglia, Norwich.

New high-spin clusters featuring transition metals

By Euan K. Brechin, Alasdair Graham, Paul E. Y. Milne,
Mark Murrie, Simon Parsons and Richard E. P. Winpenny

*Department of Chemistry, The University of Edinburgh,
West Mains Road, Edinburgh EH9 3JJ, UK*

Three possible routes to polynuclear transition metal complexes are discussed. The first route, oligomerization induced by desolvation of small cages, is exemplified by synthesis of a dodecanuclear cobalt cage, and by reactions which give octa- and dodecanuclear chromium cages. The second route involves linking cages through organic spacers, and is illustrated by use of phthalate to link together cobalt and nickel cages. For the nickel case a complex consisting of four cubanes and a sodium octahedron is found. The third route involves the use of water to introduce hydroxide bridges into cages. One method of introducing water is to use hydrated metal salts; the transformation of a Cu_6Na cage into a $Cu_{12}La_8$ illustrates this approach. Alternatively, adventitious water within solvents can be used as a source, and this approach has led to a Co_{24} cage. The structures and magnetic properties of these various cages are discussed.

Keywords: high-spin cages; cobalt cages; nickel cages;
chromium cages; pyridonates; carboxylates

1. Introduction

The discovery that a polynuclear metal cage can show magnetic hysteresis of a molecular origin is a major step forward in the study of molecular magnetism (Sessoli *et al.* 1993*a*, *b*). Since the original observation of this phenomenon in a Mn_{12} cage it has also been seen in studies of Fe_8 (Barra *et al.* 1996) and tetranuclear manganese (Aubin *et al.* 1996) and vanadium clusters (Sun *et al.* 1998). More detailed analysis has also revealed quantum tunnelling of the magnetization in the Mn_{12} (Friedman *et al.* 1996; Thomas *et al.* 1996) and Fe_8 cages (Sangregorio *et al.* 1997). The slow relaxation of magnetization within a molecular species represents the ultimate in miniaturization of a 'magnetic memory', while the observation of quantum effects within crystalline materials allows detailed analysis of the mechanism of quantum tunnelling. Thus these results have great fundamental importance, and great technological potential.

For a synthetic coordination chemist there are several challenges. The simplest is to provide more and more varied 'single-molecule magnets' for the physicists to study. Variations have included integer and non-integer spin ground states (Aubin *et al.* 1998), and spin ground states varying from $S = 3$ to $S = 10$. It would also be attractive to create cages with new shapes and examine how the topology of the cage influences the magnetic properties. A further task is to increase the spin of the ground

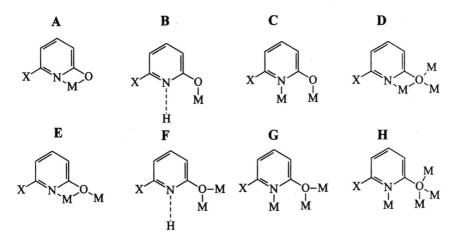

Figure 1. Possible binding modes for pyridonate ligands.
Abbreviations used: X = Cl, chp; X = Me, mhp.

states, which should lead to a higher blocking temperature below which molecules can function as magnets. The most likely route to a higher-spin ground state is via larger paramagnetic cages; therefore, coordination chemists need to learn how to make bigger cages. Underlying every challenge is the fundamental difficulty of the area, which is that designed synthesis is rarely possible. Therefore another aspect of such research is to learn to move beyond serendipity and towards designed synthesis. Here one aspect of the work taking place in Edinburgh is discussed: possible routes to larger cages.

2. Possible routes to larger cages

Compared with the massive cages made with diamagnetic metals, e.g. the polyoxometallate cages synthesized by the Müller group (Müller *et al.* 1998), or the chalcogenide clusters made by the Fenske group (Fenske *et al.* 1998), cages which contain exclusively paramagnetic metal centres are very limited in size. The most paramagnetic centres assembled in one molecule is around 20, and no group at present seems able to make and crystallize a larger compound. Our work has not got far beyond this limit, but results achieved may suggest new pathways to explore.

(a) Oligomerization induced by desolvation

It is noticeable that for many small cages growth is terminated by coordination of solvate molecules. This suggests that if the solvate molecules could be removed by heating, oligomerization may occur as remaining ligands adjust their coordination modes to satisfy the coordination number demanded by the metal sites. Ligands based on substituted pyridonates are in many ways ideal as coligands for such studies, as the pyridonates show a wide range of bonding modes (figure 1).

The initial studies we carried out were on a tetranuclear cobalt heterocubane which can be readily crystallized from MeOH (Brechin *et al.* 1996b). The cage has formula $[Co_4(\mu_3\text{-OMe})_4(chp)_4(MeOH)_7]$ **1** (where chp = 6-chloro-2-pyridonate, i.e. X = Cl in

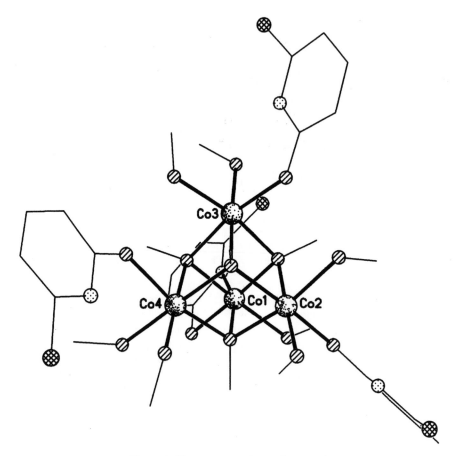

Figure 2. The structure of **1** in the crystal.

figure 1), and crystallizes with seven terminally bound MeOH ligands. Three of the four chp ligands adopt bonding mode B, while the fourth has mode A (figure 2). Heating the sample prior to crystallization, followed by crystallization from a non-coordinating solvent, drives off most of the coordinated MeOH molecules, leading to a centrosymmetric dodecanuclear cage $[\text{Co}_{12}(\text{chp})_{18}(\text{OH})_4(\text{Cl})_2(\text{Hchp})_2(\text{MeOH})_2]$ **2**. This compound has a more complex structure, and in figure 3 both the whole molecule and the metal–oxygen core are shown. As can be seen, the centrosymmetric cage contains two $[\text{Co}_4\text{O}_3\text{Cl}]$ cubes (containing Co(2), Co(3), Co(4) and Co(5)) linked by a central eight-membered ring involving four Co (Co(5) and Co(6) and symmetry equivalents) and four O atoms. The pyridonate ligands in **2** adopt bonding modes **B**, **D**, **E** and **G**. Thus, desolvating the original cage has led to a larger oligomer supported by the coordinative flexibility of the pyridonates.

The second example of this approach involves chromium carboxylates. The oxocentred triangular cages of formula $[\text{Cr}_3\text{O}(\text{O}_2\text{CR})_6\text{L}_3]^+$ can be made with a wide variety of carboxylates. When L is water or a mixture of water and hydroxide, we have found that heating the cages to between 200 and 400 °C in a stream of dry

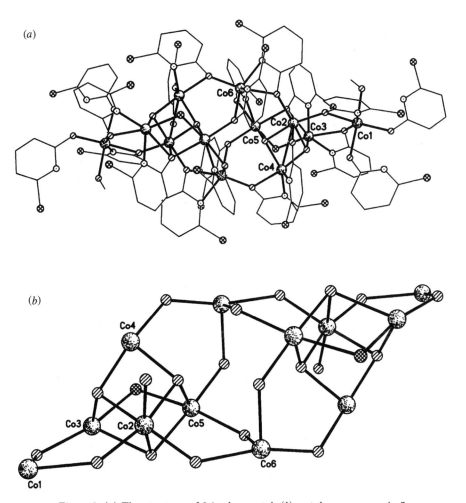

Figure 3. (a) The structure of **2** in the crystal; (b) metal–oxygen core in **2**, showing the linked cubanes.

nitrogen or under vacuum causes elimination of either water or the carboxylic acid, leading to oligomerization (Atkinson *et al.* 1999).

For example, heating $[Cr_3O(O_2CPh)_6(H_2O)_2(OH)]$ to 210 °C, followed by extraction with CH_2Cl_2 and crystallization gives an octanuclear chromium wheel $[Cr(OH)(O_2CPh)_2]_8$ **3** (figure 4). In **3** each Cr···Cr vector is bridged by one μ_2-hydroxide and two 1,3-bridging benzoates. The equivalent reaction, but heated to 400 °C, generates a more compact cage, $[Cr_8O_4(O_2CPh)_{16}]$ **4**, which contains a Cr_4O_4 heterocubane (containing Cr(1), Cr(2), Cr(3) and Cr(4)), with each oxide bound to an additional Cr centre—creating a second larger Cr tetrahedron outside the core of the cage (figure 5). Four benzoate ligands are chelating, and the remaining twelve bridge Cr···Cr vectors.

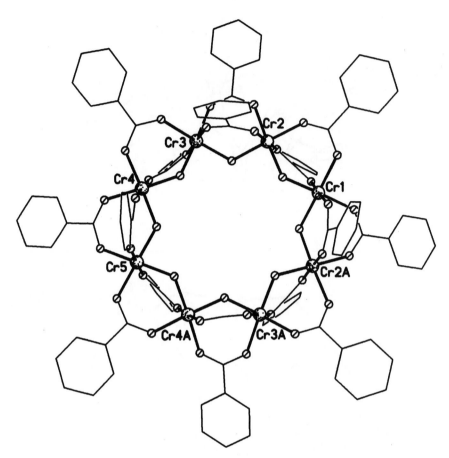

Figure 4. The structure of **3** in the crystal. A crystallographic
two-fold axis passes through Cr(1) and Cr(5).

Although the yield of **3** is low, **4** is formed almost quantitatively, and it is worth attempting a mass balance for the two reactions:

at 210°C, $8[Cr_3O(O_2CPh)_6(H_2O)_2(OH)] \rightarrow 3[Cr(\mu2\text{-}OH)(O_2CPh)_2]_8 + 8H_2O$;

at 400°C, $8[Cr_3O(O_2CPh)_6(H_2O)_2(OH)] \rightarrow 3[Cr_8(\mu4\text{-}O)_4(O_2CPh)_{16}] + 20H_2O$.

Given the low yield of **3** this is a little tendentious, but it does illustrate that the difference between the two reactions is the degree of dehydration of the original chromium triangle.

The same reaction but using pivalate (trimethylacetate) as the carboxylate was reported several years ago by the Gérbéléu group (Batsanov *et al.* 1991), giving a dodecanuclear cage. We have re-examined this synthesis and characterized the cage by both X-ray and neutron diffraction in order to find all protons and hence assign oxidation states for the chromium centres. As a result, we believe the cage should be formulated as $[Cr_{12}O_9(OH)_3(O_2CCMe_3)_{15}]$ **5**. The structure, which has crystallographic D_3 symmetry, consists of a centred-pentacapped-trigonal prism of chromium

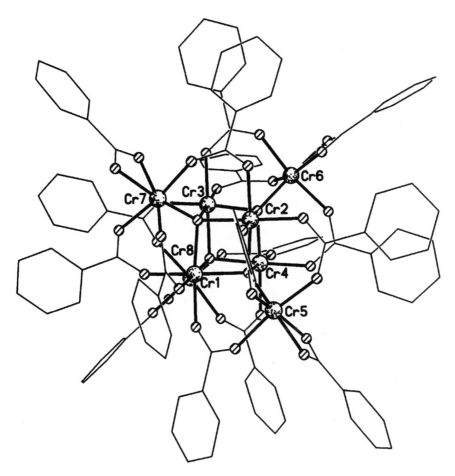

Figure 5. The structure of **4** in the crystal. A crystallographic
two-fold axis passes through Cr(1) and Cr(5).

centres (figure 6). The central chromium (Cr(1)) is bound to six μ_4-oxides, which
bridge to one vertex (e.g. Cr(4)), one triangular face-capping (e.g. Cr(2)) and one
rectangular face-capping site (e.g. Cr(3)). There are also six μ_3-oxygen sites, which
each bridge two vertex and one rectangular face-cap. Neutron studies show that a
half-weight proton is attached to each of these six sites, indicating a positional
disorder and that each position should be regarded as equal occupancy oxide/
hydroxide. The pivalate ligands 'coat' the surface of the cage.

The yield of **5** is extremely high, and carrying out a similar mass balance as for the
benzoate cages reveals a surprising difference in the synthesis of **5**:

$$4[Cr_3O(O_2CCMe_3)_6(H_2O)_2(OH)]$$
$$\rightarrow [Cr_{12}(\mu_4\text{-}O)_3(\mu_3\text{-}OH)_9(O_2CCMe_3)_{15}]\ \mathbf{5} + 9HO_2CCMe_3.$$

Here, rather than loss of water, the carboxylic acid is removed by heating to 400 °C.

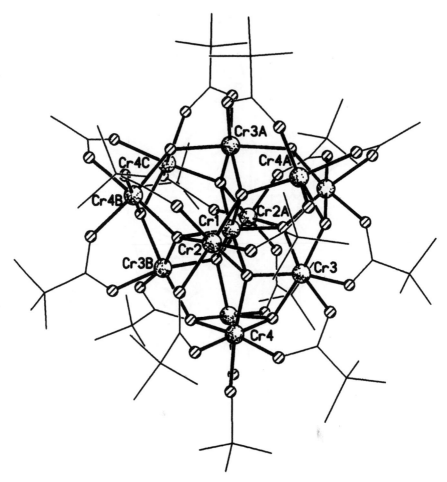

Figure 6. The structure of **5** in the crystal. A crystallographic three-fold axis passes through Cr(1) and Cr(2), and three two-fold axes pass through Cr(1) and Cr(3) and its symmetry equivalents.

This suggests that greater variation of the carboxylates present might generate further chromium cages. It is unclear why pivalic acid is removed in the latter case where water was lost in the earlier examples.

The magnetic susceptibilities of **3**, **4** and **5** have been measured from 1.8 to 300 K. For **3** the susceptibility data could be modelled using the Hamiltonian $H = JS_x S_{x+1} + JS_8 S_1$ $(1 \leqslant x \leqslant 7)$ with $J = 12.0\ \text{cm}^{-1}$. For **4** a more complex Hamiltonian is required:

$$H = J_1(S_1 S_2 + S_1 S_3 + S_1 S_4 + S_2 S_3 + S_2 S_4 + S_3 S_4) + J_2\{S_5(S_1 + S_2 + S_4)$$
$$+ S_6(S_2 + S_3 + S_4) + S_7(S_1 + S_2 + S_3) + S_8(S_1 + S_3 + 4)\},$$

where J_1 is the exchange between Cr atoms in the central heterocubane, and was

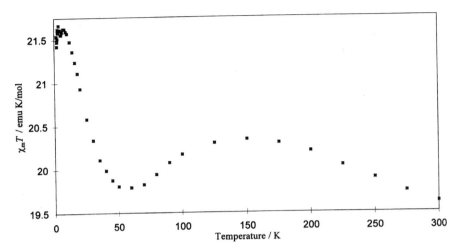

Figure 7. $\chi_m T$ (cm^3 K mol^{-1}) plotted against T (K) for **5**.

found to be 2.1 cm^{-1}, and J_2 models the exchange between the central Cr atoms and the outer tetrahedron, and was found to be 3.4 cm^{-1}. In both cases essentially diamagnetic ground states result.

For **5** the susceptibility behaviour is more interesting (figure 7). The plot of $\chi_m T$ against T shows a double maxima at *ca.* 150 and *ca.* 10 K. The value of the lower temperature maximum (21.6 cm^3 K mol^{-1}) suggests an $S = 6$ ground state ($\chi_m T$ calculated for $g = 1.99$ and $S = 6$ is 20.8 cm^3 K mol^{-1}). The complexity of the structure makes modelling this behaviour impossible at present. Confirmation of the ground state comes from EPR studies. At low temperature, and for all frequencies studied (24, 34, 90, 180 GHz), a complex multiplet is observed. The 90 GHz spectrum can be simulated with the spin-Hamiltonian parameters: $S = 6$, $g_{zz} = 1.965$, $g_{xx} = g_{yy} = 1.960$, $D = +0.088$ cm^{-1}, $E = 0$ (where D and E are the axial and rhombic zero-field splitting parameters, respectively). Unfortunately, the sign of D indicates that **5** is a high-spin cage but not a single-molecule magnet.

(b) Linking cages into supracages

Heating small cages to moderate or high temperatures is one means to cause oligomerization. A second method would be to add a linker to join the cages together. Given our success with ligands such as pyridonates and carboxylates, we examined this route using phthalate as the linker. Previous work using phthalate has been published by the Christou group (Squire *et al.* 1995). Here two examples are discussed, one with cobalt and one with nickel.

The cobalt chemistry (Brechin *et al.* 1996*b*) illustrates the new structures which can result when phthalate is included in reactions previously performed with mono-carboxylates. Reaction of cobalt chloride with sodium benzoate and Na(chp) in MeOH, followed by recrystallization from MeCN, gives a heptanuclear cobalt cage [Co$_7$(OH)$_2$(O$_2$CPh)$_4$(chp)$_8$(MeCN)] **6** (figure 8). Similar cages can be formed with trimethylacetate as the carboxylate. The polyhedron is extremely irregular, and is probably best described as based on a square-based pyramid (Co(1), Co(1A), Co(4),

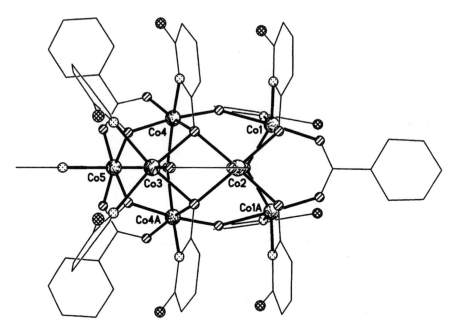

Figure 8. The structure of **6** in the crystal. Co(2), Co(3) and Co(5) lie on a mirror plane.

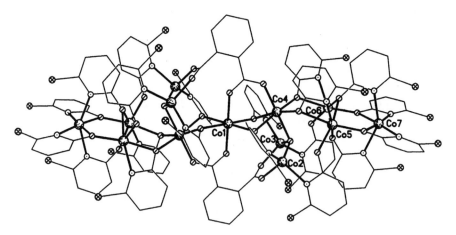

Figure 9. The structure of **7** in the crystal. Co(1) is on an inversion centre.

Co(4A), Co(2)), capped on one edge of the square-base (Co(5)) and on the neighbouring triangular face (Co(3)). While the benzoate ligands in **6** all bridge in a 1,3-manner, the chp ligands adopt bonding modes **C**, **D**, **E** and **G**.

Substituting sodium phthalate in place of Na(O$_2$CPh) in the reaction which gives **6** leads to a larger cage, [Co$_{13}$(OH)$_2$(phth)$_2$(chp)$_{20}$] **7** (where phth is phthalate) (figure 9). The structure is complex and highly irregular. The cage crystallizes with a central Co atom (Co(1)) sitting on an inversion centre. This cobalt is bound to two

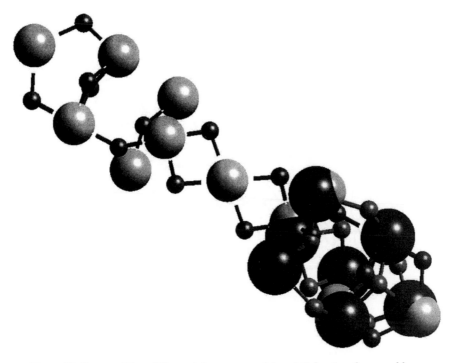

Figure 10. Superposition of the metal–oxo cores of **6** and **7** showing the resemblance between the cages: lighter balls, complex **7**; darker balls, complex **6**.

phth ligands which link to the two halves of the cage. The remainder of the cage is 'coated' with chp ligands which adopt three bonding modes: **C**, **E** and **G**. Initial examination of the structure suggests no relationship between **6** and **7**; however, matching the metal–oxo cores of the two reveals a surprising degree of overlap. Figure 10 shows a superposition of the two cores. At the upper left of the figure is one asymmetric unit of **7**. At the lower right of the figure the core of **6** is superimposed on the second-half of **7**. Six of the Co centres in **6** find an excellent fit with the six centres in **7** which are in general positions. The metal capping the triangular face of the square-based pyramid in **6** (Co(3)), and the cobalt atom on the inversion centre (Co(1)) in **7** do not find good matches. It is therefore reasonable to imagine **7** as a 'dimer' of **6**, linked through phthalates. In both structures the immense coordinative flexibility of the pyridonate ligands is crucial in allowing the formation and crystallization of these unusual structures.

While the cobalt chemistry illustrates what might be expected when a linking group is substituted for another component of a reaction, the nickel chemistry illustrates what might happen when such a ligand is simply added to a reaction. The reaction in question is a simple one. Nickel chloride reacted Na(chp) in MeOH gives a heterocubane, $[Ni_4(OMe)_4(chp)_4(MeOH)_7]$ **8**, which is isostructural with **1** (figure 2) (Blake *et al.* 1995). If sodium phthalate is added to the reaction a heterometallic complex crystallizes of formula $[Ni_{16}Na_6(OMe)_{10}(chp)_4$ $(OH)_2(phth)_8$ $(Hphth)_2(MeOH)_{20}]$ **9** (figure 11) (Brechin *et al.* 1996*a*). The compound retains the

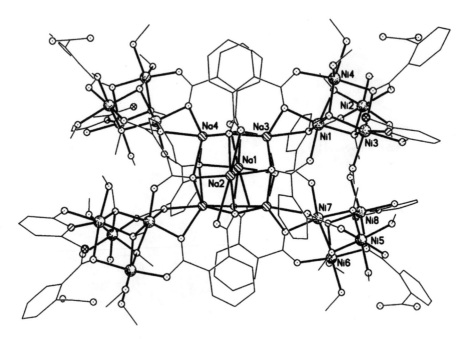

Figure 11. The structure of **9** in the crystal. A crystallographic
two-fold axis passes through Na(1) and Na(2).

nickel heterocubanes, linked through phthalate bridges, but unexpectedly an Na_6
octahedron forms at the centre of the cage. Therefore the phthalates connect five
cages—four cubanes and an octahedron.

The result is intriguing when considered in conjunction with the synthesis of
$[Co_{12}(chp)_{18}(OH)_4(Cl)_2(Hchp)_2(MeOH)_2]$ **2**, which is also formed from reaction of a
heterocubane (see §2 a). Heating a heterocubane drives off neutral solvent molecules,
causing oligomerization through modifying the bonding modes of the pyridonate
ligands, while adding a charged linker such as phthalate displaces charged terminal
pyridonate ligands, causing oligomerization in a quite different way. In principle the
cubanes are building blocks which can be assembled into different higher-nuclearity
cages depending on the method used for the assembling.

The magnetic properties of **9** are rather disappointing. Over the temperature range
300–25 K, $\chi_m T$ rises steadily from close to a value expected for 16 non-interacting
$S = 1$ centres (19.4 cm^3 K mol^{-1} for $g = 2.2$), to a maximum of 22.1 cm^3 K mol^{-1}.
Below 25 K $\chi_m T$ falls rapidly. The simplest explanation for the behaviour is to
consider each Ni$_4$ cubane separately. The spin ground state of each is probably $S = 2$,
based on correlation of coupling with bond angle (Halcrow *et al.* 1995). The rise in
$\chi_m T$ as the temperature falls to 25 K is due to depopulation of the $S = 0$ and 1 energy
levels; the fall after 25 K is probably due to depopulation of the $S = 4$ and $S = 3$
levels. There is no evidence for interaction between the four cubanes in the structure.

(c) Serendipitous solvolysis

Perhaps the best bridging ligands for synthesis of large cages are oxides and

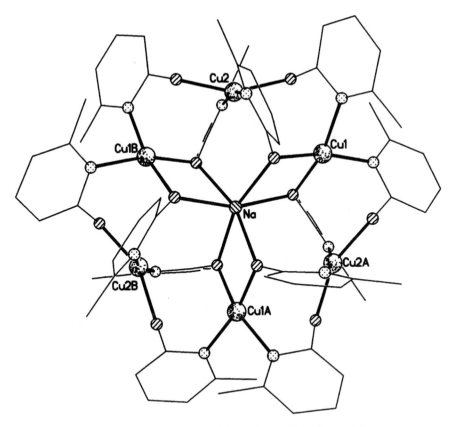

Figure 12. The structure of the cation of **10** in the crystal.
A crystallographic three-fold axis passes through Na.

hydroxides. The problem in using them in synthesis is the insolubility of the relevant metal oxides, oxy-hydroxides and hydroxides. We have been exploring a number of reactions where hydrolysis takes place during crystallization. These results suggested it might be worth examining reactions where other solvents attack complexes causing growth of larger cages.

One possible source of water to cause hydrolysis is hydrated metal salts. The only example we have of where this works is the conversion of $[Cu_6Na(mhp)_{12}][NO_3]$ **10** (where mhp is 6-methyl-2-pyridonate, X = Me in figure 1), into $[Cu_{12}La_8(OH)_{24}(NO_3)_{22}(Hmhp)_{13}(H_2O)_6][NO_3]_2$ **11** by reaction with hydrated lanthanum nitrate (Blake *et al.* 1997). **10** contains a hexanuclear copper core surrounding a sodium centre with a 'star of David' array (figure 12). This complex is soluble in CH_2Cl_2, and reacts with solid lanthanum nitrate to give blue crystals of **11** (figure 13). **11** contains a cube of La centres surrounding a Cu_{12} cuboctahedron with each $LaCu_2$ triangle containing a central hydroxide bridge. The core is thus $[Cu_{12}La_8(OH)_{24}]^{24+}$, with all other ligands attached to the remaining coordination sites of the La centres. While the core has non-crystallographic O_h symmetry the exterior ligands are extremely

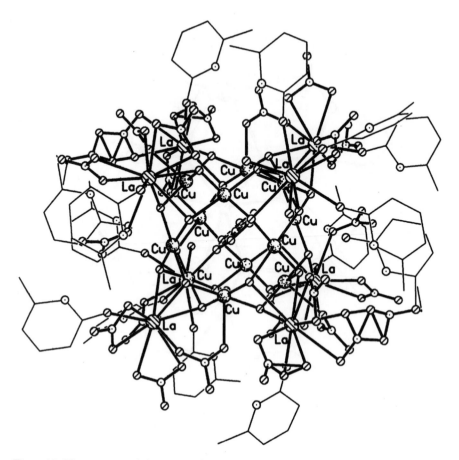

Figure 13. The structure of the cation of **11** in the crystal. The structure is centrosymmetric.

disordered. Although this reaction looks potentially very exciting, it has proved difficult to extend to other systems.

A second, and more productive source of adventitious water, is solvent. Solvents such as ethyl acetate contain *ca.* 1–3% water unless dried carefully. This amount of moisture seems suitable for solvolysis reactions. For example, reaction of anhydrous cobalt chloride with Na(mhp) in MeOH gives a purple solution which can be evaporated to dryness and redissolved in ethyl acetate. From the resulting purple solution, large purple crystals form after a period of days. The resulting cage, $[Co_{24}(OH)_{18}(mhp)_{22}(OMe)_2(Cl)_6]$ **12** (figure 14), resembles a fragment of cobalt hydroxide (Brechin *et al.* 1997). The cage contains sixteen cobalt centres which are based on octahedral coordination geometries (Co(1), Co(2), Co(3), Co(4), Co(5), Co(6), Co(7), Co(8) and symmetry equivalents), with varying amounts of distortion, while the remaining eight cobalt sites have distorted tetrahedral geometries. The central region of the cage contains the octahedral sites, bridged mainly by μ_3-hydroxides and some μ_3-methoxides and chlorides. The circumference of the disc has mhp ligands attached, showing bonding modes **C**, **E** and **G**. Again, the flexibility of

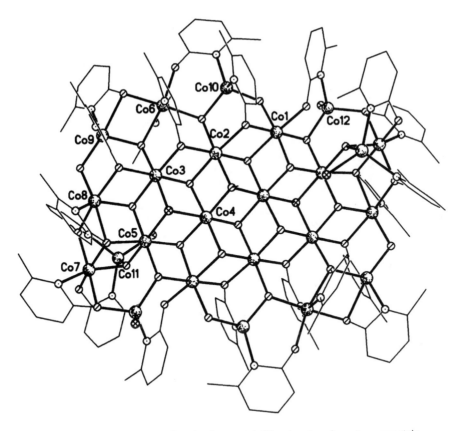

Figure 14. The structure of **12** in the crystal. The structure is centrosymmetric.

the ligand helps support formation of the cage. The magnetic behaviour of **12** seems unusual, with a divergence between field-cooled and zero-field-cooled susceptibility at 4.5 K. Whether this is of a molecular origin, or is evidence of spin-glass behaviour is open to question. The molecules of **12** pack efficiently in the crystal, and intermolecular interactions are likely to be important in determining the bulk magnetic properties.

12 contains 24 paramagnetic centres, which makes it one of the largest cages crystallographically characterized to date. We therefore spent some time examining the reaction which gives **12**, and in attempting to make analogous cages for nickel. The amount of moisture present is crucial in forming and crystallizing **12**: too much moisture and an insoluble product forms which appears to be, in the main, cobalt hydroxide. We also attempted to exclude moisture, and although we failed we found another new cage, $[Co_9(OH)_4(mhp)_8(Cl)_6(Hmhp)_4]$ **13** (figure 15). The structure does not resemble **12** closely, although there are common features. **13** crystallizes on a two-fold axis, with a central cobalt atom (Co(1)) lying on the symmetry element. Co(1) is part of two vertex-sharing cubanes which are both distorted as one cobalt (Co(5)) is slightly out of position. Co(5) and Co(4) have tetrahedral coordination geometries, while Co(1) and Co(2) have octahedral geometries and Co(3) is extremely irregular.

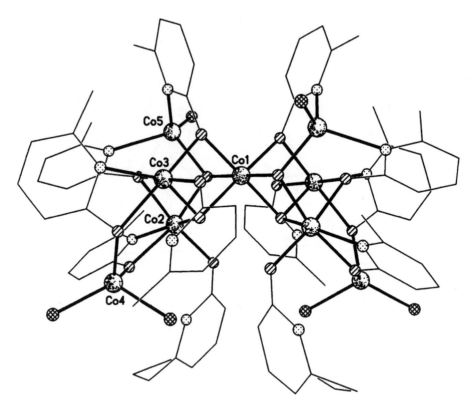

Figure 15. The structure of **13** in the crystal. A crystallographic
two-fold axis passes through Co(1).

This mixture of coordination environments is also found in **12**. The mhp ligands in **13**
adopt bonding modes **C**, **E** and **G**.

Attempting to make the nickel analogue of **12** failed entirely, generating a quite
different cage $[Ni_4Na_4(mhp)_{12}(Hmhp)_2]$ **14** (figure 16) (Brechin *et al.* 1998*b*). This
centrosymmetric structure features four chemically identical $[Ni(mhp)_3]^-$ units
surrounding a central sodium 'chair'. Each mhp therefore chelates to the nickel site,
and bridges on to either one or two sodium atoms. No hydroxide is found within
the cage indicating that the nickel and cobalt chemistry is quite different in this
system.

We have managed to incorporate hydroxide into a nickel-mhp cage, but by starting
with nickel hydroxide and reacting that directly with 6-methyl-2-pyridone (Hmhp).
The resulting cage is probably best written as $[Ni_6(OH)_6\{Ni(mhp)_3\}_5(Hmhp)(Cl)$
$(H_2O)_2]$ **15** (figure 17) (Brechin *et al.* 1998*a*). As for **14**, $[Ni(mhp)_3]^-$ units are found
attached to a central cage, but here the central cage is a $Ni_6(OH)_6$ face-sharing
double-cubane (containing Ni(1), Ni(2), Ni(3), Ni(4), Ni(5) and Ni(6)). Two of the five
$[Ni(mhp)_3]^-$ 'complex ligands' use all three O atoms to bind to nickel centres, while
the remaining three units use two O atoms to bind nickel and the final oxygen to
hydrogen-bond to hydroxide within the double-cubane core. This core resembles a

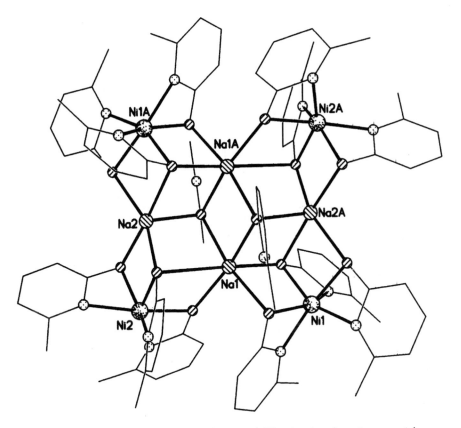

Figure 16. The structure of **14** in the crystal. The structure is centrosymmetric.

small fragment of nickel oxide. The magnetic properties of **15** are disappointing; susceptibility studies indicate a low-spin ground state.

3. Discussion

Each of the three possible routes to high-nuclearity cages has advantages and disadvantages, which are worth considering.

Oligomerization induced by desolvation has the advantage that very compact cages result, with many short superexchange paths between metal centres. It appears to be applicable to a broad range of metal precursors and generates new structures. The main drawback with the approach is the lack of control of the structure of the resulting product. While some control seems possible by choice of temperature, this is likely to remain, in reality, rationalization of observations rather than a predictive tool.

Linking cages together through ligands such as phthalate is considerably more elegant, and offers more potential for control and prediction. Unfortunately, the disadvantage here is that the long superexchange path between cages can limit the magnetic communication as in **9**, and therefore although structurally a cage may be

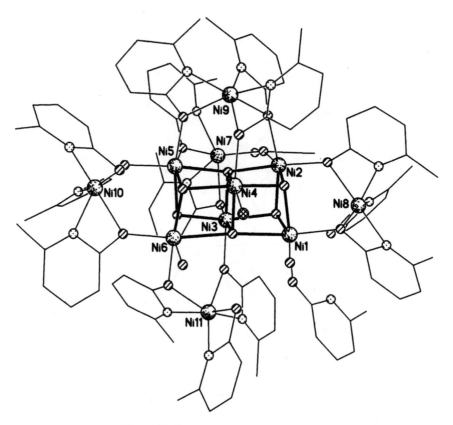

Figure 17. The structure of **15** in the crystal.

exciting, magnetically it may be rather dull. There are many obvious ways around this problem, including use of ligands which are stable radicals.

Serendipitous solvolysis produces the largest paramagnetic cages. It is related to the method by which very large polyoxometallates are produced, but is much less well developed at present. The main drawback here is reproducibility of conditions. Dependence on adventitious moisture can be extremely unreliable, but this problem should be comparatively straightforward to solve. There is a further drawback that the resulting structures are not controlled, although perhaps the resemblance of both **12** and **15** to minerals—cobalt hydroxide and nickel oxide, respectively—suggests that fragments of extended structures will be made through this route. On the positive side, in addition to the large cages which result, the superexchange pathways are short and lead to reasonably strong communications between spin centres.

This work was supported by the EPSRC(UK). Some of the crystallography was carried out by Sandy Blake, Robert Gould and Steven Harris (all Edinburgh). The EPR studies were performed by Eric McInnes and Frank Mabbs (Manchester), and Graham Smith (St. Andrews). Magnetic measurements were carried out in collaboration with Cristiano Benelli (Florence); this collaboration is supported by NATO.

References

Atkinson, I. M., Benelli, C., Murrie, M., Parsons, S. & Winpenny, R. E. P. 1999 Turning up the heat: synthesis of octanuclear chromium(III) carboxylates. *Chem. Commun.*, pp. 285–286.

Aubin, S. M. J., Wemple, M. W., Adams, D. M., Tsai, H.-L., Christou, G. & Hendrickson, D. N. 1996 Distorted $Mn^{IV}Mn_3^{III}$ cubane complexes as single-molecule magnets. *J. Am. Chem. Soc.* **118**, 7746–7754.

Aubin, S. M. J., Spagna, S., Eppley, H. J., Sager, R. E., Christou, G. & Hendrickson, D. N. 1998 Resonant magnetization tunnelling in the half-integer-spin single-molecule magnet $[PPh_4][Mn_{12}O_{12}(O_2CEt)_{16}(H_2O)_4]$. *Chem. Commun.*, pp. 803–804.

Barra, A. L., Debrunner, P., Gatteschi, D., Schulz, C. E. & Sessoli, R. 1996 Superparamagnetic-like behaviour in an octanuclear iron cluster. *Europhys. Lett.* **35**, 133–138.

Batsanov, A. S., Timko, G. A., Struchkov, Y. T., Gérbéléu, N. V. & Indrichan, K. M. 1991 Dodecanuclear chromium oxopivalate as the representative of new types of metal carboxylates: synthesis and structure. *Koord. Khim.* **17**, 662–669.

Blake, A. J., Brechin, E. K., Codron, A., Gould, R. O., Grant, C. M., Parsons, S., Rawson, J. M. & Winpenny, R. E. P. 1995 New polynuclear nickel complexes with a variety of pyridonate and carboxylate ligands. *J. Chem. Soc. Chem. Commun.*, pp. 1983–1985.

Blake, A. J., Gould, R. O., Grant, C. M., Milne, P. E. Y., Parsons, S. & Winpenny, R. E. P. 1997 Reactions of copper pyridonate complexes with hydrated lanthanoid nitrates. *J. Chem. Soc. Dalton Trans.* 485–495.

Brechin, E. K., Gould, R. O., Harris, S. G., Parsons, S. & Winpenny, R. E. P. 1996*a* Four cubes and an octahedron: a nickel–sodium supracage assembly. *J. Am. Chem. Soc.* **118**, 11 293–11 294.

Brechin, E. K., Harris, S. G., Parsons, S. & Winpenny, R. E. P. 1996*b* Desolvating cubes and linking prisms: routes to high nuclearity cobalt coordination assemblies. *Chem. Commun.*, pp. 1439–1440.

Brechin, E. K., Harris, S. G., Harrison, A., Parsons, S., Whittaker, A. G. & Winpenny, R. E. P. 1997 The synthesis and structural characterisation of a tetraicosanuclear cobalt complex. *Chem. Commun.*, pp. 653–654.

Brechin, E. K., Clegg, W., Murrie, M., Parsons, S., Teat, S. J. & Winpenny, R. E. P. 1998*a* Nanoscale cages of manganese and nickel with 'rock salt' cores. *J. Am. Chem. Soc.* **120**, 7365–7366.

Brechin, E. K., Gilby, L. M., Gould, R. O., Harris, S. G., Parsons, S. & Winpenny, R. E. P. 1998*b* Heterobimetallic nickel–sodium and cobalt–sodium complexes of pyridonate ligands. *Dalton Trans.* 2657–2664.

Fenske, D., Zhu, N. Y. & Langetege, T. 1998 Synthesis and structure of new Ag-Se clusters: $[Ag_{30}Se_8(Se^tBu)_{14}(P^nPr_3)_8]$, $[Ag_{90}Se_{38}(Se^tBu)_{14}(PEt_3)_{22}]$, $[Ag_{114}Se_{34}(Se^nBu)_{46}(P^tBu_3)_{14}]$, $[Ag_{112}Se_{32}(Se^nBu)_{48}(P^tBu_3)_{12}]$ and $[Ag_{172}Se_{40}(Se^nBu)_{92}(dppp)_4]$. *Angew. Chem. Int. Ed. Engl.* **37**, 2640–2644.

Friedman, J. R., Sarachik, M. P., Tejada, J. & Ziolo, R. 1996 Macroscopic measurement of resonant magnetisation tunnelling in high-spin molecules. *Phys. Rev. Lett.* **76**, 3830–3833.

Halcrow, M. A., Sun, J. S., Huffman, J. C. & Christou, G. 1995 Structural and magnetic properties of $[Ni_4(\mu_3\text{-}OMe)_4(DBM)_4(MeOH)_4$ and $[Ni_4(\mu_3\text{-}N_3)_4(DBM)_4(EtOH)_4]$: magneto-structural correlations for $[Ni_4X_4]^{4+}$ cubane complexes. *Inorg. Chem.* **34**, 4167–4177.

Müller, A., Krickemeyer, E., Bögge, H., Schmidtmann, M., Beugholt, C., Kögerler, P. & Lu, C. 1998 Formation of a ring-shaped reduced 'metal oxide' with the simple composition $[(MoO_3)_{176}(H_2O)_{80}H_{32}]$. *Angew. Chem. Int. Ed. Engl.* **37**, 1220–1223.

Sangregorio, C., Ohm, T., Paulsen, C., Sessoli, R. & Gatteschi, D. 1997 Quantum tunnelling of the magnetisation in an iron cluster nanomagnet. *Phys. Rev. Lett.* **78**, 4645–4648.

Sessoli, R., Gatteschi, D., Caneschi, A. & Novak, M. A. 1993*a* Magnetic bistability in a metal–ion cluster. *Nature* **365**, 141–143.

Sessoli, R., Tsai, H.-L., Schake, A. R., Wang, S., Vincent, J. B., Folting, K., Gatteschi, D., Christou, G. & Hendrickson, D. N. 1993*b* High-spin molecules: $[Mn_{12}O_{12}(O_2CR)_{16}(H_2O)_4]$. *J. Am. Chem. Soc.* **115**, 1804–1816.

Squire, R. C., Aubin, S. M. J., Folting, K., Streib, W. E., Hendrickson, D. N. & Christou, G. 1995 Octadecanuclearity in manganese carboxylate chemistry $K_4[Mn_{18}O_{16}(O_2CPh)_{22}$ $(phth)_2(H_2O)_4]$ ($phthH_2$ = phthalic acid). *Angew. Chem. Int. Ed. Engl.* **34**, 887–889.

Sun, Z., Grant, C. M., Castro, S. L., Hendrickson, D. N. & Christou, G. 1998 Single-molecule magnets: out-of-phase AC susceptibility signals from tetranuclear vanadium(III) complexes with an $S = 3$ ground state. *Chem. Commun.*, pp. 721–722.

Thomas, L., Lionti, F., Ballou, R., Gatteschi, D., Sessoli, R. & Barbara, B. 1996 Macroscopic quantum tunnelling of magnetisation in a single crystal of nanomagnets. *Nature* **383**, 145–147.

From ferromagnets to high-spin molecules: the role of the organic ligands

By Talal Mallah, Arnaud Marvilliers and Eric Rivière

Laboratoire de Chimie Inorganique UMR CNRS 8613,
Université Paris-Sud, 91405 Orsay, France

The nature of the organic ligands is crucial in influencing the dimensionality of bi-metallic cyanide-bridged systems. By controlling the number of coordination sites around a metal ion, we show that systems with new structure and original architecture may be obtained. Blocking four of the coordination sites around Ni^{II} and Mn^{II}, new structures with new architectures can be obtained depending on the position of the remaining two available sites: in the *trans* or *cis* position.

Keywords: molecular magnetism; ferromagnet; metamagnet; high-spin molecules;
cyanide-bridged systems; low-dimensional systems

1. Introduction

The design of new systems with expected dimensionality and architecture is one of the challenges in molecular magnetism. Magnetic properties are intimately related to dimensionality. Extended systems possessing three-dimensional structure are expected to order magnetically below a critical temperature T_c. Polynuclear molecules with a high-spin ground state and large magnetic anisotropy may behave as nanomagnets.

We have been involved during the last few years in the design of molecular systems where dimensionality and, thus, magnetic properties may be addressed chemically by means of organic chelating ligands. In this paper, we will focus on the synthesis of cyanide-bridged systems obtained from the reaction of hexacyanochromate(III) and mononuclear complexes of general formula $[M^{II}L_n(H_2O)_{6-n}]^{2+}$, where M may be Ni or Mn, and L_n is a chelating ligand that blocks two, four or five of the coordination sites around the metal. Depending on the nature of the chelating ligand—(i) the number of chelating sites, (ii) the size of the ligand, (iii) the isomer used for a given number of chelating sites (for tetradentate ligands cis and trans isomers)—an infinite number of possibilities can be expected and making prediction is hazardous. Our endeavour is directed towards the study of different parameters that influence the dimensionality and the architecture for a given dimension in order to establish some rules that enable a control of the properties of the final material.

In this paper, we present the crystallographic structures and the magnetic properties of a series of Cr^{III}–M^{II} cyanide-bridged compounds; our aim is to present our work more from the chemical design point of view, without going deeply into the subtleties of the magnetic properties. The paper is organized as follows: §2 is an introduction to the nature of the exchange interaction through the cyanide bridge between Cr^{III} and Ni^{II} on one hand and Cr^{III} and Mn^{II} on the other hand. A

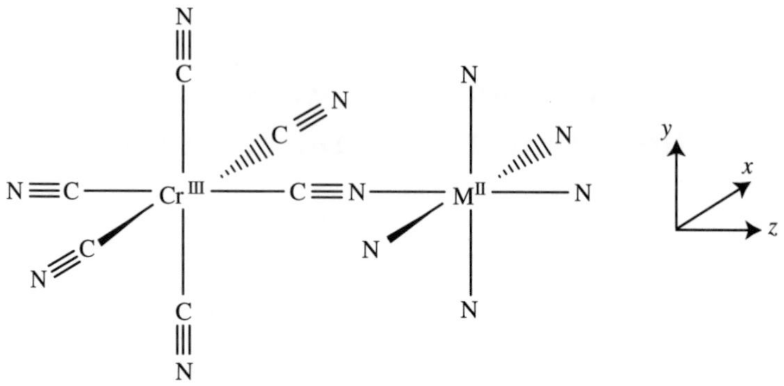

$$N$$
$$\|\|\|$$
$$C$$
$$|$$

$$N\!\equiv\!C\!-\!\!-\!\!Cr^{III}\!\cdots\!\!\!\!\!\!\!\!\!\!^{||||}C\!\equiv\!N\!-\!\!-\!C\!\equiv\!N\!-\!\!-\!M^{II}\!\cdots\!\!\!\!^{||||}N$$

Scheme 1.

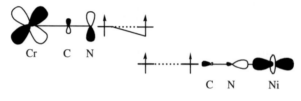

Scheme 2.

description of the structure and the magnetic properties of several compounds according to the nature of the organic ligand L_n can be found in §3, and a brief discussion mainly centred on the possibilities of predicting the dimensionality and the architecture of new cyanide-bridged materials is included in this section. A general conclusion is given in §4.

2. Short-range exchange interaction through the cyanide bridge

Let us consider the molecular fragment depicted in scheme 1, in which the local symmetry around each metal ion is octahedral.

When M is Ni, the three single electrons of Cr^{III} are described by t_{2g} orbitals strictly orthogonal to the e_g orbitals of the two Ni^{II} single electrons. This is shown in scheme 2 between the Cr^{III} d_{yz} and the Ni^{II} d_{z^2} orbitals. Ginsberg (1971) was the first to predict that ferromagnetic interaction is expected from the interaction between Cr^{III} and Ni^{II} ions through a linear bridge. We find, experimentally, that the exchange coupling interaction between Cr^{III} and Ni^{II} through the cyanide bridge is, indeed, ferromagnetic in nature. Kahn (1985) showed that the amplitude of the ferromagnetic interaction depends on the overlap density defined as $\rho(1) = a(1)b(1)/r_{12}$. When the bridge is a single atom, the overlap density concentrated on a very small region of space between the two metal ions is expected to have a large local value, leading to an appreciable ferromagnetic exchange interaction. In Cr^{III}–Ni^{II} cyanide-bridged systems, the metal–metal distance is *ca.* 5.2 Å, the overlap density is expected to be spread over the carbon and nitrogen atoms leading to a small amplitude for the interaction. Gadet *et al.* (1992) have shown, using semi-empirical calculations, that this was not the case, because despite the presence of a two-atom bridge, the overlap density is mainly

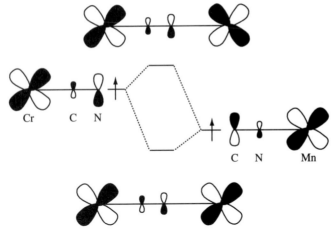

Scheme 3.

localized on the nitrogen atom, giving a large local value for the overlap density. Scheme 2 shows that the contribution of the nitrogen atom to the Cr^{III} d_{yz} and the Ni^{II} d_{z^2} magnetic orbitals is larger than that of carbon.

When M is Mn the situation is different, since Mn^{II} has five single electrons: three on t_{2g} orbitals and two on e_g orbitals. The overall exchange coupling constant J can be expressed as

$$J_{CrMn} = \frac{1}{n_{Cr} n_{Mn}} \sum j_{\mu\nu}, \qquad (2.1)$$

where n is the number of single electrons for a given metal ion and $j_{\mu\nu}$ are the different exchange pathways between pairs of magnetic orbitals. Fifteen pathways are possible, corresponding to $n_{Cr} n_{Mn} = 15$, but only a few are effective. Effective pathways that contribute to the interaction are those corresponding to magnetic orbitals possessing a contribution to the cyanide bridge between the metal ions. Of the possible 15 pathways, only four should be taken into account, i.e. j_{xz-xz}, j_{yz-yz}, j_{xz-xz}, j_{xz-z^2} and j_{yz-z^2}. The former two pathways are built from orbitals with non-zero overlap integral (scheme 3), while the latter correspond to orthogonal orbitals, as in the case of Ni^{II}.

For $Cr^{III} Mn^{II}$ cyanide-bridged compounds, the interaction will be the result of two contributions: one antiferromagnetic, due to the overlap between Cr^{III} and Mn^{II} t_{2g} magnetic orbitals; and the other ferromagnetic, due to t_{2g}–e_g orthogonal orbitals. In such a case, the overall interaction depends on the nature of the dominant interaction. Experimental results presented below clearly show that the antiferromagnetic contribution dominates and that the interaction between Cr^{III} and Mn^{II} through cyanide is antiferromagnetic. Our prediction is based on the assumption that the Cr–C–N–M sequences are linear. This is the case in most of the compounds, but if the CNM angle deviates from linearity by more than $10°$, it is possible to observe a Fe(III)–Ni(II) antiferromagnetic interaction, as was postulated by El Fallah *et al.* (1996) to explain the ferrimagnetic behaviour in $[Ni(tren)]_3[Fe(CN)_6]_2$.

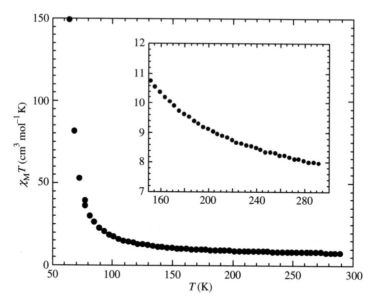

Figure 4. $\chi_M T = f(T)$ at $H = 100$ Oe for CrNien.

3. From three-dimensional systems to polynuclear complexes

(a) *Three-dimensional* $[Cr(CN)_6]_2[Ni(1,2\text{-diaminoethane})]_3 \cdot 2H_2O$: *a ferromagnet*

$[Cr(CN)_6]_2[Ni(1,2\text{-diaminoethane})]_3 \cdot 12H_2O$, referred to simply as CrNien below, is obtained by reacting $[Ni(1,2\text{-diaminoethane})(H_2O)_4]Cl_2$ with $K_3[Cr(CN)_6]\cdot 2H_2O$ in water. A yellow precipitate appears immediately; it is filtered, thoroughly washed with water and dried under vacuum. The crystal structure is determined on a powder sample by refining the X-ray diffraction diagram using the Rietvelt method. The refinement is done using the *Fm3m* space group of Prussian blue. The unit cell is face-centred cubic with a parameter a equal to 10.46 Å corresponding to the Cr^{III}–Cr^{III} distance. Each Cr atom is surrounded by six Ni atoms through the cyanide bridge, and each Ni is surrounded by only four Cr and one 1,2-diaminoethane bidentate ligand.

When decreasing temperature, the $\chi_M T$ product increases from a value that corresponds to non-interacting spins at room temperature (7.29 cm^3 mol^{-1} K) and diverges *ca.* $T = 60$ K (figure 1). The increase in $\chi_M T$ is the signature of a short-range ferromagnetic interaction between Cr^{III} and Ni^{II} through the cyanide bridge.

In order to study the behaviour at low temperature, magnetization versus temperature measurements were performed in three different regimes (figure 2). The starting temperature is $T = 70$ K and the field is set at 30 Oe, the temperature is decreased down to 3 K, which affords the field cooled magnetization (FCM). At $T = 3$ K, the field is cut-down and the magnetization measured with increasing temperature up to $T = 70$ K, which gives the remnant magnetization (REM) curve that goes to zero at $T = 40$ K. The zero-field cooled magnetization (ZFCM) curve is obtained by cooling the sample in zero field down to $T = 3$ K; the field is set again at 30 Oe and the magnetization measured from 3 K to 70 K. These three curves are indicative of a three-dimensional magnetic order, and since the short-range interaction

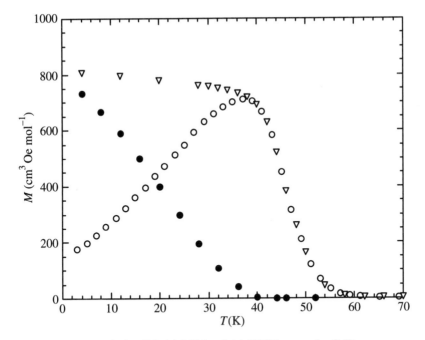

Figure 2. (∇) FCM, ($\bullet$) REM and ($\circ$) ZFCM curves for CrNien.

is ferromagnetic in nature, the order is a ferromagnetic one. The ordering temperature T_c is defined as the temperature where the REM curve goes to zero, which should correspond to the temperature where the FCM and the ZFCM curves merge. For CrNien, T_c is equal to 40 K.

The cubic structure of CrNien allows us to use the molecular-field approach to estimate the value of the exchange coupling constant J_{CrNi} between Cr^{III} and Ni^{II}. The relationship between T_c and J_{CrNi} is given by (Ferlay 1999)

$$T_c = \frac{2}{k} y \sqrt{y} |J_{CrNi}| \sqrt{S_{Cr}(S_{Cr} + 1) S_{Ni}(S_{Ni} + 1)}, \tag{3.1}$$

where k is the Boltzmann constant, S_{Cr} ($= 3/2$) and S_{Ni} ($= 1$) are the Cr^{III} and Ni^{II} spins, respectively, and y is the chromium-to-nickel atomic ratio ($2/3$) within the compound. Applying equation (3.1) with $T_c = 40$ K leads to a value of 9.3 cm^{-1} for the exchange coupling constant J_{CrNi}.

The introduction of the bidentate ligand in the coordination sphere of Ni^{II} does not alter the overall structure, which is still the same as Prussian blue, this is because the ligand used is not bulky enough to destroy the cubic arrangement. We have increased the size of the bidentate ligand by substituting the four hydrogen atoms on the amine groups by four methyl groups in the mononuclear Ni^{II} precursor; this leads to the formation of a new system possessing a non-cubic structure, as was shown from the powder diffraction pattern. Unfortunately, no single crystals were obtained, and, hence, an analysis of the new structure could not be undertaken.

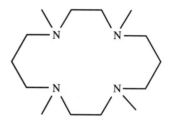

Scheme 4.

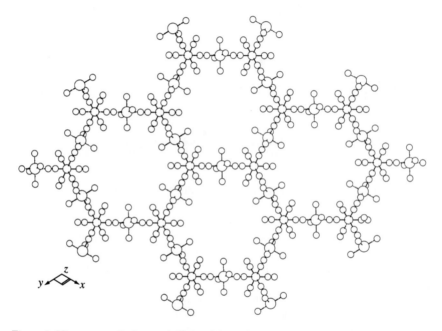

Figure 3. View perpendicular to the layer (along the c-axis) for CrNitmc; the tmc's carbon atoms were removed for clarity.

(b) *Two-dimensional* $[Cr(CN)_6]_2[Ni(tmc)]_3 \cdot 12H_2O$: *a metamagnet*

(i) *Description and discussion of the structure*

When the mononuclear complex $[Ni^{II}(tmc)](ClO_4)_2$, where tmc is a tetradentate ligand (scheme 4), is reacted with hexacyanochromate(III) in water or in acetonitrile, a neutral 3/2 stoichiometric compound of formula $[Cr(CN)_6]_2[Ni(tmc)]_3 \cdot 12H_2O$ (referred to simply as CrNitmc below) is obtained independently of the relative stoichiometry of the reactants.

X-ray diffraction on single crystals reveals a structure built up from corrugated layers, as shown in figures 3 and 4. The layers are stacked along the c-axis (figure 5), the interlayer metal–metal distance is 9.25 Å. The 12 water molecules in the interlayer space are disordered.

The Cr–C distances for the three bridging cyanides (2.059(11) Å) are slightly shorter than those of non-bridging ones (2.098(11) Å). The Ni surrounding is more

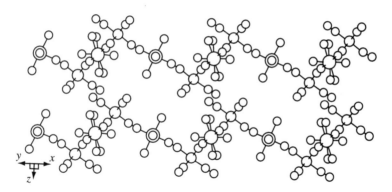

Figure 4. View (two layers) perpendicular to the *c*-axis for CrNitmc.

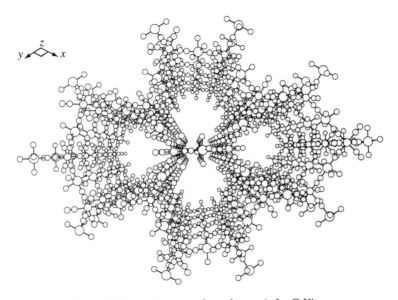

Figure 5. Perspective view along the *c*-axis for CrNitmc.

distorted: within the tmc plane, the Ni–N distances have a difference of *ca.* 0.1 Å (Ni–N(1) = 2.24(2) Å, Ni–N(4) = 2.313(15) Å). The Ni–N axial distances are much shorter (Ni–N(10) = 2.086(11) Å) (figure 6). The local Ni^{II} environment may then be considered as a compressed octahedron along the Cr–CN–Ni–NC–Cr connection perpendicular to the tmc plane.

Each Cr(CN)$_6$ unit is surrounded in a facial manner by three Ni(tmc) complexes; the Ni(tmc) units bridge two Cr(CN)$_6$ in a *trans* fashion. The arrangement of the Cr(CN)$_6$ moieties around the Ni was expected because [Ni(tmc)]$^{2+}$ has its two available sites in the *trans* position. The number of Ni atoms around Cr is dictated by the neutrality of the compound, which imposes a Ni/Cr:3/2 stoichiometry. Let us analyse the possible structures that may grow by considering different starting local arrangements around the metal ions. Before doing that we assume that

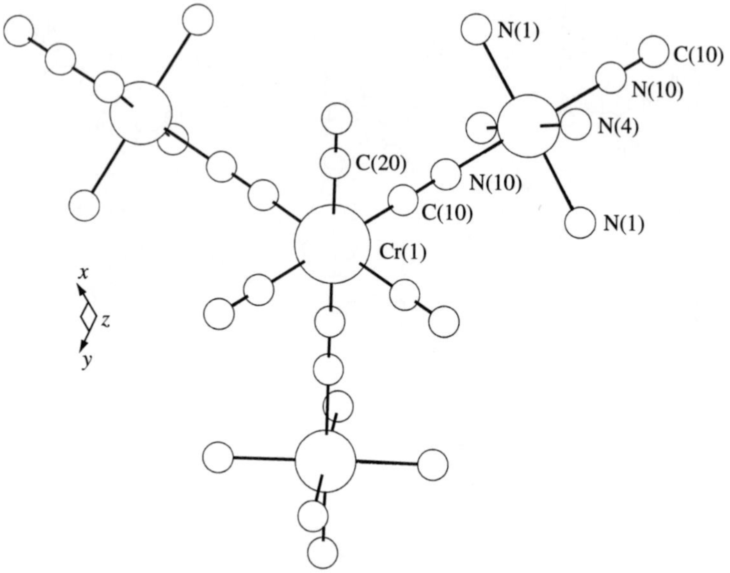

Figure 6. View of a CrNi$_3$ unit along the C$_3$ axis of the Cr(CN)$_6$ complex for CrNitmc.

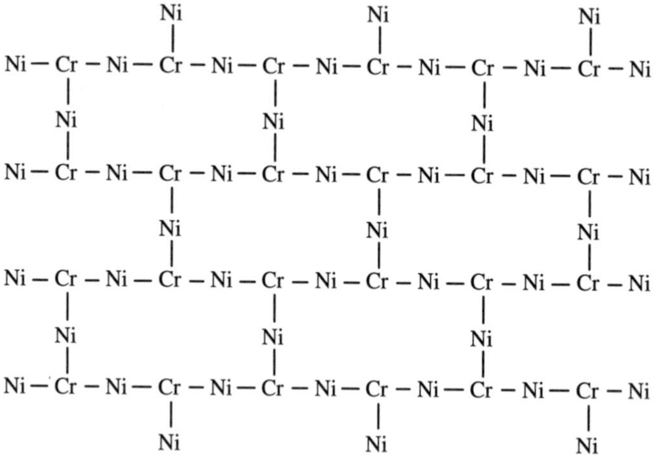

Scheme 5.

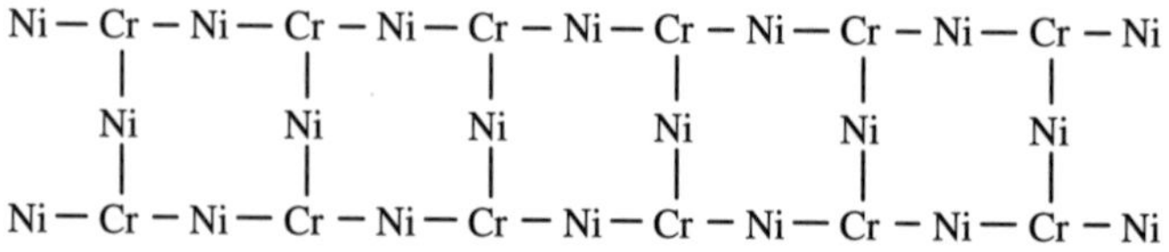

Scheme 6.

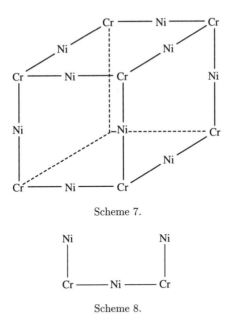

Scheme 7.

Scheme 8.

(i) the nature of tmc imposes a *trans* arrangement of two $Cr(CN)_6$ units around each Ni^{II} ion; and

(ii) the neutrality of the compound allows each Cr to be surrounded by three Ni: meridional and facial spatial $CrNi_3$ arrangements will then be considered.

A meridional $CrNi_3$ arrangement leads either to a bidimensional flat layer or a double chain stair-like structure (schemes 5 and 6).

For the facial arrangement, one expects the formation of a discrete cube-like molecule (scheme 7) or a bidimensional corrugated layered compound. Experiment shows that the latter possibility has been obtained. One may think that within the powder there is a mixture of different compounds; this hypothesis is ruled out because the magnetic behaviour of a powder sample obtained by rapid precipitation in water and that of a powder sample obtained by grinding a collection of single crystals is identical.

It is not easy to rationalize the selectivity of the reaction that leads to the observed structure (the reaction gives *ca.* 80% yield of powder when performed in water), especially because the cube-like discrete molecule arrangement should, on an entropy basis, be thermodynamically favoured. The entropy contribution to the total free energy favours the formation of a compound built up from discrete entities, rather than a polymeric one. It is possible, however, to invoke one argument in favour of the corrugated layer structure. A common spatial arrangement found in the three theoretical structures is shown in scheme 8.

This arrangement of the five complexes in the same plane (two $Cr(CN)_6$ and three Ni(tmc)) imposes a steric hindrance between the three tmc ligands. In order to preclude hindrance between the organic ligands, the compound locally adopts the only other arrangement (scheme 9) that, when translated in two dimensions, leads to the observed structure.

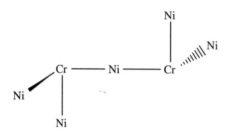

Scheme 9.

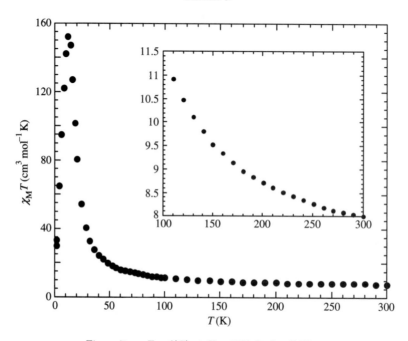

Figure 7. $\chi_M T = f(T)$ at $H = 1000$ Oe for CrNitmc.

(ii) *Magnetic properties*

The behaviour of the $\chi_M T = f(T)$ curve (figure 7) between room temperature and $T = 2$ K is similar to that of the preceding compound. As expected, the short-range interaction between Ni^{II} and Cr^{III} through the cyanide bridge is ferromagnetic within the bidimensional layer. The maximum in the curve corresponds to a $\chi_M T$ value of 155 cm^3 mol^{-1} K and occurs at $T = 12$ K. The corresponding spin value S is about 17. This is a rough measure of the number of correlated spins within the plane; it corresponds to nine Ni^{II} and six Cr^{III} ions having their spins in the same direction.

In order to get insight into the compound properties at low temperature, a powder sample was blocked with parafilm in order to preclude particle motion at low temperature, and the magnetization was measured between 50 and 2 K in an applied field of 30 Oe (figure 8). Below 20 K, the magnetization rises abruptly, attains a maximum of 523 cm^3 Oe mol^{-1} at $T = 14$ K, decreases and then levels off at 5 K with

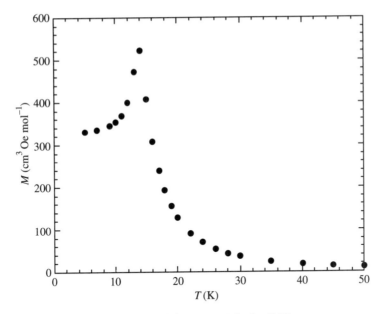

Figure 8. $M = f(T)$ at $H = 30$ Oe for CrNitmc.

a value of 332 cm^3 Oe mol^{-1}. The presence of a maximum in the FCM curve indicates the presence of an antiferromagnetic three-dimensional order between the ferromagnetic planes. At low temperature, the magnetization does not go to zero but to a value that corresponds to about two-thirds of the one observed at $T = 14$ K. This is the expected behaviour of an antiferromagnet with non-negligible anisotropy, as shown by de Jongh & Miedema (1974). In such compounds, if the magnetization is measured on a single crystal with the applied field parallel to the easy axis, the magnetization falls to zero at the ordering temperature T_N. When the field is perpendicular to the easy axis, the magnetization saturates at T_N. For a powder, the magnetization below T_N should, thus, correspond to the mean of the perpendicular and the parallel magnetization, $(2M_\perp + M_\parallel)/3$, which is what we observe experimentally.

Layered antiferromagnets with non-negligible magnetic anisotropy are expected to behave in a different manner from antiferromagnets with extremely small anisotropy. Below T_N, the magnetization can be reversed abruptly when a magnetic field is applied, providing that the interlayer antiferromagnetic coupling responsible for the bulk behaviour is of the same order of magnitude as the applied magnetic field. Figure 9 shows that, at $T = 2$ K, it is possible to reverse the magnetization by applying a field of *ca.* 1200 Oe and to reach saturation with only 2000 Oe. The compound behaves as a metamagnet with a critical field H_C of 1200 Oe at $T = 2$ K. The magnetization at saturation is equal to 11.5μ_B; this is the expected value for the paramagnetic aligned phase (or the field induced ferromagnetic phase) that corresponds to the sum of the spins of three Ni^{II} and two Cr^{III} metal ions $(M/N\beta = gS = 2 \times (3 \times 1 + 2 \times 3/2) = 12)$. The interlayer antiferromagnetic interaction is probably due to water molecules, which may link two different layers by hydrogen bonds between the nitrogen atoms of non-bridging cyanide groups. Van Langenberg *et al.* (1997) have shown that ferromagnetic three-dimensional order

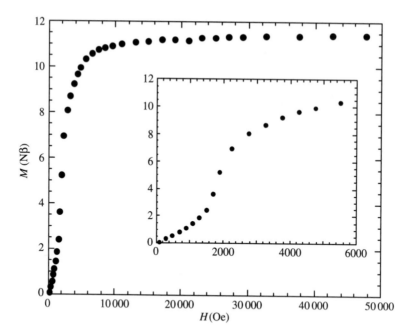

Figure 9. $M = f(H)$ at $T = 2$ K for CrNitmc.

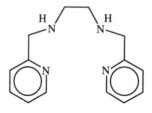

Scheme 10.

occurs because of the presence of a three-dimensional hydrogen-bond network in $[Fe(CN)_6]_3[Ni(bpm)_2]_2 \cdot H_2O$. Unfortunately, the water molecules occupying the interlayer space are disordered; it is not possible to visualize the bond network responsible for the interaction between the layers.

(c) One-dimensional $[Mn(bispicen)]_3[Cr(CN)_6]_2 \cdot 6H_2O$

The reaction of *cis* $[Mn(bispicen)Cl_2]$, where bispicen is a tetradentate ligand (see scheme 10), with $((C_4H_9)_4N)_3[Cr(CN)_6]$ in methanol affords a yellow powder. Elemental analysis fits with the general formula $[Mn(bispicen)]_3[Cr(CN)_6]_2 \cdot 6H_2O$, referred to below as CrMnbispic.

Yellow block-shaped single crystals are obtained by letting two methanolic solutions of the two reactants diffuse for two months in an H-tube. The structure shows the formation of a one-dimensional system (figures 10 and 11). Two kinds of $Cr(CN)_6$ molecules are present: within the body of the chain one chromicyanide is surrounded by five Mn(bispicen) molecules: four in equatorial positions and one in an apical position leaving one non-bridging cyanide group; the second $Cr(CN)_6$ molecule

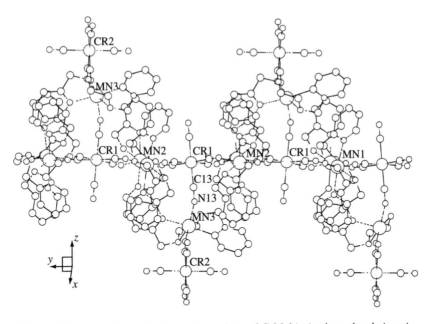

Figure 10. View of the architecture of one chain of CrMnbispic along the chain axis.

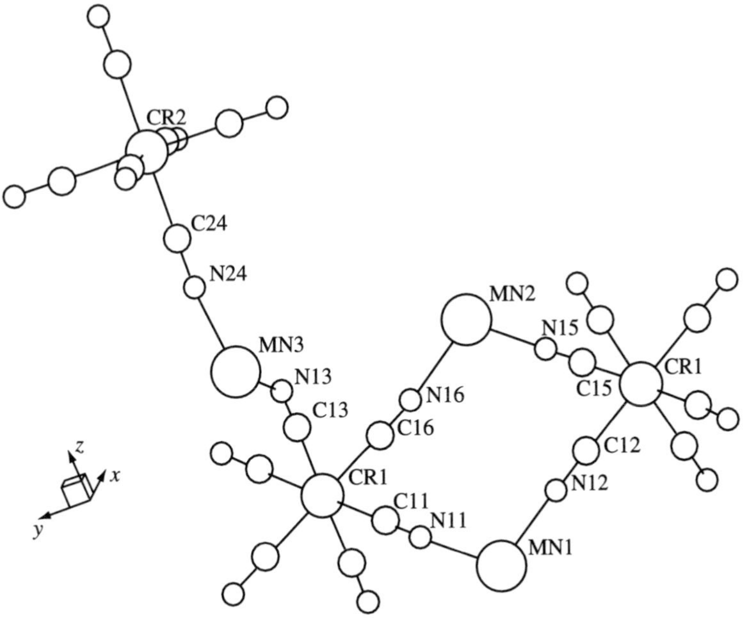

Figure 11. View of a hexanuclear fragment of CrMnbispic.

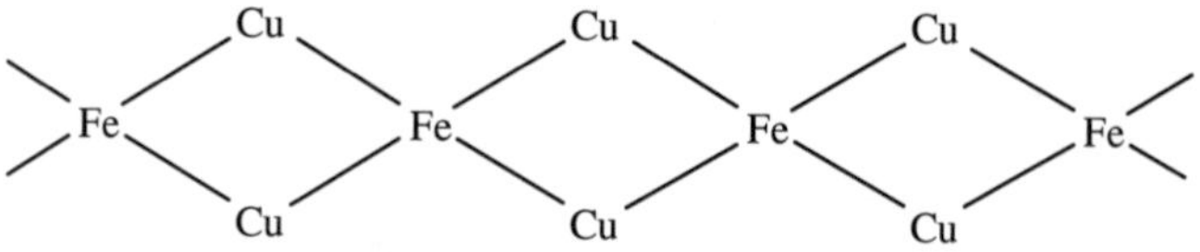

Scheme 11.

is not bridging, it is linked only to the apical Mn unit. Each of the equatorial manganese molecules bridges two $Cr(CN)_6$ within the chain, and the apical manganese molecule bridges two $Cr(CN)_6$, one inside the chain and one outside.

The architecture of the body of the one-dimensional chain was observed, for the first time, by Morpugo *et al.* (1980) in $[Fe(CN)_6]_2[Cu(dien)]_3$. This compound is made from the association, within the same crystal, of a one-dimensional anionic chain $[Fe(CN)_6][Cu(dien)]_2^-$ with the same structure as the body structure of our compound and a cationic dimer $[Fe(CN)_6][Cu(dien)]^+$. Scheme 11 is a view of the chain architecture, the tridentate ligand diethylenetriamine (dien) capping the Cu^{II} ions and the bridging cyanide groups were removed for clarity. The reason for such similarity is the nature of the complex linking the metallocyanides in the two compounds, i.e. possessing two potential available sites for cyanide nitrogen in the *cis* position. The difference between the two structures is that in Morpugo's compound the binuclear unit is not linked to the chain, while in CrMnbispic it occupies one apical position on each in-chain chromium atom in an alternating way.

The Cr–C distances are all *ca.* 2.05 Å and the Cr–C–N angles are very close to 180° (176° on average). The surrounding of the Mn^{II} ions is more distorted. The distances between Mn and the nitrogen atoms belonging to the tetradentate ligands vary between 2.27 and 2.34 Å, while the distances between the manganese ions and the cyanide's nitrogen atoms are slightly shorter, 2.17 Å. The Mn–N–C angles deviate well from linearity:

$$Mn(1)\text{–}N(12)\text{–}C(12) = 177.61°; \quad Mn(1)\text{–}N(11)\text{–}C(11) = 163.75°;$$
$$Mn(2)\text{–}N(15)\text{–}C(15) = 168.32°; \quad Mn(2)\text{–}N(16)\text{–}C(16) = 161.43°;$$
$$Mn(3)\text{–}N(24)\text{–}C(24) = 167.18°; \quad Mn(2)\text{–}N(13)\text{–}C(13) = 146.75°.$$

The smaller angle corresponds to a cyanide bridge between the apical Mn ion and the chromium ion belonging to the body of the chain.

We present here the preliminary magnetic properties. At room temperature, the $\chi_M T$ value (13.9 cm^3 mol^{-1} K) is lower than that corresponding to the three non-interacting Mn^{II} ($S = 5/2$) and two non-interacting Cr^{III} ($S = 3/2$) ions (which have a $\chi_M T$ value of 16.875 cm^3 mol^{-1} K). On decreasing the temperature, $\chi_M T$ decreases to a minimum at $T = 84$ K and then increases, as shown in figure 12. This behaviour indicates the presence of a short-range antiferromagnetic interaction between the Cr^{III} and the Mn^{II} ions as expected. Below $T = 30$ K, $\chi_M T$ increases abruptly to a value of 308 cm^3 mol^{-1} K and then decreases at low temperature, which suggests the onset of a three-dimensional magnetic ordering. Further studies are necessary in order to understand the magnetic behaviour at low temperature. Measurements will be made on single crystals as well in order to check the magnetic properties along the parallel and the perpendicular axes of the chain.

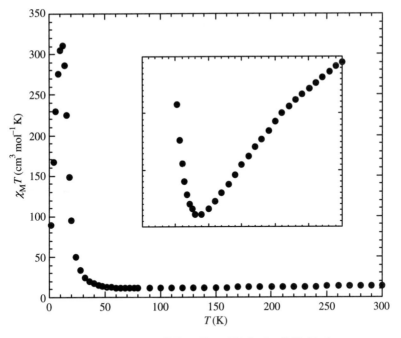

Figure 12. $\chi_M T = f(T)$ at $H = 1000$ Oe for CrMnbispic.

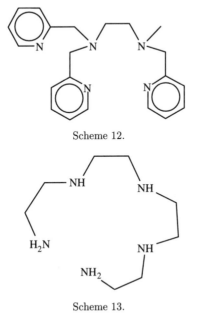

Scheme 12.

Scheme 13.

(d) *Zero-dimensional systems*, CrMn$_6$ *and* CrNi$_6$: *high-spin molecules*

In order to drastically reduce the dimensionality of the systems, it is possible to block all but one of the coordination sites on a metal ion (see Scuiller *et al.* 1996;

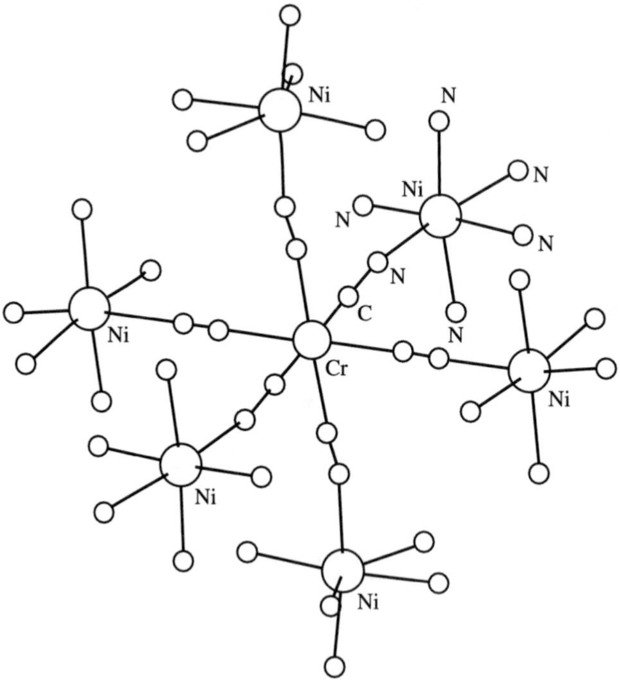

Figure 13. View of the heptanuclear molecule CrNi$_6$;
the carbon skeleton of tetren is not shown here.

Mallah *et al.* 1995). Pentadentate ligands may be used to do this. We prepared two mononuclear complexes,

$$[Mn(trispicmeen)(H_2O)](ClO_4)_2 \quad \text{and} \quad [Ni(tetren)(H_2O)](ClO_4)_2,$$

where trispicmeen and tetren are depicted in schemes 12 and 13, respectively.

The synthesis of the heptanuclear compounds consists of adding a salt of chromicyanide to a solution containing an excess of the mononuclear compound. For

$$[Cr(CNMn(tripicmeen)_6](ClO_4)_9$$

(referred to simply as CrMn$_6$ below), the terabutylammonium salt and the acetonitrile solvent were used, the compound is isolated as powder by adding tetrahydrofuran to the mixture. [Cr(CNNi(tetren)$_6$](ClO$_4$)$_9$ (CrNi$_6$, see figure 13) is obtained by using the potassium salt in a 1/1 mixture of acetonitrile and water. Single crystals grow within a week if the mother solution is left to evaporate.

The $\chi_M T = f(T)$ plot for CrMn$_6$ (figure 14) presents a minimum $T = 210$ K, while that of CrNi$_6$ increases on cooling down and reaches, at $T = 4$ K, a value (32.4 cm^3 mol^{-1} K) corresponding to a spin $S = 15/2$. This indicates the occurrence of an antiferromagnetic interaction between the central Cr^{III} and the six peripheral Mn^{II} in CrMn$_6$ and a ferromagnetic interaction for CrNi$_6$. The experimental $\chi_M T = f(T)$ data can be fitted to the $\chi_M T$ theoretical expression derived from the

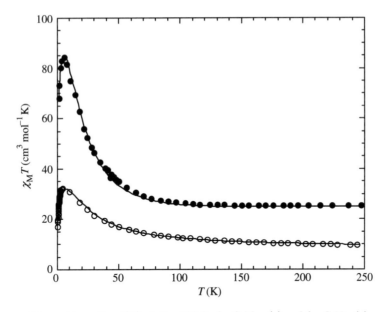

Figure 14. $\chi_M T = f(T)$ at $H = 100$ Oe for CrMn$_6$ ($\bullet$) and for CrNi$_6$ ($\circ$). The lines correspond to the fit results.

following spin Hamiltonian:

$$H = -J_{CrM} S_{Cr} \sum_{i=1}^{6} S_{M_i} + \beta g \left[\sum_{i=1}^{6} S_{M_i} + S_{Cr} \right] H + DS_z^2, \qquad (3.2)$$

where J_{CrM} is the exchange coupling constant between Cr^{III} and M^{II}, M is Mn or Ni, g is the g-factor, taken to be the same for the metal ions, S_{Cr} and S_M are Cr^{III} and M^{II} spin operators, and D is the zero-field splitting parameter. The fit gives the following results: for CrMn$_6$, $J_{CrMn} = -8$ cm^{-1}, $g = 1.98$ and $D = 0.006$ cm^{-1}; for CrNi$_6$, $J_{CrNi} = +15$ cm^{-1}, $g = 2.04$ and $D = 0.008$ cm^{-1}.

The magnetization versus field plots at $T = 2$ K confirm the nature of the ground state for the two compounds. Figure 15 shows that, for CrMn$_6$, saturation magnetization ($27\mu_B$) corresponds to $S = 27/2$, where the six peripheral $S = 5/2$ spins are aligned in an anti-parallel fashion to the central $S = 3/2$ spin. For CrNi$_6$, the observed value of $15\mu_B$ indicates that the $S = 15/2$ ground state is the result of parallel alignment of the local spins within the ground state.

The dynamic magnetic properties of CrNi$_6$ show the occurrence of a blocking of the $S = 15/2$ moment at $T_B = 0.3$ K. Figure 16 shows the $\ln(\tau) = f(1/T)$ curve, where τ is the inverse of the frequency of the oscillating field in the AC susceptibility experiment, and T is the temperature of the maximum of the imaginary component of the susceptibility for a given frequency. The energy barrier corresponds to the slope of the $\ln(\tau) = f(1/T)$ curve and is found to be equal to 4 cm^{-1}.

Assuming an axial anisotropy, the energy barrier is given by DS_z^2, where $S_z = 15/2$. It is possible to compute the value of the zero-field splitting parameter D to be 0.07 cm^{-1}. This value is an order of magnitude more than the value extracted from fitting the $\chi_M T = f(T)$ data (see above). One reasonable explanation is that our

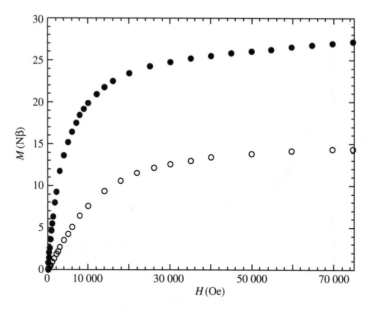

Figure 15. $M = f(H)$ at $T = 2$ K for CrMn$_6$ ($\bullet$) and for CrNi$_6$ ($\circ$).

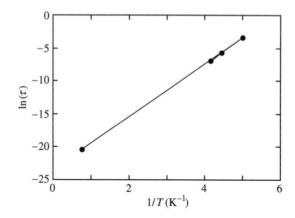

Figure 16. $\ln(t) = f(1/T)$ for CrNi$_6$: experimental data ($\bullet$); and linear fit result (——).

assumption of an axial anisotropy as responsible for the blocking of the moment at low temperature is wrong, especially the notion that the CrNi$_6$ heptanuclear complex has an octahedral symmetry. However, the blocking of the magnetic moment is due to the presence of magnetic anisotropy, which is probably cubic in nature and not axial in this case.

4. Conclusion

We have shown how, everything being equal, the nature of the organic ligand can influence the dimensionality and, thus, the magnetic properties of cyanide-bridged systems. When one 1,2-diaminoethane, a bidentate ligand, is used as chelate for

Ni^{II}, the three-dimensional structure of Prussian blue analogues is not altered and a bulk magnetic order is observed. On the other hand, when pentadentate ligands are used as chelates, it is possible to preclude polymerization and stabilize heptanuclear high-spin molecules. The role of the organic ligand is clearly determinant. It is then possible to explore new systems possessing new architecture, such as the $[Mn(bispicen)]_3[Cr(CN)_6]_2 \cdot 6H_2O$ compound presented above. On the other hand, polynuclear complexes with very high-spin ground states may be designed as well. A nice example of a pentanuclear compound was reported by van Langenberg *et al.* (1997). This compound is obtained by using a Ni^{II} complex containing two bulky bidentate ligands, leaving two coordination sites available in the cis position. The bulkiness of the bidentate ligands precluded the formation of an extended network because of steric hindrance, and thus allowed the stabilization of a molecular complex. Another nice example is reported by Heinrich *et al.* (1998), who showed that substituting three cyanide ligands with a tridentate ligand leads to the formation of an octanuclear box-like molecule. In Long's compound, diamagnetic ions are used but similar cyanide-bridged compounds that contains twenty paramagnetic metal ions (scheme 7) may be designed providing a judicious choice of the chelating organic ligands.

References

de Jongh, L. J. & Miedema, A. R. 1974 Experiments on simple magnetic systems. *Adv. Phys.* **23**, 1–239.

El Fallah, S. M., Rentschler, E., Caneschi, A., Sessoli, R. & Gatteschi, D. 1996 A three dimensional molecular ferrimagnet based on ferricyanide and $[Ni(tren)]^{2+}$ building blocks. *Angew. Chem. Int. Ed. Engl.* **35**, 1947–1949.

Ferlay, S., Mallah, T., Ouahès, R., Veillet, P. & Verdaguer, M. 1999 A chromium–vanadyl ferrimagnetic molecule-based magnet: structure, magnetism and orbital interpretation. *Inorg. Chem.* **38**, 229–234.

Gadet, V., Mallah, T., Castro, I. & Verdaguer, M. 1992 High T_c molecular-based magnets: a ferromagnetic bimetallic chromium(III)–nickel(II) cyanide with $T_c = 90$ K. *J. Am. Chem. Soc.* **114**, 9213–9214.

Ginsberg, A. P. 1971 *Inorg. Chim. Acta Rev.* **5**, 45.

Heinrich, J. L., Berseth, P. A. & Long, J. R. 1998 Molecular prussian blue analogues: synthesis and structure of cubic $Cr_4Co_4(CN)_{12}$ and $Co_8(CN)_{12}$. *J. Chem. Soc. Chem. Commun.*, pp. 1231–1232.

Kahn, O. 1985 Dinuclear complexes with predictable magnetic properties. *Angew. Chem. Int. Ed. Engl.* **24**, 834–850.

Mallah, T., Auberger, C., Verdaguer, M. & Veillet, P. 1995 A heptanuclear $Cr^{III}Ni_6^{II}$ complex with a low-lying $S = 15/2$ ground state. *J. Chem. Soc. Chem. Commun.*, pp. 61–62.

Morpugo, G. O., Mosini, V. & Porta, P. 1980 Crystal structure and spectroscopic properties of a polynuclear complex between [bis(2-aminoethyl)amine]copper(II) and hexacyanoferrate(III). *J. Chem. Soc. Dalton Trans.*, pp. 111–117.

Scuiller, A., Mallah, T., Verdaguer, M., Nivorozkhin, A., Tholence, J.-L. & Veillet, P. 1996 A rationale route to high spin molecules via hexacyanometallates: a new μ-cyano $Cr^{III}Mn_6^{II}$ heptanuclear complex with a low-lying $S = 27/2$ ground state. *New J. Chem.* **20**, 1–3.

van Langenberg, K., Batten, S. R., Berry, K. J., Hockless, D. C. R., Moubaraki, B. & Murray, K. S. 1997 Structure and magnetism of a bimetallic pentanuclear cluster $[(Ni(bpm)_2)_3Fe(CN)_6)_2] \cdot 7H_2O$ (bpm = bis(1-pyrazolyl)methane)). The role of the hydrogen-bonded $7H_2O$ 'cluster' in long-range magnetic ordering. *Inorg. Chem.* **36**, 5006–5015.

Discussion

P. DAY (*The Royal Institution, London, UK*). Has Professor Mallah made a cluster containing a central ferrocyanide ion with six surrounding high-spin Fe(III)? Such a species would mimic the local structure of Prussian blue, for which we proposed a mechanism to explain the ferromagnetic exchange between Fe(III) involving the intervalence charge-transfer configuration many years ago (Mayoh & Day 1976).

T. MALLAH. We have made a pentanuclear cluster containing a central ferrocyanide with four surrounding high-spin Fe(III) chelated by a pentadentate ligand containing two phenolato groups. The magnetic properties clearly show a ferromagnetic interaction between the high-spin Fe(III) ions through the diamagnetic low-spin Fe(II), as observed in Prussian blue. A mechanism involving the intervalence charge transfer configuration proposed by Mayoh & Day (1976) explains the ferromagnetic exchange interaction. Unfortunately, we were not able to observe the intervalence optical band in our compound because of the presence of an intense band centred at 580 nm due to the charge transfer between the phenolato groups and Fe(III).

Additional reference

Mayoh, B. & Day, P. 1976 Contribution of valence delocalisation to the ferromagnetism of Prussian blue. *J. Chem. Soc. Dalton Trans.*, pp. 1483–1486.

Molecular-based magnets: an epilogue

By JOEL S. MILLER

Department of Chemistry, University of Utah, 315 S. 1400 E. RM Dock,
Salt Lake City, UT 84112-0850, USA

This Royal Society Discussion Meeting provided an overview and sampling of some of the exciting current developments in the rapidly growing area of molecule-based magnets. Clearly, as only a few of the multitude of groups studying aspects of the technologically important topic were represented, and time, as it is always in short supply, was not available for the speakers to cover many topics, the audience received a taste of the field. Hence, a variety of overview references to reviews are provided (Gatteschi 1994; Kahn 1993; Kinoshita 1994; Miller & Epstein 1994, 1995a, b; Coronado et al. 1996; Turnbull et al. 1996).

Magnetic materials prepared from molecules, not atoms or ions, have enabled the establishment of a diverse new class of magnets with magnetic ordering temperatures well above room temperature. As a class, molecule-based magnets are anticipated to exhibit a plethora of technologically important attributes that include modulation/tuning of properties via organic chemistry methodologies, compatibility with polymers for composites, low density, flexibility, transparency, low temperature processibility, insulating, solubility, high coercivity, high strength, low environmental contamination, biocompatibility, high magnetic susceptibilities and permeabilities, high magnetizations, low magnetic anisotropy, semiconducting behaviour, etc. Many of these non-magnetic characteristics are not available with conventional atom-based magnets.

Molecular-based magnets have indeed been prepared with critical temperatures ranging from 0.2 to 400 K, with the purely organic magnets, i.e. those with spins solely residing in the p-orbital, having the lowest T_c, i.e. ca. $0.2 < T_c < 1.48$ K. Compounds with spins only residing in d-orbitals have a broad range of T_c up to 372 K for a new material reported first reported at this Discussion Meeting (Hatlevik et al. 1999). Materials with spins residing in both p- and d-orbitals also have a broad range of T_c up to 400 K. Likewise, examples of molecular-based magnets exhibit a broad range of coercivity spanning from low values (ca. 0 Oe) associated with soft magnets to several teslas associated with hard magnets (Kurmoo, this issue; Miller et al. , unpublished research).

To achieve magnetic ordering, it is mandatory that unpaired electrons are present and their number and coupling dictate the resultant magnetic behaviour. As emphasized by several speakers, the couplings are a consequence of the three-dimensional structure motif and although an a priori understanding of the structure–magnetic property relationship is strongly desired, it is increasingly clear that this is complex and so far elusive. Nonetheless, it is a high priority for researchers in the area. Structural motifs documented to stabilize magnetic ordering can be zero (isolated molecules and ions), one (linear chain), two (layers), or three dimensional (networks).

In addition to complex structures, the diversity of magnetic behaviours is bewildering. Within this broad class of materials, ferromagnets, ferrimagnets, antiferromagnets, canted antiferromagnets (or weak ferromagnets), metamagnets, spin glasses, spin frustration, spin flops, as well as the exciting new area of high spin 'small' clusters, some of which are 'single' molecule magnets have been reported. In the area of high spin clusters, the one with the largest spin state, $S = 39/2$, was reported for the first time at this Discussion Meeting (Hashimoto & Ohkoshi, this issue).

Materials that combine magnetic properties with other properties are envisioned to lead to new technologically important materials. In this regard, photomagnetic behaviour was discussed (Hashimoto & Ohkoshi, this issue) and spin crossover materials were touched upon. The former is a rapidly growing embryonic area, while the more mature latter area still has many exciting aspects that will fuel continued studies worldwide.

All emerging research areas have issues that need to be addressed. As a consequence of growth by diverse research groups worldwide (each with differing backgrounds, e.g. synthetic organic, organometallic, main group, and coordination chemistry, as well as physical organic and inorganic chemistry, not to ignore physical chemistry, condensed matter physics, materials science, etc.), language problems arise that impede communication and progress. It behoves all participants in this field to devote some time both to define and clarify the terms they use as well as learn the terms of others.

Issues from a more technical perspective include three-dimensional structure control, identification of structure function relationships, and control of anisotropy. Identification of structure cannot be overemphasized. Akin to the study of biologically important systems, it is necessary to understand and control the primary, secondary, and tertiary structures. Structure determination, particularly by neutron diffraction, at the temperatures at which the magnetic phenomena and transitions occur is the Holy Grail. As challenging as this is, progress in this area should be targeted. The availability of CCD detectors has enabled the structure determination of increasingly smaller crystals and this should provide an impetus to the development of structure–function relationships. Nonetheless, disorder persists as a problem and improved methods need to be developed to unravel complex disorder. The growth of larger crystals will simplify, in some cases, the structure determination by X-ray diffraction, but more importantly will enable important single crystal neutron diffraction studies as well as anisotropic magnetic (EPR and magnetization) and optical measurements. Probing of the local magnetic properties should be a highly placed agenda item and muon spin resonance is being used more and more to provide information in this regard.

Perhaps in a decade, another Royal Society Discussion Meeting on this topic can be convened, and I trust enormous progress will have been made with many fascinating new and totally unexpected results being reported.

References

Coronado, E., Delhaès, P., Gatteschi, D. & Miller, J. S. (eds) 1996 *Molecular assemblies to the devices*. NATO Advanced Studies Workshop, E321. Kluwer.

Gatteschi, D. 1994 *Adv. Mat.* **6**, 635.

Hatlevik, Ø., Buschmann, W. E., Zhang, J., Manson, J. L. & Miller, J. S. 1999 *Adv. Mat.* **11**, 914.

Kahn, O. 1993 *Molecular magnetism*. New York: VCH.

Kinoshita, M. 1994 *Jap. J. Appl. Phys.* **33**, 5718.

Miller, J. S. & Epstein, A. J. 1994 *Angew. Chem. Int. Ed.* **33**, 385.

Miller, J. S. & Epstein, A. J. 1995*a* *Chem. Eng. News* **73** (40), 30.

Miller, J. S. & Epstein, A. J. (eds) 1995*b* Molecule-based magnets. *Mol. Cryst. Liq. Cryst.* **271–274**.

Turnbull, M. M., Sugimoto, T. & Thompson, L. K. (eds) 1996 *Molecule-based magnetic materials*. ACS Symp. Ser., 644.

The Bakerian Lecture, 1999
The molecular chemistry of
magnets and superconductors

By Peter Day

*Davy Faraday Laboratory, The Royal Institution of Great Britain,
21 Albemarle Street, London W1X 4BS, UK*

*Lecture held at The Royal Institution 10 February 1999 and
University of Oxford 15 February 1999*

Only recently have magnetism or superconductivity been associated with molecular compounds. Observing these properties in molecular-based lattices has brought chemistry to bear on the relation between crystal structure and physical properties. Recent work from The Royal Institution is used to identify features associated with the molecular-based materials that distinguish them from conventional continuous lattice solids, and which are unique to molecular solid state.

**Keywords: magnetic materials; organic magnets; metal–organic solids;
superconductors; molecular materials**

1. Introduction: conducting and magnetic molecular compounds

In the 224 years since its inauguration, the Bakerian Lecture has quite frequently taken as its subject matter work performed at The Royal Institution. Indeed, in the more expansive days of the early 19th century, Humphrey Davy gave no fewer than five such lectures, exceeded only by our Institution's most famous son, Michael Faraday, who gave six! In the year of its bicentenary, it is therefore particularly appropriate to survey a topic which, while its gestation goes back more than 20 years, has formed the principal activity of my own research group at The Royal Institution. Especially apposite, too, is the fact that it concerns both magnetism and conductivity, subjects that have long pedigrees in the Albemarle Street laboratories. For example, by his discovery of potassium, set out in one of his Bakerian Lectures (Davy 1808), Davy overturned at a stroke the conventional view of metals, since the material he described was both low melting and chemically extremely reactive. Likewise, the metals described in the present lecture are quite unconventional, being soft molecular crystalline compounds obtained from organic solvents at room temperature. Some contain solvent molecules, including water, trapped within their crystal lattices, or have chiral or magnetic centres as an integral part of their structures.

(a) Conductivity in molecular crystals

Of all the physical properties characterizing condensed matter, electrical conductivity is the one spanning the largest number of orders of magnitude, from the

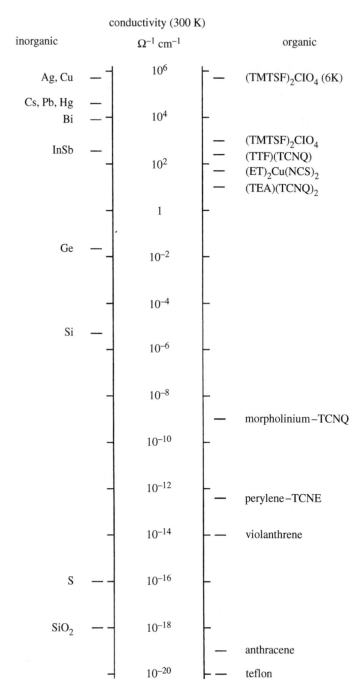

Figure 1. Room temperature specific conductivities (S cm^{-1}) of representative inorganic (left hand) and molecular organic (right hand) solids.

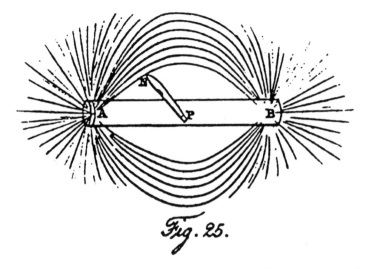

Figure 2. Lines of magnetic flux around a bar magnet (from Faraday 1832).

most highly conducting metals to the most insulating solids. It may therefore appear surprising at first sight that, over the last two decades, chemical synthesis has led to the discovery of many molecular solids with room temperature electrical conductivities spanning as wide a range as those of more conventional metals, semiconductors and insulators. The left-hand side of figure 1 indicates specific conductivities at room temperature for various examples of structurally simple solids, most of them elements, whose properties have long provided paradigms of the metallic, semiconducting and insulating states. Nearly all are continuous lattice materials. In contrast, on the right-hand side of figure 1 we see a series of chemical formulae whose initial characteristic, compared with those on the left, is that they are much more complicated. Indeed most of the formulae would not be recognizable immediately by chemists, since they are based on abbreviated acronyms invented largely by physicists interested by their properties but unable to handle the complexities of the correct chemical nomenclature. Thus we have a parable of the circumstance, quite often found in science, that a topic may start by being multidisciplinary, but has further steps to take before it can properly be said to be interdisciplinary. Compounds with the highest conductivities are of two types: charge transfer complexes, consisting of an electron donor and acceptor, and charge transfer salts, typically containing an electron donor molecule in the form of a cation and an inorganic anion. The examples that we have prepared and characterized are all of the latter type. Uniting all these compounds is the observable characterizing the metallic state first pointed out by Davy in another of his Bakerian Lectures (Davy 1821), namely a negative temperature coefficient of conductivity. As Davy himself put it, 'The most remarkable general result that I obtained by these researches ... was that the conducting power of metallic bodies varied with the temperature and was lower in some inverse ratio as the temperature was higher'.

(b) *Magnetism in molecular crystals*

A quite analogous evolution, in which the role of solid state chemistry has become progressively more marked, can be seen when we examine the history of magnetic materials. It is also a story in which The Royal Institution played a key part in the 19th century. When Michael Faraday (1832) published the first illustration of the lines of magnetic flux around a magnet (figure 2), there can be no doubt at all that the cylindrical bar in the picture was of iron, since, apart from the discovery of lodestone millennia previously, that was the only practical magnetic material available at the time. The 160 years since Faraday's picture have seen a veritable explosion in the number and variety of solids showing spontaneous magnetization below some critical temperature, from the ferro- and ferrimagnetic oxides whose synthesis created the challenge taken up in Néel's molecular field model to the most recent ferromagnets that contain no metallic elements at all, and whose ordering mechanism remains a subject of theoretical debate.

Most of the magnetic solids synthesized by chemists of the past 40 years have not been metallic conductors (certainly not the metal-organic and organic materials that constitute the burgeoning subject of 'molecule-based magnetism' surveyed recently at a Royal Society Discussion Meeting (this issue) and a series of biennial international conferences (*Proc. Int. Conf. Mol. Magnets, Mol. Cryst. Liq. Cryst.*)). Here, the presence of ordered localized moments in an insulating lattice gives rise to numerous properties quite distinct from those of metallic (or itinerant) magnets. Not only are such materials frequently soluble in organic solvents, or even water, but their optical properties are quite unusual: it is over 20 years since we showed that the transparent ferromagnet $(CH_3NH_3)_2CrCl_4$, soluble in ethanol, changed colour visibly from green to yellow on passing through the Curie temperature near 40 K, because excitons arising from spin-forbidden ligand field transitions gained intensity by coupling to spin-waves propagating in the two-dimensional ferromagnetic lattice (Bellitto & Day (1978); for a review see Bellitto & Day (1992)).

(c) Chemically constructed multilayers

The present brief account cannot hope to survey all the remarkable variety of conducting and magnetic compounds synthesized by chemists (or even by our own group) over the past 20 years. Rather, by selecting a small number of examples prepared and studied at The Royal Institution in the last few years, my aim will be to illustrate a few of the unusual structures and unlooked for properties emerging from the application of supramolecular chemistry to collective electronic properties. To focus matters still further, most of the compounds chosen have strongly developed low-dimensional character, usually expressed by the presence of alternating layers of organic and inorganic material. What we have called 'organic–inorganic composites' (Day 1985) or 'chemically constructed multilayers' (Day 1990) (to distinguish them from the physicists' multilayers laid down by molecular beam epitaxy) combine many of the physical properties of both classes of solid, with the added subtlety that the organization of molecules within the layers (and hence of the interactions determining the collective electronic behaviour) depends crucially on the interfaces with the neighbouring layers. Quite small variations in the chemistry result in major changes in physical properties.

2. Metal–organic and organic molecular magnetic compounds

In this section, three distinct chemical systems are described, in each of which a spontaneous magnetization is found arising from three quite different physical mechanisms. In each case, too, we identify features of the magnetic behaviour peculiar to the molecular nature of the lattice.

(a) *Organophosphonato-metal(II) salts: canted antiferromagnets*

It is far from easy to construct insulating lattices from molecular building blocks that exhibit ferromagnetic exchange between neighbouring localized moments because, following the classical models of Anderson (1963), Kanamori (1959) and Goodenough (1955), it is necessary to engineer the alignment of adjacent units so that the orbitals bearing the unpaired electrons are orthogonal. Based on antiferromagnetic superexchange between localized moments, mediated through polyatomic bridging ligands, an easier task is to organize a lattice in which adjacent moments are not exactly antiparallel, but make a small angle to one another so that the resultant is a weak ferromagnetic moment (sometimes also called canted antiferromagnetism). This situation arises when the site symmetry of the ions carrying the magnetic moments is low, and there is no centre of inversion between neighbouring sites (the so-called Dzialoshinskii–Moriya mechanism (Dzialoshinskii 1958; Moriya 1960)). An excellent example of this phenomenon is provided by the organophosphonate salts of divalent 3d metals. They also exemplify one straightforward method of introducing molecular organic groups into a purely inorganic lattice.

The extensive series of ternary transition metal phosphate salts with general formula $A^I M^{II} PO_4 \cdot H_2O$ (A^I is either a Group 1 element or NH_4^+) already have well-defined layers of M^{II} coordinated to PO_4^{3-} and H_2O, forming approximately square arrays of M^{II} (Carling *et al.* 1995). By replacing one of the PO_4 atoms with an organic moiety, we arrive at the equally extensive series of organophosphonato-salts $RPO_3 M^{II} \cdot H_2O$. The R groups, which can be *n*-alkyl, aryl or other functionalized groups, extend outwards from the layers of M^{II}, $-PO_3$ and H_2O, replacing the A^I, and forming purely organic layers bound only by van der Waals interactions. Like the purely inorganic ternary phosphates, the organophosphonates are also weak ferromagnets, but with two significant differences (Carling *et al.* 1993). First, the magnitude of the ferromagnetic moment varies greatly with the organic group, increasing monotonically in the *n*-alkyl-sequence from C_1 to C_4. Thus, we have a first instance where a change in the organic part of the lattice remote from the site of the atomic moments changes the bulk magnetic properties of the crystal. However, the second difference is even more surprising.

Naively, one might imagine that increasing the *n*-alkyl chain length, which increases the distance between the weakly ferromagnetic metal phosphonate layers, would bring about a corresponding decrease in the three-dimensional magnetic ordering temperature T_N which must be induced by interlayer interaction. On the contrary, however, T_N alternates; it *increases* from CH_4 to C_2H_5 decreasing again to C_3H_7 and increasing to C_4H_9 (Carling *et al.* 1993)! What have been called 'alternation effects' are well known in the structural chemistry of aliphatic molecular crystals, but this appears to be the first time such an effect has been detected in the bulk magnetism of such a solid. The most likely explanation is in the orientation of the terminal $-CH_3$ of the alkyl chain. Given that the $P-C$ bond is almost perpendicular to the $Mn-O$ plane, *n*-alkyl phosphonates containing odd numbers of C atoms in the alkyl group have terminal $C-C$ bonds likewise perpendicular to that plane. On the other hand, the *n*-alkyl groups with even numbers of C atoms have terminal $C-C$

nearly parallel to the Mn—O plane, because of the tetrahedral C—C—C bond angles. Consequently, the $-CH_3$ groups mesh together differently in the two cases, transmitting a small variation in C—P—O angle to the plane of Mn, with consequent small modulation of the Mn—O—P—O—Mn superexchange pathway and hence the exchange constant between Mn neighbours. It is this exchange constant, which in turn determines the magnetic correlation length within each layer, whose build up with decreasing temperature provides the extended magnetic dipoles that interact across the organic molecular layer to bring about three-dimensional long-range magnetic ordering at T_N (Kurmoo *et al.* 1999). Thus, once again, a small change in an organic group remote from the locus of the moment has a significant influence on the bulk magnetic properties.

(b) Nitronylnitroxides: purely organic ferro- and antiferromagnets

A similar sensitivity of the electronic ground state to fine details in the molecular arrangement is found among the class of purely organic magnetic solids based on the nitronylnitroxide moiety. Until the early 1990s, all magnetic materials since the dawn of history, whether they were metallic conductors or not, had owed their magnetic properties to the presence of metal atoms within the structure, specifically to partly occupied d- or f-shells. Then Kinoshita and his colleagues (Tamamura *et al.* 1991) discovered ferromagnetism in a series of solids based on the stable, organic, nitronyl–nitroxide radicals containing only C, H, N and O, the first examples of ferromagnetism due to p-electrons. Although their Curie temperatures are extremely low (mostly below 1 K), in their bulk magnetic behaviour these molecular crystals are indistinguishable from other ferromagnets, for example, having internal fields increasing from zero to T_c to reach a saturation value as the temperature decreases. A very convenient and direct way of measuring the internal field is to record the zero field muon spin rotation (μSR), whereby positive muons are implanted in a sample and the angular asymmetry of their decay is observed using detectors in the forward and backward direction relative to the muon beam. The asymmetry arises because the magnetic moment of the muon precesses around the internal field, the precession frequency being a measure of the field strength. Typical μSR oscillations measured at the ISIS pulsed muon facility at the Rutherford Appleton Laboratory in a collaboration with colleagues in Oxford and Yokohama are shown in figure 3a, and the resulting temperature variation of the internal field in a nitronyl–nitroxide ferromagnet in figure 3b (Pratt *et al.* 1993).

Since Kinoshita's original discovery (Tamamura *et al.* 1991), many nitronylnitroxides have been synthesized to explore structure–property relations in this unusual class of magnet by systematically changing the organic substituent R attached to the nitronylnitroxide moiety. One example will suffice: when R = 3-quinolyl (3QNNN) we find a ferromagnet, but for R = 4-quinolyl (4QNNN) it is an antiferromagnet. The reason for the striking difference becomes apparent from the crystal structures (Sugano *et al.* 1993). Although the two substituents are very similar from an electronic point of view, the difference in the point of attachment of the quinolyl group to the nitronylnitroxide means that the molecules have quite different shapes and hence pack differently in their respective crystals. In 3-QNNN the axes of the O—N—C—N—O groups that carry the unpaired spins are approximately perpendicular suggesting the orthogonality between magnetic orbitals necessary for ferromagnetism, while in

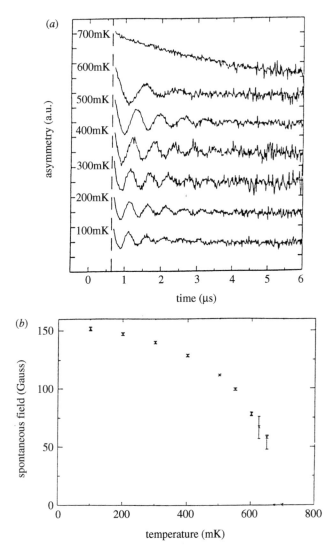

Figure 3. (*a*) Zero field muon spin rotation in *p*-nitrophenyl–nitronylnitroxide. (*b*) Temperature variation of internal field determined by muon spin rotation.

4-QNNN the N—O groups on neighbouring molecules are brought close enough for electron pairing to take place.

(*c*) *Bimetallic tris-oxalato-salts: honeycomb layer ferrimagnets*

A distinguishing feature of molecular-based magnetic materials is that the vast majority are insulators, and the microscopic magnetic moments are therefore localized. Maximum connectivity between centres carrying the moments is therefore desirable to maximize ordering temperatures, so ambidentate ligands are most effective. The bimetallic *tris*-oxalato series $AM^{II}Fe^{III}(C_2O_4)_3$ is particularly interesting from

Table 1. M^{II}–$Fe^{III}(d_1)$ and interlayer separation (d_2) in two-dimensional $AM^{II}Fe^{III}(C_2O_4)_3$

	M^{II} = Fe		M^{II} = Mn	
	d_1 (Å)	d_2 (Å)	d_1 (Å)	d_2 (Å)
$N(n\text{-}C_3H_7)_4^+$	4.667	8.218	4.686	8.185
$N(n\text{-}C_4H_9)_4^+$	4.701	8.980	4.731	8.937
$N(n\text{-}C_5H_{11})_4^+$	4.703	10.233	4.728	10.158
$P(n\text{-}C_4H_9)_4^+$	4.735	9.317	4.760	9.525
$As(C_6H_5)_4^+$	4.683	9.655	4.722	9.567
$N(C_6H_5CH_2)(n\text{-}C_4H_9)_3^+$	4.690	9.633	4.735	9.433
$(C_6H_5)_3PNP(C_6H_5)_3^+$	4.690	14.433	4.707	14.517

this point of view, since a wide range of organic cations A^+ stabilize a hexagonal layer structure and the details of magnetic ordering are highly sensitive to changes in the packing of A between the layers.

One strategy for achieving finite zero-field magnetization in a molecular-based array, without the need for ferromagnetic near-neighbour exchange, is to exploit ferrimagnetism. Compounds with general formula $AM^{II}M^{III}(C_2O_4)_3^{3-}$ constitute a very extensive series, formed by a wide range of organic cations A^+, as well as divalent and trivalent M both from transition metal and B-subgroup ions (Tamaki *et al.* 1992). Depending on the connectivity of the $M^{III}(C_2O_4)_3^{3-}$ units effected by the M^{II} one can have either a two- or three-dimensional array, with the build up of the long-range order state being 'templated' by the organic A^+ (Decurtins *et al.* 1994*b*). Connection between $M^{III}(C_2O_4)_3^{3-}$ by M^{II} in two dimensions produces a honeycomb structure in which both metal ions occupy sites of trigonally distorted octahedral geometry, with all near-neighbour M^{II}, M^{III} pairs bridged by oxalate ions (figure 4). Many compounds in this series therefore have crystal structures that are approximately hexagonal, with basal plane unit cell constants that vary only slightly with A^+, though with strongly varying interlayer separations. Some unit cell constants are listed in table 1, which show that a factor of 2 in interlayer separation is easily achievable.

The set of compounds we have studied in detail have M^{II} = Mn or Fe and M^{III} = Fe (Mathionière *et al.* 1996). Averaged over the whole group, the spacing between the metal ions in the plane decreases from Mn to Fe by 0.03 Å, in line with the decrease in ionic radius expected from ligand field considerations. In contrast, though, the interplanar spacing *increases* by an average of 0.08 Å, most probably because the organic groups which enter the hexagonal cavities are slightly extended as the cavity becomes smaller.

One example studied in detail is $N(n\text{-}C_5H_{11})_4MnFe(C_2O_4)_3$, whose crystal structure consists of alternate layers of $[MnFe(C_2O_4)_3]^-$ and $N(n\text{-}C_5H_{11})_4^+$ (Carling *et al.* 1996). The former comprise honeycomb networks of alternating Mn and Fe bridged by $C_2O_4^{2-}$ (figure 5). Thus, both metal ions are coordinated by six O originating from three bidentate oxalate ions forming trigonally distorted octahedra. Similar networks have been observed in $P(C_6H_5)_4MnCr(C_2O_4)_3$ (Decurtins *et al.* 1994*a*) and in the $P(C_6H_5)_4^+$ compound, one P—C bond lies parallel to a threefold axis and the unit cell is rhombohedral.

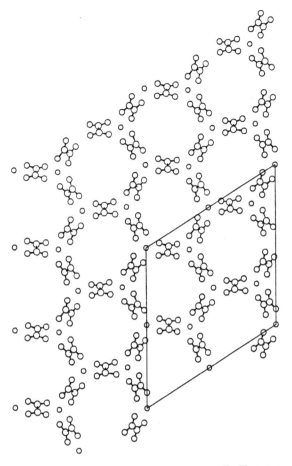

Figure 4. Layer honeycomb structure of $AM^{II}M^{III}(C_2O_4)_3$.

As far as the bimetallic *tris*-oxalato layer is concerned (figure 5), the deviation of the 3d ions from a hexagonal array is implicit in the orthorhombic space group: the angles Fe–Mn–Fe and Mn–Fe–Mn are, respectively, 112° and 138° instead of 120°. The site symmetry of the metal ions, which would be D_3 if the cell were rhombohedral, is reduced to C_2 and the metal–oxygen bond lengths are not all equal. One index of the distortion of the MO_6 units from regular octahedra is the deviation of the *trans* O—M—O bond angles from 180°: at the Fe site two such angles are 163° and one 170°. As expected for bidentate chelating oxalate groups the 'bite angle' O—M—O averages 78.0° around the Fe site and 79° around the Mn, similar to those found in other oxalato-complexes of Fe(III) and Mn(II) (Julve *et al.* 1984). Since the mean O—M—O angle for O atoms on adjacent oxalate groups exceeds 90° (99.4° at the Fe site and 98.6° at Mn), both MO_6 octahedra may be considered as slightly elongated perpendicular to the plane of the $[MnFe(C_2O_4)_3]^-$ layer. Alternate layers have opposite chirality (i.e. Mn(Λ) and Fe(Δ) in one layer, and Mn(Δ), Fe(Λ) in the next).

The N atoms of the $N(n\text{-}C_5H_{11})_4^+$ form rectangular planar arrays, interleaving the

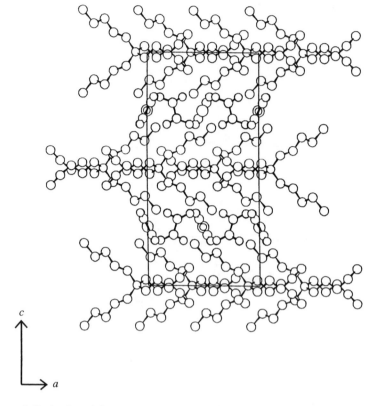

c

a

Figure 5. Projection of the crystal structure of $N(n\text{-}C_5H_{11})_4MnFe(C_2O_4)_3$ along the b-axis
(Carling *et al.* 1996).

$MnFe(C_2O_4)_3^-$ layers, with the four attached alkyl chains extended: two in the plane
of the N and two perpendicular. The first four C atoms in each chain are fully
extended but the terminal CH_3 is twisted away from the plane of the four CH_2
towards the *gauche* configuration. Attempts to synthesize compounds in the series
$N(n\text{-}C_nH_{2n+1})_4MnFe(C_2O_4)_3$ have only been successful for $n = 3\text{-}5$, no C_6H_{14}
derivative being obtained (Nuttall 1997). It would appear that the steric requirement
for accommodating the alkyl chains within the hexagonal Mn_3Fe_3 cavities cannot be
satisfied for $n > 5$, though in the series with $A = (n\text{-}C_nH_{2n+1})P(C_6H_5)_3^+$, compounds
with $n = 6,7$ have been prepared (I. D. Watts, unpublished work).

   The $M^{II} = Mn$ compounds constitute a rather unusual kind of ferrimagnetism in
that the electronic ground states of the two metal ions are the same: $^6A_1$ in D_3
symmetry. The near-neighbour exchange interaction is strongly antiferromagnetic,
as indicated by the large negative Weiss constants, which do not vary much with
A since the exchange pathway is only slightly affected by changing the organic
group. Further evidence of strong antiferromagnetic spin correlations within the
layers is provided by the existence of a broad maximum in the susceptibility at
55 K, again independent of A. The short-range magnetic order therefore mimics
that of a two-dimensional antiferromagnet. However, with the onset of long-range

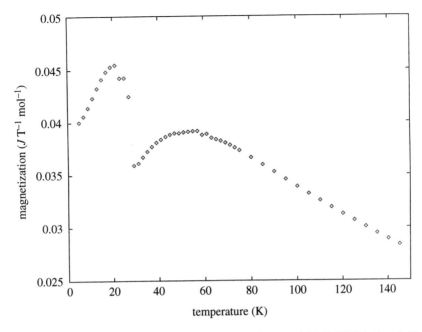

Figure 6. Temperature-dependent magnetization of $N(n\text{-}C_5H_{11})_4MnFe(C_2O_4)_3$ in a field of 10^{-2} T (Mathionière *et al.* 1996).

order around 27 K (nearly independent of A), the susceptibility increases abruptly (figure 6), to reach a value which does vary strongly with A, being smallest for $N(n\text{-}C_4H_9)_4^+$ and largest for $(C_6H_5)_3PNP(C_6H_5)_3^+$ at 5 K (Mathionière *et al.* 1996). As in the Mn alkyl-phosphonates (Carling *et al.* 1993), the magnitude of the uncompensated moment is determined by an organic group which is not only not implicated in the exchange mechanism, but is spatially remote from the site of the magnetic moment, a phenomenon which has no analogue among conventional magnetic materials.

When M^{II} = Fe in the bimetallic *tris*-oxalato-Fe(III) series, we have the interesting situation of a mixed valency compound (Day 1980). The physical properties are those of Class II in the Robin–Day classification (Robin & Day 1966), but unexpectedly a bizarre magnetic phenomenon is seen. The two magnetic ions being $S = 2$ and $S = 5/2$, the resulting behaviour is that of a conventional ferrimagnet. However, depending on the nature of the organic cation A, one either has a conventional magnetization at low temperature, increasing monotonically from zero at T_c to a limiting value at $T \rightarrow 0$ (figure 7a), or a magnetization that increases at first from zero below T_c but then reaches a maximum. At lower temperatures, the magnetization then falls again, passing though zero and becoming strongly negative (Mathionière *et al.* 1994; Nuttall & Day 1998)(figure 7b). This behaviour is extremely rare among molecular-based magnetic materials but finds a precedent among continuous lattice oxides (Goodenough 1963). As long ago as the 1940s there was great interest in ferrimagnetic mixed-valency Fe oxides with spinel and garnet structure because they were among the first materials discovered with non-zero spontaneous magnetization which were not metallic.

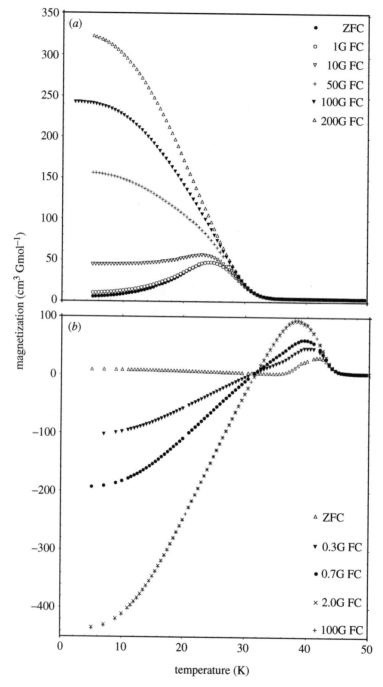

Figure 7. Temperature-dependent magnetization of ferrimagnets $AFe^{II}Fe^{III}(C_2O_4)_3$ measured in a 100 G field after cooling in different fields: (*a*) A = P $(C_6H_5)_4$; 'normal' behaviour; (*b*) A = $N(n\text{-}C_4H_9)_4$, negative magnetization (Nuttall & Day 1998).

The origin of the apparently bizarre situation that the net magnetization of a sample could be antiparallel to the applied measurement field was actually described as long ago as 1948 by Néel (1948). In a ferrimagnet the net magnetization at a given temperature is the vector sum of the magnetizations of each sublattice. Should the temperature dependence of the magnetizations of each sublattice be similar, the resultant will be a monotonic increase from T_c to absolute zero as shown in figure 7 for $P(C_6H_5)_4Fe^{II}Fe^{III}(C_2O_4)_3$. On the other hand if the temperature derivatives of sublattice magnetization $dM_{Fe(II)}/dT$ and $dM_{Fe(III)}/dT$ have a different dependence on temperature, then the temperature derivative of the resultant $d(M_{Fe(II)} - M_{Fe(III)})/dT$ can change sign. It is also feasible for the magnetizations of the two sublattices to cancel at a 'compensation temperature'. However, the feature distinguishing the bimetallic *tris*-oxalato compounds from the oxides is that once again the drastically varying magnetic behaviour comes about by changing organic groups situated quite far away in the lattice from the magnetic centres.

3. Superconducting and semiconducting molecular charge transfer salts

Superconductivity, certainly an unlooked-for phenomenon when it was first discovered in 1910, was first found in an organic molecular solid in the early 1980s (Jerome *et al.* 1980), bringing the topic into the realm of synthetic chemistry. The compounds in question are molecular charge transfer salts, defined as substances derived from an electron donor molecule in the form of a cation and an inorganic anion of defined charge, such as I_3^-, ClO_4^-. (A much smaller number of salts arise from anionic molecular acceptors and small cations such as NH_4^+.) At first sight, superconductivity might appear as unlikely a property to find in a molecular material as ferromagnetism, and for the same reason, namely that by definition intramolecular electronic interactions are much stronger than intermolecular ones in such a solid. Nevertheless, the past 15 years have seen many such compounds synthesized, and broad structure–property relationships emerge (see, for example, Williams *et al.* 1992). Critical temperatures for the onset of superconductivity have risen from 1.5 to 13 K (33 K in fulleride salts) which, although far lower than the high T_c cuprates, lies well within the upper quartile of all known superconductors with the exception of the cuprates, which still remain a distinct and enigmatic group.

Given that the T_c of the molecular superconductors are relatively low, it is pertinent to ask why it is worth devoting much effort to them. The question can be answered in several ways. First, from the straightforward standpoint of synthesis and processing, it is undoubtedly appealing to be able to envisage making superconductors from solution at room temperature, in contrast to conventional metallurgical or ceramic methods. Second, it is attractive to be in a position to modify (and in some cases completely transform) the electronic ground state by small chemical changes. Third, because of their molecular make-up, they are highly compressible, so that the electronic ground state can be altered drastically by quite modest applied pressures. Fourth, and perhaps most significant for scientific novelty, is the opportunity presented by the molecular nature of the lattice to introduce properties that would scarcely be feasible in a simple close-packed continuous lattice. Two examples taken up below are structural chirality and magnetism.

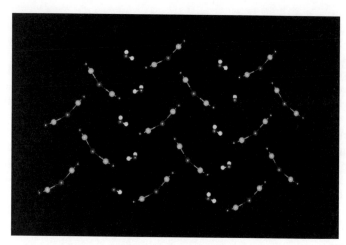

Figure 8. The anion layer in κ-(BEDT-TTF)$_2$Ag(CN)$_2 \cdot$ H$_2$O (Kurmoo *et al.* 1990). Note the cavities which contain the BEDT-TTF –CH$_2$CH$_2$– groups.

(a) Bis-ethylenedithio-tetrathiafulvalene (BEDT-TTF) salts: structures and properties

The first molecular superconductors were salts of tetramethyl-tetraselenofulvalene (TMTSF), but in the mid-1980s the organo-sulphur donor BEDT-TTF came on the scene (Mizuno *et al.* 1978) and its salts now provide the physicist with examples of virtually every collective electronic ground state known: semiconductors, metals, superconductors, charge density waves, spin-density waves, spin–Peierls transitions, etc. Such richness comes from the fact that the collective properties emanate from two main parameters: the mean charge on the BEDT-TTF molecules, which determines the Fermi energy, and the molecular packing (i.e. the distances and angles between neighbouring molecules), which determines the overlap between frontier molecular orbitals and hence the intermolecular transfer integrals.

BEDT-TTF charge transfer salts are known with mean cation charges of 0.5, 0.67, 1.0 and 2.0 and a precise, if empirical, correlation has been found with the C — C and C — S bond lengths in the central tetra-thiafulvalene part of the molecule, so that effective cation charges can now be estimated reliably from crystallographic data (Guionneau *et al.* 1997). This is especially valuable in the not infrequent cases where inequivalent molecules in the unit cell bear significantly different charges. For example (BEDT-TTF)$_4$ReCl$_6 \cdot$ C$_6$H$_5$CN, with a mean cation charge of +0.5, contains alternate stacks of donor molecules with charges of 0 and +1; the former lying close to the neutral solvent molecules and the latter to the ReCl$_6^{2-}$. Physically, the compound is a semiconductor with (apart from the 5d moment on the Re) a one-dimensional antiferromagnetic interaction between the BEDT-TTF$^+$ ($S = 1/2$) (Kepert *et al.* 1997).

Among the superconducting BEDT-TTF salts two main structure types have emerged, labelled β and κ, in both of which (as is commonly the case with salts of this donor) the organic donor cations and inorganic anions form separate alternate layers. We illustrate each by examples from our earlier work. Salts of both categories have mean cation charges of +0.5, so that in β-(BEDT-TTF)$_2$AuBr$_2$, for instance, the donor molecules are organized into stacks with the mean planes of the TTF groups parallel (Talham *et al.* 1987; Chasseau *et al.* 1993). However, because of numerous interstack S ... S contacts below the van der Waals distance, the Fermi surface is closer to being two dimensional than one, so that the ratio of the normal-state metallic conductivity within the donor layer, parallel and perpendicular to the stacks, is only about 3:1. In contrast, in κ-phase salts, exemplified by κ-(BEDT-TTF)$_2$Ag(CN)$_2 \cdot$ H$_2$O (Kurmoo *et al.* 1990), the BEDT-TTF are not arranged in stacks but as discrete plane–plane dimers, with the planes of adjacent dimers orthogonal. Conductivity parallel to the donor layer is then isotropic. This compound, which has a T_c of 5.5 K, is the first ambient pressure superconductor of any kind to have water in the crystal lattice. Another, (BEDT-TTF)$_3$Cl$_2 \cdot$ 2H$_2$O, prepared earlier, only becomes superconducting under pressure (Rosseinsky *et al.* 1988; Kurmoo *et al.* 1988).

At once the question arises as to why salts with two such similar anions, both linear and with the same charge, should induce such different packing motifs in the donor layers. The answer lies in the arrangement of the Ag(CN)$_2^-$ in the anion layer, linked together by H-bonds to the H$_2$O (figure 8). The resulting planar array contains cavities into which the terminal $-$CH$_2$CH$_2-$ groups 'dock'. As we shall see below, such interactions between the donor and anion layers are a crucial factor determining the orientation of the donors, and hence the bulk physical properties of BEDT-TTF salts.

(b) *BEDT-TTF salts with tris-oxalato-metallate(III) anions:*
magnetism, chirality and solvent templating

Superconductivity and long-range magnetic order have long been thought to be ground states incapable of coexistence because the internal field generated by magnetic centres would break up the superconducting Cooper pairs. It is therefore of interest to use the chemical flexibility available in a molecular-based compound to try and synthesize superconductors containing magnetic centres. In the charge transfer salts an obvious approach is to make salts with paramagnetic anionic transition metal complexes, and over a period of years we pursued this strategy by preparing and characterizing a large number of BEDT-TTF salts with anions ranging from simple halogeno-species such as FeCl$_4^-$ (Mallah *et al.* 1990) and ReCl$_6^{2-}$ (Kepert *et al.* 1997) to large metal clusters such as Mo$_6$Cl$_{14}$ (Kepert *et al.* 1998) and PW$_{12}$O$_{40}^{6-}$ (Bellitto *et al.* 1995). This work has been reviewed (Kurmoo & Day 1997). Most compounds of this kind are semiconductors because of charge disproportionation within the donor sublattices, but one, (BEDT-TTF)$_4$CuCl$_4 \cdot$ H$_2$O (Day *et al.* 1992), is a metal from ambient temperature down to 400 mK, though without any transition to superconductivity, and also shows evidence for short-range ferromagnetic correlations between the Cu moments. However, more recently we have prepared an extensive series of compounds with the general formula (BEDT-TTF)$_4$[A^{I}M^{III}(C$_2$O$_4$)$_3$] (solvent), among which are the first ever superconductors containing paramagnetic 3d ions (Graham *et al.* 1995). As with the bimetallic *tris*-oxalato-salts described in §3 *a*, the chirality of the M(C$_2$O$_4$)$_3^{3-}$ is a crucial factor in determining the crystal structure and indeed, there are two different ways of incorporating Λ and Δ stereoisomers into the ultimately racemic lattice. The two phases are also strikingly different in their physical properties, one being superconducting and the other semiconducting.

The compound β''-(BEDT-TTF)$_4$[(H$_3$O)Fe(C$_2$O$_4$)$_3$]C$_6$H$_5$CN was the first molecular superconductor containing paramagnetic metal centres, an important precedent

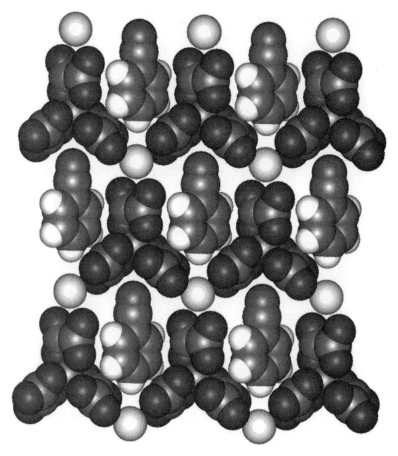

Figure 9. The anion layer in β''-(BEDT-TTF)$_4$[(H$_3$O)M(C$_2$O$_4$)$_3$]C$_6$H$_5$CN; M = Cr, Fe
(Kurmoo *et al.* 1995; Martin *et al.* 1997).

because of the conflict between superconductivity and magnetism in solids (Fisher &
Maple 1983). The crystal consists of alternating layers of BEDT-TTF within which
the cations form stacks in the β''-packing motif and layers of anions and solvent
molecules very similar to that found in the bimetallic *tris*-oxalate-salts, except that
one of the metal sites is occupied by H$_3$O$^+$ and the hexagonal cavity between the
oxalate ions contains the C$_6$H$_5$CN rather than the organic cation (Kurmoo *et al.*
1995). If, however, the H$_3$O$^+$ is replaced by K$^+$ or NH$_4^+$, the resulting compounds are
semiconductors, although superficially at least, the hexagonal arrangement of the
anion layer is almost the same.

Close inspection reveals a most significant difference. Both compounds are race-
mates: the lattices contain equal proportions of Δ and Λ enantiomers of Fe(C$_2$O$_4$)$_3^{3-}$,
but distributed in different ways. In the superconducting compound, which is mono-
clinic ($C2/c$), we find alternate layers consisting of only Λ or Δ molecules (figure 9);
in the semiconducting phase, on the other hand, which is orthorhombic (Pbcn), rows
of Δ and Λ molecules alternate within each layer (figure 10). In consequence, the site

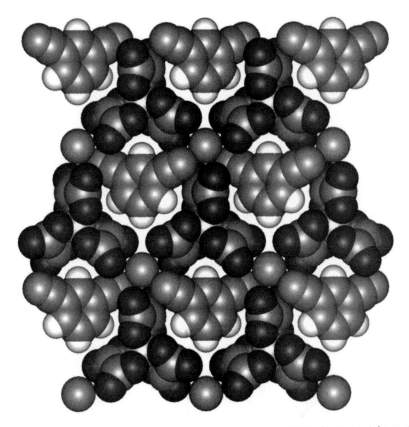

Figure 10. The anion layer in semiconducting (Pbcn): $(BEDT\text{-}TTF)_4[AM(C_2O_4)_3]C_6H_5CN$;
A = K, NH_4, H_3O; M = Cr, Fe, Co, Al (Kurmoo *et al.* 1995; Martin *et al.* 1997).

occupied by the A^+, though surrounded by six O atoms from oxalate anions in both cases, has quite a different symmetry from that found in the monoclinic phase, and in fact has a cavity on one side which accommodates the –CN group of the C_6H_5CN, thus rendering the site seven coordinate.

Thus, we have the apparently paradoxical situation of two remarkably similar *tris*-oxalato-metallate(III) anion sublattices leading to two quite different BEDT-TTF packing types, and hence to totally different properties.

In the series with general formula $(BEDT\text{-}TTF)_4[AM(C_2O_4)_3]C_6H_5CN$, five examples are known to date which crystallize in the Pbcn space group, with the anion sublattice illustrated in figure 9 (Kurmoo *et al.* 1995). They are listed in table 2 (Martin 1999), where it should be noted in particular that the $(H_3O)Cr$ phase is polymorphic (Martin *et al.* 1997).

The arrangement of BEDT-TTF molecules in the Pbcn phase is illustrated in figure 11. Analysis of the C—C and C—S bond lengths using the algorithm of Guionneau *et al.* (1997) reveals that half the BEDT-TTF are uncharged, while the other half, which form face-to-face dimers, carry a charge of +1. From magnetic susceptibility measurements we find there is no contribution from the donor layer so the single unpaired electrons expected on each $(BEDT\text{-}TTF)^+$ are paired. The result,

Table 2. *Phases of* $(BEDT\text{-}TTF)_4[AM(C_2O_4)_3]C_6H_5CN$

M =	Cr	Fe	Co	Al
A = K	—	Pbcn	—	—
NH$_4$	—	Pbcn	—	Pbcn
H$_3$O	Pbcn	C2/c	Pbcn	—
	C2/c			

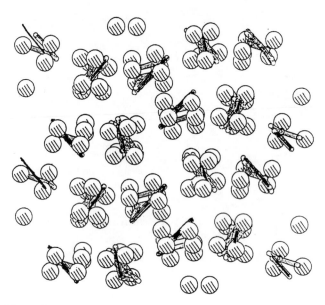

Figure 11. The BEDT-TTF sublattice in $(BEDT\text{-}TTF)_4[AM(C_2O_4)_3]C_6H_5CN$, Pbcn phase. Shaded spheres are H atoms.

confirmed by band structure calculation (Kurmoo *et al.* 1995), is a semiconductor. In seeking the reason for this unusual packing motif we focus on the interface between the donor and anion sublattices, and in particular on the terminal $-CH_2CH_2-$ moieties of the BEDT-TTF, together with the 'top' layer of O coming from the oxalate anions.

The second phase of $(BEDT\text{-}TTF)_4[AM(C_2O_4)_3]C_6H_5CN$, which crystallizes in the $C2/c$ space group, has been identified in $(H_3O)M$ examples with M = Fe, Cr (table 2). The difference between the distribution of chiral anions compared with the semiconducting phase transforms the packing of the donor layer, inducing a β'' arrangement of BEDT-TTF (Kurmoo *et al.* 1995). As mentioned earlier, the β-stacking arrangement leads to superconductivity in numerous salts with mean BEDT-TTF changes of $+0.5$, so it is not a surprise that the *tris*-oxalato-metallate(III) salts with this structure are also superconducting. The Fe salt has $T_c = 8$ K while the Cr one has $T_c = 6$ K and a smaller critical field (Martin *et al.* 1999). All the anions in each layer now have the same chirality (figure 9) so the network of terminal O atoms on the oxalate ions is quite different from that of the orthombic phase. The temperature dependent resistance (typical of a metal) and the

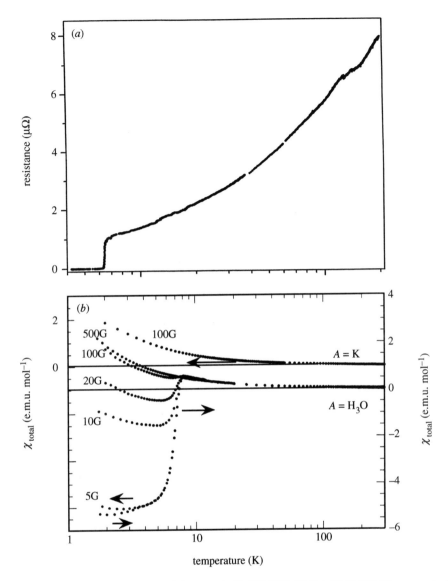

Figure 12. Superconductivity of β''-$(BEDT\text{-}TTF)_4[(H_3O)Fe(C_2O_4)_3]C_6H_5CN$ (Kurmoo *et al.* 1995). (*a*) Temperature dependence of resistance ($\mu\Omega$). (*b*) Temperature dependence of magnetic susceptibility measured after cooling in different fields $(BEDT\text{-}TTF)_4[AFe(C_2O_4)_3]C_6H_5CN$ (left-hand scale, A = K; right-hand scale, A = H_3O).

superconductivity transition in the $(H_3O)Fe$ salt are shown in figure 12*a* while the magnetic susceptibility (figure 12*b*) shows typical Curie–Weiss behaviour at high temperature, because of the Fe moment, with sudden onset of diamagnetism at low temperature when the sample becomes superconducting.

Other molecules may be substituted for C_6H_5CN in the hexagonal cavity formed by A^+ and $M(C_2O_4)_3^{3-}$. In particular, $C_6H_5NO_2$ and C_5H_5N form β''-BEDT-TTF

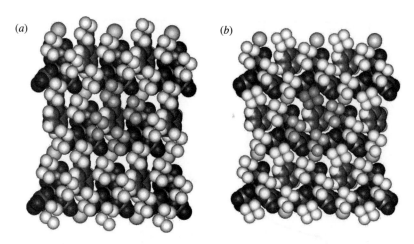

Figure 13. Terminal H atoms of BEDT-TTF in β''-(BEDT-TTF)$_4$[(H$_3$O)Fe(C$_2$O$_4$)$_3$] (solvent): (*a*) solvent = C$_6$H$_5$CN; (*b*) solvent = C$_5$H$_5$N (Turner *et al.* 1999). Light spheres are H atoms.

salts with H$_3$O$^+$ and M = Cr and Fe, which crystallize in the $C2/c$ space group with unit cell parameters very similar to the corresponding C$_6$H$_5$CN compounds. The C$_6$H$_5$NO$_2$ salts are superconducting, albeit with lower T_c and critical fields (S. Rashid, unpublished work), but the pyridine solvate presents an even more fascinating aspect. At room temperature it has metallic conductivity similar to the C$_6$H$_5$CN compound, but instead of becoming superconducting at low temperature it undergoes a sharp metal-to-insulator transition at 116 K (Turner *et al.* 1999). At 150 K, when both materials are in the metallic regime, they have β''-BEDT-TTF stacking, but in the pyridine solvate one-quarter of the BEDT-TTF have an unresolved–twisted conformation in contrast with the C$_6$H$_5$CN one, where they are all twisted–twisted and eclipsed with low thermal parameters. All bond distances and angles within the anionic layers are closely similar in the two (H$_3$O)Fe compounds except that in the pyridine solvate there is a void in each hexagonal cavity, corresponding to the position of the unresolved –CH$_2$CH$_2$– grouping in the BEDT-TTF layers on each side. Below the metal–insulator transition temperature the molecular conformation is resolved into two distinct sites, so that all the BEDT-TTF molecules have twisted–twisted conformations, but one-quarter are staggered and the remainder are eclipsed, as indicated in figure 13 by the positions of the H-atoms. At corresponding temperatures the C=C and C—S bond lengths within the TTF moieties are almost the same in both compounds, so we can rule out charge localization as a reason for the contrasting physical behaviour. This appears to be the first example of a change in molecular conformation bringing about such large change in electrical properties in a molecular conductor, illustrating once again how subtle is the relationship between supramolecular organization and physical behaviour in the molecular solid state.

4. Conclusion: supramolecular chemistry and solid state physics

Let me end where I began, with Humphry Davy. His Bakerian Lecture introduced a wider audience to what must have seemed at the time some highly unusual metals. It is to be hoped that the present brief survey of some examples of molecular-based superconductors and magnets has achieved the same objective. In one small respect, at least, it has surpassed Davy's endeavour: in his lecture on the properties of potassium Davy refers to it as being 'a perfect conductor of electricity' (Davy 1808). Now, of course, we know that it is not; good, yes, but perfect, no. At low temperature, potassium is not among the metals that becomes superconducting, the only category of conductor that can strictly be called 'perfect'. However, as we have seen, there are numerous organic and metal–organic molecular solids that are superconductors, and others that have long-range-ordered magnetic ground states. The molecular nature of these materials has not only provided a wealth of opportunities for synthetic chemists, but served to introduce completely new aspects into this area of physical and materials science: solubilities, compressibility, chirality and intercalation, to name just a few. No doubt in the years to come this new interface between supramolecular chemistry and materials science will uncover many more.

The examples cited in this article are taken from the work of my own research group, first at Oxford and more recently at The Royal Institution. They have been made possible by the ingenuity and enthusiasm of many colleagues, principally Dr Mohamedally Kurmoo, Dr Simon Carling and Dr Scott Turner, and talented graduate students Anthony Graham, Cameron Kepert, Chris Nuttall, Lee Martin, Ian Watts, Samina Rashid and Justin Bradley. We have also benefited from many invaluable collaborations with crystallography and physics groups, especially Professor Michael Hursthouse and Professor Daniel Chasseau, and Dr John Singleton, Dr Stephen Blundell and Dr Bill Hayes. Our group has been supported by the Engineering and Physical Sciences Research Council, the European Commission Human Capital and Mobility, and Training and Mobility of Researchers Programmes and the British Council.

References

Anderson, P. W. 1963 *Solid state physics* (ed. F. Deitz & M. Turnbull), vol. 14, p. 99. Wiley.

Bellitto, C. & Day, P. 1978 *J. Chem. Soc. Chem. Commun.*, p. 511.

Bellitto, C. & Day, P. 1992 *J. Mater. Chem.* **2**, 265.

Bellitto, C., Bonamico, M., Fares, V., Federici, F., Riglini, G., Kurmoo, M. & Day, P. 1995 *Chem Mat.* **7**, 1475.

Carling, S. G., Visser, D. & Day, P. 1993 *J. Sol. State Chem.* **106**, 11.

Carling, S. G., Day, P. & Visser, D. 1995 *Inorg. Chem.* **34**, 3917.

Carling, S. G., Mathionière, C., Day, P., Malik, K. M. A., Coles, S. J. & Hursthouse, M. B. 1996 *J. Chem. Soc. Dalton Trans.*, p. 1839.

Chasseau, D., Gaultier, J., Bravic, G., Ducasse, L., Kurmoo, M. & Day, P. 1993 *Proc. R. Soc. Lond.* A **442**, 207.

Davy, H. 1808 *Phil. Trans. R. Soc. Lond.* **98**, 10.

Davy, H. 1821 *Phil. Trans. R. Soc. Lond.* **111**, 431.

Day, P. 1980 *Chem. Brit.* **16**, 217.

Day, P. 1985 *Phil. Trans. R. Soc. Lond.* A **314**, 145.

Day, P. 1990 In *Lower dimensional systems and molecular devices* (ed. R. M. Metzger, P. Day & G. Pappavassilliou). New York: Plenum Press.

Day, P. (and 10 others) 1992 *J. Am. Chem. Soc.* **114**, 10722.

Decurtins, S., Schmalle, H. R., Oswald, A., Linden, T., Ensling, J., Gütlich, P. & Hauser, A. 1994*a* *Inorg. Chem. Acta.* **216**, 65.

Decurtins, S., Schmalle, H. W., Schneuwly, P., Ensling, J. & Gütlich, P. 1994*b* *J. Am. Chem. Soc.* **116**, 9521.

Dzialoshinskii, I. E. 1958 *J. Phys. Chem. Solids* **4**, 241.

Faraday, M. 1832 *Phil. Trans. R. Soc. Lond.* **122**, 132.

Fisher, E. G. & Maple, M. B. (eds) 1983 *Topics in current physics*, vols 32, 34. Springer.

Goodenough, J. B. 1955 *Phys. Rev.* **100**, 564.

Goodenough, J. B. 1963 *Magnetism and the chemical bond.* New York: Interscience.

Graham, A. W., Kurmoo, M. & Day, P. 1995 *J. Chem. Soc. Chem. Commun.*, p. 2061.

Guionneau, P., Kepert, C. J., Chasseau, D., Truter, M. R. & Day, P. 1997 *Synth. Met.* **86**, 1973.

Jerome, D., Mazard, A., Ribault, M. & Bechgaard, K. 1980 *J. Physique Lett.* **41**, L95–L98.

Julve, M., Frans. J., Verdaguer, M. & Gleizes, L. 1984 *J. Am. Chem. Soc.* **106**, 8306.

Kanamori, J. 1959 *J. Phys. Chem. Solids* **10**, 87.

Kepert, C. J., Kurmoo, M. & Day, P. 1997 *J. Mater. Chem.* **7**, 221.

Kepert, C. J., Kurmoo, M. & Day, P. 1998 *Proc. R. Soc. Lond.* A **454**, 487.

Kurmoo, M. & Day, P. 1997 *J. Mater. Chem.* **7**, 1291.

Kurmoo, M., Rossensky, M. J., Day, P., Autan, P., Kang, W., Jerome, D. & Batail, P. 1988 *Synth. Met.* A **27**, 425.

Kurmoo, M., Pritchard, K., Talham, D., Day, P., Stringer, A. M. & Howard, J. A. K. 1990 *Acta Crystallogr.* B **46**, 348.

Kurmoo, M. (and 10 others) 1995 *J. Am. Chem. Soc.* **117**, 12 209.

Kurmoo, M., Day, P., Estouves, C., Drillon M., Deroy A., Poinsot, R. & Kepert, C. J. 1999 *J. Sol. State Chem.* **145**, 452.

Mallah, T., Hollis, C., Bott, S., Kurmoo, M. & Day, P. 1990 *J. Chem. Soc. Dalton Trans.*, p. 859.

Martin, L. 1999 PhD thesis, University of London.

Martin, L., Turner, S. S., Day, P., Mabbs, F. E. & McInnes, F. I. L. 1997 *J. Chem. Soc. Chem. Commun.*, p. 1367.

Martin, L., Turner, S. S., Day, P., Hursthouse, M. B. & Malik, K. M. A. 1999 *J. Chem. Soc. Chem. Commun.*, p. 513.

Mathionière, C., Carling, S. G., Dou, Y. & Day, P. 1994 *J. Chem. Soc. Chem. Commun.*, p. 1551.

Mathionière, C., Nuttall, C. J., Carling, S. G. & Day, P. 1996 *Inorg. Chem.* **35**, 1201.

Mizuno, M., Garito, A. F. & Cava, M. P. 1978 *J. Chem. Soc. Chem. Commun.*, p. 18.

Moriya, T. 1960 *Phys. Rev.* **120**, 91.

Néel, L. 1948 *Ann. Physique* **1**, 137.

Nuttall, C. J. 1997 PhD thesis, University of London.

Nuttall, C. & Day, P. 1998 *Chem. Mater.* **10**, 3050.

Pratt, F. L., Valladares, R., Caulfield, J., Dekkers, I., Singleton, J., Fisher, A. J., Hayes, W., Kurmoo, M. & Day, P. 1993 *Synth. Met.* **61**, 171.

Robin, M. B. & Day, P. 1966 *Adv. Inorg. Chem. Radiochem.* **10**, 247.

Rosseinsky, M. J., Kurmoo, M., Talham, D. R., Day, P., Chasseau, D. & Watkin, D. 1988 *J. Chem. Soc. Chem. Commun.*, p. 88.

Sugano, T., Tamura, M., Goto, T., Kato, R., Kinoshita, M., Sakai, Y., Ohashi, Y., Kurmoo, M. & Day, P. 1993 *Mol. Cryst. Liq. Cryst.* **232**, 61.

Talham, D., Kurmoo, M., Day, P., Obertelli, A., Parker, I., Friend, R. H., Stringer, A. & Howard, J. A. K. 1987 *Sol. State Commun.* **61**, 459.

Tamaki, H., Zhong, J., Matsumoto, M., Kida, S., Koikawa, N., Achiwa, Y., Hashimoto, Y. & Okawa, H. 1992 *J. Am. Chem. Soc.* **114**, 6974.

Tamamura, M., Nakazawa, Y., Shiomi, D., Nozawa, K., Hosokoshi, Y., Ishakawa, M. Takahashi, M. & Kinoshita, M. 1991 *Chem. Phys. Lett.* **186**, 401.

Turner, S. S., Day, P., Malik, K. M. A., Hursthouse, M. B., Teat, S. J., MacLean, E. J. & Martin, L. 1999 *Inorg. Chem.* **38**, 3543.

Williams, J. M., Ferraro, J. R., Thorn, R. J., Carlson, K. D., Geiser, U., Wang, H. H., Kini, A. M. & Whangbo, M.-H. 1992 *Organic superconductors (including fullerenes).* Englewood Cliffs, NJ: Prentice Hall.